HVAC Controls and Systems

HVAC Controls and Systems

John I. Levenhagen, P.E.

Donald H. Spethmann, P.E.

McGraw-Hill, Inc.

New York San Francisco Washington, D.C. Auckland Bogotá
Caracas Lisbon London Madrid Mexico City Milan
Montreal New Delhi San Juan Singapore
Sydney Tokyo Toronto

Library of Congress Cataloging-in-Publication Data

Levenhagen, John I.
HVAC controls and systems / John I. Levenhagen, Donald H. Spethmann.
p. cm.
ISBN 0-07-037509-7
1. Heating—Control. 2 Air conditioning—Control.
I. Spethmann, Donald H. II. Title.
TH7466.5.L48 1992
697—dc20 92-25594
CIP

2 3 4 5 6 7 8 9 0 DOC/DOC 9 8 7 6 5 4

ISBN 0-07-037509-7

The sponsoring editor for this book was Robert W. Hauserman, the editing supervisor was Kimberly A. Goff, and the production supervisor was Pamela A. Pelton. This book was set in Century Schoolbook. It was composed by McGraw-Hill's Professional Book Group composition unit.

Printed and bound by R. R. Donnelley & Sons Company.

I dedicate this book first and foremost to my wife Theresa whose patience and strength over the months and years in the writing of this book was conspicuous.

I also dedicate this book to the many friends and associates at ASHRAE (The American Society of Heating, Refrigerating and Air Conditioning Engineers), as well as the ones at some of the major control companies who helped with the research, in particular people like: Ron Caffery, Robert Stahl, Geno Strehlow, Kirk Drees, Jim Grevers, Bob Weeks, Mark Maduza, George Huhnkie, Jerry Kubiak, Dan Krajna, John Meyer, Trish Woolfer, John Trabor, and Dave Podeszwa. My thanks go out to them as well as others who helped in the final composition of this book.

John I. Levenhagen, P.E.

I would like to dedicate my portion of this book to control application engineers, those people who strive for an understanding of total system operation to provide proper control of each unique total HVAC system.

I would also like to thank those who helped me write my portion of this book: My wife, Rose, for her encouragement and understanding; Bill Pienta for his helpful review of building automation and DDC explanations; Anil Saigal for sharing his considerable knowledge of the proposed ASHRAE standard, "BACnet: A Data Communications Protocol for Building Automation and Control Networks"; and Dick Poey of Honeywell and Bob Parsons of ASHRAE for arranging for the use of drawings.

Donald H. Spethmann, P.E.

Contents

Preface xiii

Chapter 1. Overview 1

HVAC Design Considerations 3
HVAC Systems 4
Basic Control 5
Supervisory Control (Building Automation) 6
Optimizing Control 7
Sources of Supply 7
History of Supply 8

Chapter 2. Thermostats 11

Room Thermostats 11
Unit Thermostats 19
Special Thermostats 26

Chapter 3. Dampers and Damper Motors 29

Dampers 29
Parallel Blade Versus Opposed Blade Dampers 30
Fire and Smoke Dampers 34
Manual Dampers 35
Static Control Dampers 36
Sizing of Automatic Dampers 36
Damper Motors 41
Pneumatic Damper Motors 41
Electric Damper Motors 46

Chapter 4. Automatic Valves 51

Types of Automatic Valves 51
Two-Way Automatic Valves 52
Three-Way Automatic Valves 54
Valve Actuators 60

Valve Sizing 66
Butterfly Valves 67

Chapter 5. Pneumatic Transmitters, Indicating Receivers, and Receiver Controllers 71

Transmitters 71
Indicating Receivers 78
Receiver–Controllers 80

Chapter 6. Auxiliary Devices 83

Pneumatic Relays and Cumulators 83
Electric Relays 87
Pneumatic Switches 90
Electric Switches 92
Transducers 93
Pressure Controllers 96
Clocks and Miscellaneous Devices 97

Chapter 7. Construction Systems and Devices 101

Air Compressors 101
Control Air Systems for Pneumatic Controls 111
Wiring Systems for Electric and Electronic Controls 118
Troubleshooting Systems 122

Chapter 8. Electric and Electronic Control Products 125

Electric Control Components and Systems 125
Oil Furnace 126
Natural Gas Furnace 126
Room Thermostats 127
Commercial Electric Controls 130
Electronic Control Components and Systems 133
Residential Electronic Controls 135
Commercial Electronic Controls 135
Specifying an Appropriate System 139

Chapter 9. Direct Digital Control 143

Definition and Historical Evolution 143
Functionality with Examples of Applications 144
Sequence of Control for Fig. 9.1 144
Variations in Programming Methods 147
Energy Management Functions and Interfaces to DDC 150
Stand-Alone DDC Controllers Versus DDC as Part of a BAS 151
Functional Considerations 151
Communication Considerations 151

Specifying Automatic Control by DDC 153
Checkout, Commissioning, and Acceptance 155

Chapter 10. Air-Handling Units 157

Evolution of Air-Handling Units 157
Packaged Units 162
Single-Path Units 163
Dual-Path Units 178
General Comments 182
Troubleshooting 186
Maintenance of Air-Handling Units 187
Design Considerations with Air-Handling Units 188

Chapter 11. Terminal Units and Systems 191

Unit Ventilators 193
Basic Cycles of Operation 194
Fan-Coil Units 197
Double-Duct Mixing Boxes 203
Induction Units 205
Variable Air Volume Terminal Boxes 205
Other Terminal Units 211

Chapter 12. Primary Supply Systems 213

Heating Supply Systems 213
Boilers 213
Heating Converters 216
Chiller–Condenser Rejected Heat 217
Cogeneration Heat Source 218
Chilled Water Supply Systems 221
Chiller Plant Optimization 222
Centrifugal Chillers 225
Positive Displacement Chillers 227
Absorption Chillers 227
Free Cooling Cycles 228
Thermal Storage 228
Cool Storage 228
Heat Storage 233

Chapter 13. Heat Pumps and Heat Pump Controls 235

Air-to-Air Heat Pumps 238
Earth-to-Air Heat Pumps 240
Water-to-Air and Air-to-Water Heat Pumps 241
Water-to-Water Heat Pumps 242
Other Commercial Heat Pump Systems 243
Heat Pump Components 243

Heat Recovery/Heat Pump Systems 247
Heat Recovery 249

Chapter 14. Distribution Systems of All Types 255

Water Distribution Systems 255
Two-Pipe Systems 255
Three-Pipe Systems 257
Four-Pipe Systems 259
Pumping Control and Differential Pressure Regulation 260
Primary–Secondary Pumping Variations 266
Steam Distribution Systems 269

Chapter 15. Supervisory Control and Total System Optimization 273

Definition and Historical Background 273
System Configurations and Communications 275
Proprietary Versus Open Systems 277
Possible Supervisory Functions 280
Customized Global Control Functions 280
Standard Energy Management Functions 281
Optimum Start and Stop 281
Demand Control 282
Duty Cycle 283
Load Reset and Zero Energy Band 283
Night Cycle and Night Purge 283
Enthalpy Control 284
Interface to Local Loop Control 284
Operator Interface Functions 285
Custom Programming and Time Event Programming 287
Application Engineering Functions and Tools 287
Specifying Building Automation Needs 288
Check Out Commissioning and Acceptance 289
Documentation 289

Chapter 16. Operating and Maintaining Control and HVAC Systems 297

Operating and Maintaining Primary Systems and Controls 297
Pumps 299
Boilers 299
Chillers 301
Steam Distribution Systems 305
Primary Air Systems 305
Operating and Maintaining Terminal Equipment 307
Radiators and Convectors 308
Operating and Maintaining Controls 309
Pneumatic Controls 309
Electric Controls 311

Calibrating Controls 312
Operating and Maintaining Building Automation Systems 314
Summary 315

Chapter 17. Total Facility Approach to Planning Controls 317

Building Usage and Zoning 317
Building Management Method 318
Total Mechanical System Design 320
Selection of Types of Control Systems 322
Methods of Specifying and Procuring a Control System 323

Index 327

Preface

During the past 40 years HVAC systems in nonresidential buildings in North America have been tried with varying degrees of success. As these new systems were tried, new and innovative control systems were designed to allow them to work satisfactorily. As with anything new, problems developed and the solutions were not always quickly forthcoming. But, in time, solutions did surface and the owners and operators of the building were in most cases satisfied with the results.

Some of the solutions to those problems were never published; others were published as in-house papers for employees only. The need therefore existed to uncover the practical and unknown solutions.

To respond to that need, *HVAC Controls and Systems* covers the major components of HVAC commercial control systems in as much detail as necessary to provide the solutions to the problems that existed. These problems involve everything from automatic dampers, automatic valves, unit thermostats, and electronic controls to the problems caused by faulty or poorly designed controls in the systems. The book, therefore, covers room thermostats, room humidistats, and all of the devices they as well as other instruments control. It also includes pneumatic controls, analog electronic controls, digital electronic controls, and standard electric controls. Residential controls are covered, although the emphasis is on nonresidential building controls. Some mention is made of industrial controls, where that field comes close to the commercial field. The information presented will enable you to work on problems not mentioned in other books and solve those problems based upon the practical knowledge of the authors.

A discussion of digital controls, as well as analog electronic controls brings out the disadvantages and advantages of all of the systems, from unit ventilators to today's VAV systems. These systems are the ones used in all types of commercial buildings from schools and hospitals to high-rise office buildings.

To understand the complexities of today's modern systems and the controls used to maintain the conditions in them, *HVAC Controls and Systems* presents the history of systems as they were developed.

John I. Levenhagen, P.E.
Donald H. Spethmann, P.E.

HVAC Controls and Systems

Chapter

1

Overview

Controls for heating ventilating and air conditioning (HVAC) cover a broad range of products, functions, and sources of supply. *Control* can be defined as the starting and stopping or modulation of a process to regulate the condition being changed by the process. In this book the process involved is heating, ventilating and air conditioning. HVAC commonly refers to buildings of public, commercial, or institutional usage that require a provision for ventilation as well as heating and cooling. This book deals with the common practices in using control devices and systems to control HVAC systems. It does not deal with the control of residential heating and cooling, except in some special cases where residential controls crossover into light commercial controls. The application of controls starts with an understanding of the building and HVAC systems and the use of the spaces to be conditioned and controlled. The type of HVAC system determines the control sequence. The basic control sequence can then be done by several types of control products such as pneumatic, electric, analog electronic, or electronic direct digital control (DDC).

The use of building spaces and the general strategy of managing and running the building(s) determines the benefits to be obtained from additional controls for centralization, automation, and/or optimization. The sources of supply vary with the simplicity or complexity of the products planned for the project. This chapter will discuss the general range of choices and the considerations in making these choices. Later chapters will give more detail.

A brief historical review of HVAC control can help us understand the subject. Back when heating was controlled by draft dampers, the thermostat was invented to control the dampers. The use of mechanical stokers for coal firing required another step in the use of control. When oil burners were introduced, the concept of combustion safety control became necessary. This involved the sensing and proof of flame in the proper time sequence of introducing draft, fuel, and ignition.

The use of steam and hot water radiators led to the concept of zone control and eventually individual room control (IRC). Forms of zone control included closed loop control using zone thermostats and open loop control with outside conditions setting the rate of heat delivery to the zone. Both of these forms of control were used to regulate the delivery of heat. The means of regulation included the following: valves to control the flow of steam or hot water, controlling pumps to circulate hot water, and controlling boiler operation. When IRC was used, the central supply was maintained and radiator valves were controlled by room thermostats.

The use of fans to deliver ventilation as well as heated air was controlled by dampers, which varied the source and volume of air. This included unit ventilators in classrooms. The typical control of unit ventilators was by pneumatic controls and included the following features: minimum outside air, discharge air, low-temperature limits, and thermostats with lower night settings activated by compressed air supply pressure level. The increased usage of air conditioning led to more complex control sequences and in larger systems to centralized monitoring and control. The increased complexity of sequences to accommodate air conditioning included the following: control of cooling in sequence with heating and ventilation, control of dehumidification by subcooling and reheat, and control by economizer sequence that used outdoor air for cooling when it was a suitable source of cooling. In larger systems, centralized monitoring and control was typically done with electronic sensing and some form of multiplexed wiring to the centralized monitoring point. The operator interface at the central location initially used gauges, meters, and switches mounted on panels that included a graphic representation of the piping schematic for the system being controlled. The use of photographic slide projectors increased the graphic capacity and flexibility. The shared wiring schemes initially had control cables that switched the connection of a data cable from one system's wires to another. The use of digital codes and time-shared transmission allowed communication of many channels of information over a single pair of wires.

The industrywide development and use of computers and microprocessors has caused great changes in the HVAC controls industry. First, minicomputers were installed on very large jobs to collect data and provide centralized control. Then, microprocessors were used in remote data-gathering panels to gather data and provide direct digital control. Computers are used both as on-site central controllers with operator interface and as computer assisted engineering (CAE) tools in the design and generation of system programs, databases, and documentation. Microprocessors are used in remote data gathering panels with DDC, as well as in smaller unit controllers and smart thermostats.

The scope of HVAC control has grown to include building automation. Building automation systems can include fire alarm or life safety systems and security systems. This book will not deal with these systems, however, but will confine coverage to HVAC control.

From this brief historical review we see that the evolution of HVAC equipment and systems caused the evolution of the basic controls for HVAC and that the evolution of control technology finds new ways to improve the operation and maintenance of mechanical and electrical systems by what is broadly called building automation. Also, technology has led to the use of direct digital control and is pushing the application of expert systems in control of HVAC.

In reviewing the scope of HVAC control, this chapter will briefly examine each of the following subjects:

HVAC Systems

Basic control

Supervisory control

Optimizing control

Sources of supply

HVAC Design Considerations

Before considering the description of controls or HVAC systems, it would be wise to take a long look at what is involved in the design of HVAC for commercial buildings. Although this book is about controls, it is also about systems, and understanding systems is far more important than understanding controls.

A smart engineer once said, "All the controls in the world will not correct the problems with a poorly designed HVAC system." Truer words have never been spoken. Engineers must be responsible and design systems that can be controlled and can work at all levels of load. This involves knowing load profiles for all hours of operation for all zones of usage and also understanding part load operating characteristics of the HVAC equipment to be used.

There is another phrase used by knowledgeable control engineers, the kiss principle, which means keep it simple stupid. Controls piled on top of controls will only accentuate a problem if the problem is with the system itself. A high percentage of the troubleshooting done in the past where the indicated problem was said to be a control problem later proved to be a system problem. An example that has come up time and time again is the matter of ductwork and hydronic system balancing. The state of the art has progressed greatly in the past 15 years, and balancing is now almost an exact science, so much so that a system that is properly designed with no scrimping on the duct work or piping and that has adequate balancing dampers or valves can be balanced to the satisfaction of all concerned.

If a project is straightforward, like an office building with no special conditions, a normal variable air volume (VAV) or similar system is in order. If, however, the project is not straightforward and has special requirements such as a hospital, a smart engineer does not try to do the job with one all-purpose HVAC unit. Whatever type of system is designed, it should be able to be operated properly at all loads before the controls are added.

HVAC Systems

HVAC system types can be classified as either self-contained unitary packages or as central systems. With unitary packages the one package converts a primary energy source (such as electricity or gas) and provides final heating and cooling to the space to be conditioned. Examples of self-contained unitary packages are rooftop HVAC systems, packaged terminal air conditioning (ac) units for rooms, and air-to-air heat pumps.

The typical uses of packaged unitary systems are in single-story structures where there is easy access to outside air or rooftop mounting. They are also used in situations where first cost is more important than operating costs.

Central systems are a combination of central supply subsystem and multiple end use subsystems. End use subsystems can be fan systems or terminal units. If the end use subsystems are fan systems, they can be single or multiple zone type.

Whatever controlled HVAC unit maintains space temperature, it can be referred to as the end use zone system. The central supply can be packaged boilers and chillers in smaller sizes or custom applied boilers and chillers in larger sizes. In addition to size, the difference between a package and a custom-applied unit is the extent of control and auxiliary equipment furnished with the unit. The safety controls associated with chillers and boilers are applied by the unit manufacturer because their locations and control actions are directly dependent on the mechanical equipment design. These are original equipment manufacturer (OEM) furnished controls that come with the unit whether it is a packaged or a custom-applied unit. Additional controls for capacity to maintain supply temperature levels and auxiliary equipment such as pumps are included in a package but not necessarily with a custom-applied chiller or boiler.

With central systems, the primary conversion from fuel such as gas or electricity takes place in a central location, with some form of thermal energy distributed throughout the building or facility.

There are many variations of combined central supply and end use zone systems. The most frequently used combination is central hot and chilled water distributed to multiple fan systems. The fan systems use water to air heat exchangers called coils to provide hot and/or cold air to condition the controlled spaces. Another combination central supply and end use zone system is a central chiller and boiler for the conversion of primary energy, as well as a central fan system to delivery hot and/or cold air. The multiple end use zone systems are mixing or VAV boxes. The typical uses of central systems are in larger, multistoried buildings where access to outside air is more restricted. Typically central systems have lower operating costs.

Besides packaged unitary and central systems, there are a variety of special-purpose systems. These include

1. Heat pump cycles on chillers that use rejected heat or tower free cooling
2. Thermal storage
3. Cogeneration of electricity and heat

Each special-purpose system has unique control strategy requirements. These are discussed in the chapters on primary supply (Chap. 12) and heat pump cycles (Chap. 13).

Basic Control

Basic control regulates the amount of heating or cooling necessary to meet the load in conditioned spaces. Minimum outside air needed for ventilation is provided whenever a space is occupied. When outside air temperature is a suitable source for free cooling, it is controlled as needed at values greater than minimum.

The approach in packaged unitary equipment is to control the generation of heating or cooling by space thermostats. The approach in central systems is to control the delivery of heating and cooling by the end use zones to match the load in the space. The supply (or rate of primary conversion) is controlled to match the load imposed by the sum of the end use zones. A typical method of doing this is for end use zones to be controlled by room thermostats and central supplies to be controlled by discharge controllers. Generally, discharge temperature controllers control the rate of primary conversion (chillers or boilers), and pressure controls determine the delivery rate of the pumps or fans distributing the central supply. In many cases there are multiple boilers and/or chillers and pumps put on or off line as necessary to provide proper capacity. Those on line are modulated as necessary to meet load needs. The controls to put units on and off line would normally be job applied to meet the system needs.

Supervisory Control (Building Automation)

The role of *supervisory control,* as a generic class of control, is to control scheduling and interaction of all the subsystems to meet building needs with appropriate operator input. Supervisory control systems have had many names, each used for a particular emphasis. Among the names and their acronyms are the following: building automation system (BAS), energy monitoring and control system (EMCS), facility management system (FMS), and energy management system (EMS). In the context of this book on HVAC control, the most appropriate name is building automation system. The important thing is that there is a clear understanding of what a system is expected to do and that this is clearly communicated to the people who will make it happen.

The term *direct digital control* is sometimes used to describe everything done by a computer- or microprocessor-based control system. The original use of direct digital control referred to providing closed loop control of local loops by a digital computer or microprocessor. These local loops were originally controlled by a hardware set composed of a sensor, a controller, and a controlled device such as a damper or valve actuator. With DDC the controller was replaced by a calculation in a digital computer or microprocessor. One benefit was that a single computer or microprocessor could do many such calculations and replace many hardware controllers. Building automation was also being done by centrally located computers or microprocessors. The functions of these systems were general monitoring, scheduled starting and stopping, and changing local loop set points. These programs were generally described as energy management applications.

Direct digital control is implemented both in stand-alone panels and in intelligent data-gathering panels that are the remote panels of a building automation system. Energy management programs that were originally in the central computer of a building automation system are sometimes placed in the remote data-gathering panels or even in stand-alone DDC controllers. This has led some people to use DDC to describe all functions of control done by microprocessor-based control systems. This book, however, uses the narrower definition, where DDC refers to local loop control.

Energy management application programs are different than local loop control and are named for their specific function, such as optimum start or demand control. The considerations of which energy management application programs should be used rely upon the type of building and of HVAC system. For instance, the optimum start–stop programs are not appropriate for a hospital that has 24-h operation and use of spaces. Also load reset of supply temperatures is appropriate for systems that supply both heating and cooling simultaneously, such as reheat systems or hot and cold deck mixing box systems. These guidelines are given in the chapters discussing the type of system being controlled, such as Chap. 10 "Air-Handling Units" and Chap. 12 "Primary Supply Systems."

Optimizing Control

The concept of optimizing control is not only to control space conditions but also to do it in a manner that minimizes the energy and costs when different forms of energy are available. An optimizing strategy is generally to improve the efficiency of primary supply equipment or to reduce the losses of energy in end use systems. The sizing of equipment is to meet maximum loads, but the majority of time equipment is run at much less than maximum load. This means that the part load characteristics of the equipment determines the efficiency in meeting a given load. When there are multiple chillers or boilers, an optimizing strategy would be to choose the most efficient equipment that has the capacity to meet the load at any given time. Also, with some types of end use systems, energy wasted by bucking heating against cooling can be minimized by resetting supply temperature levels to be no more than is necessary to meet a given load condition. Another way to optimize is to use the thermal storage of a building mass or of a storage facility to make use of energy stored at low cost and used when needed and when it would be high cost to generate it. Also, moving heat from one area of a building to another can be an optimizing opportunity.

These optimizing principles are used for specific types of HVAC systems or building load circumstances. The measured variable in all of these circumstances is the amount of heating or cooling load and the control action to make some change in the way a load is supplied. This process has led to the use of the terms *load reset* and *dynamic load control* to describe this general approach to optimizing control. The selection of the most efficient combination of chillers to supply a cooling load has been called *optimized chiller selection.* These strategies are covered in more detail in the chapters pertinent to the systems and equipment involved.

Sources of Supply

The primary source of supply is the manufacturer of the control equipment involved. The majority of HVAC control is supplied as systems installed as a part of the construction process when making a new building. Adding controls to existing buildings is another way to improve building performance. Frequently, energy retrofit is an economically viable project that involves updating a control system with energy management strategies and DDC. The marketing of controls can be through different channels, which include the following:

Full-line control companies that provide turnkey installed systems and comprehensive service for all types of control, including the most complex building automation. Their marketing is typically through multiple channels, which include OEM sales and distributor sales of control products in addition to direct sale of installed systems.

Control distributors selling and, in some cases, installing and servicing controls. At times, simple energy management systems for light commercial type buildings are also installed by distributors with specialized training.

Full-line HVAC equipment companies that include a limited line of controls and simple building automation in their product line. They typically offer packages that integrate the HVAC equipment and controls in popular configurations.

Specialized control companies that have a limited product line dealing with some niche in the overall HVAC control field. Their marketing may be through direct sales, installing contractors, or manufacturer's representatives.

As stated at the beginning of this chapter, the application of controls for HVAC starts with an understanding of the HVAC system and the use of the spaces to be conditioned. The use of the spaces determines the desired environment, the time schedules involved, and to a great extent the ventilation and air-conditioning load levels and timing. The HVAC system characteristics determine the sequence of control required and levels of costs and results that will be achieved. The selection of the type of HVAC system is beyond the scope of this book. The characteristics of basic control and optimizing control for different types of HVAC systems will, however, be covered. The organization of this book is into two sections, one dealing with control devices and one with HVAC and supervisory control or building automation systems.

History of Supply

A brief history of the evolution of controls suppliers will help you understand the present situation. Before World War II, the main suppliers of HVAC controls in commercial buildings were companies that primarily promoted pneumatic controls. Of the three major companies in this field, two started as electric control companies; one of them even manufactured electric cars before seriously getting in the control business. The second one purchased a pneumatic controls company so as to participate in that business as well. The third company started in business by selling controls and other plumbing items to the industry.

The one thing that was common among the original big three was the concept that controls for commercial buildings were too complicated to sell over the counter and had to be installed and supervised by the controls manufacturer. This concept included having branch offices with installers and servicepeople.

Electric control systems for commercial buildings were modulating type controls and were sold on a supervised basis. When several other companies entered the commercial controls market with electric and electronic controls, some of their distribution was through distributors as well as branches. Some of the newcomers who started with electric and electronic controls expanded into pneumatic controls either by their own development or by association with foreign companies.

When computer-based supervisory control systems came to market, some larger companies with computer-based products entered the HVAC controls market but eventually gave up. As international business developed and companies became multinational, some foreign-based controls companies expanded into the U.S. markets directly or through associations with smaller U.S. control companies. During the period when energy conservation was a hot button and application of microprocessors became commonplace, some small companies evolved with limited product lines for energy management functions. When DDC became accepted, some small companies developed microprocessor-based DDC controllers and supervisory systems.

With all this activity, there has been considerable change in the sources of supply at any given time. A number of companies have come and gone from the HVAC control scene. Presently the full line control companies that started out as major players remain as major players but with a few more competitors with limited systems if not full lines of controls. Some major HVAC systems manufacturers have acquired or developed control capabilities. They market packaged HVAC systems with controls and supervisory control systems.

Some small companies provide products for specific applications. The understanding of the HVAC system and its control needs will always be a key element in the successful purchasing and ownership of such a facility. The goal of this book is to help in that understanding. The selection of a source of supply should consider the life cycle needs and costs as well as the track record of potential suppliers.

Chapter

2

Thermostats

This chapter will discuss room thermostats, room humidistats, unit capillary thermostats, insertion thermostats, insertion humidistats, safety thermostats, and special thermostats and their uses in the various systems. It will also deal with problems involving thermostats and present solutions based upon experience and tests performed over the years. The discussions will refer mostly to pneumatic controls, only because pneumatic control systems have been the standard of the commercial industry for many years. Some of the principles also apply to electronic and/or electric controls. Self-contained controls are also a factor to be reckoned with, but usage of those types of controls is minimal in the field of commercial controls and systems.

Room Thermostats

The mounting of room thermostats and room humidistats has been the subject of much discussion in the past, and for many years the industry standard has been for the thermostat to be mounted near the door of a room 5 ft from the floor. The only problem is that if the room is full of short people, such as school children, the thermostat is not controlling the temperature where the occupants are. Although many rooms are not filled just with children, the point is that too often the conditions that are or will be in the space are forgotten. Thus, the placement of a room thermostat or humidistat is important enough that we need to study its location with the view of trying to keep the occupants comfortable *where they are*.

Further, it is important to study the location of the room thermostat or humidistat as to the affect of conditions *at the thermostat*. Remember, the thermostat knows only what is going on at its location. If there is a ceiling diffuser blowing air at the location where the thermostat is mounted, there is going to be cycling of the system. It is, therefore, important to think of a thermostat on the wall as a device that knows only what is going on in its immediate vicinity. The thermostat cannot think; it can only do what it is set to do. These principles apply to all types of thermostats and humidistats (pneumatic, electric, and electronic).

Sometimes installers and others are concerned about the way thermostats and humidistats are mounted on the wall; that is, whether they should be mounted in a horizontal or a vertical position. Generally speaking, except for the fact that the words on the cover may read wrong or the thermometer might be upside down, there is no reason that thermostats and humidistats cannot be mounted either horizontally or vertically. There is, however, one important exception to that rule—when the thermostat is electric and has a mercury bulb for the switch contact (Fig. 2.1). This type is common in residential and some commercial buildings and requires the thermostat to be mounted a certain way. Some of these require the installer to use a level to mount it on the wall.

Generally, however, manufacturers design and sell thermostats that can be mounted either way. They also sometimes make kits to allow a change in the mounting method. All of the above not withstanding, the majority of the wall thermostats and humidistats are vertical mount. When in doubt, it is advisable to check with the manufacturer to see what is provided and what is available.

Room *thermostats and humidistats* are devices that control automatic valves and dampers in a control system. These devices have built in *sensors* as well as moving parts that control the device being controlled. An example is a pneumatic thermostat that has a bimetallic sensor and a relay to do the work required on the device being con-

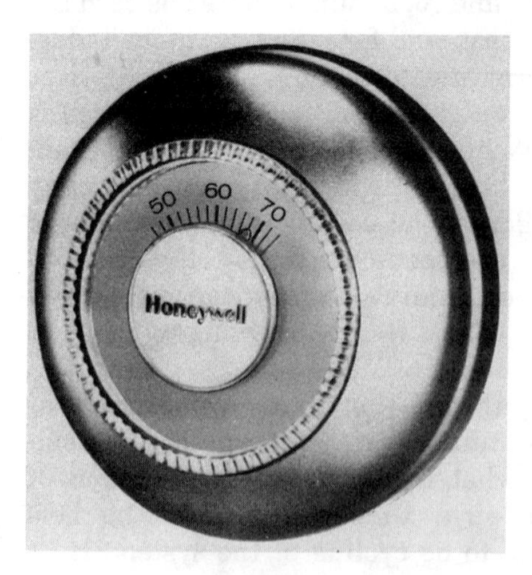

Figure 2.1 Typical electric room thermostat. (*Courtesy of Honeywell Inc.*)

trolled. Usually the complete package is under one cover on the wall, and all action takes place at the thermostat. There are, however, sensors that are mounted under the cover in the room that have no action devices (e.g., relays) under the same cover. They usually transmit the temperature information to another device at a remote location that does the controlling with relays, and so on. Normally this principle is used in electronic control systems involving a wire wound resistance (RTD) (Fig. 2.2) mounted under a cover that reads the temperature in the space and transmits that information to an electronic controller in an equipment room.

Often, there is confusion with the terms thermostat and sensor. The concept of a sensor under a cover in the room is new and came about because of the advent of electronic control systems. Room sensors are used with other control systems and are sometimes called transmitters. In the case of pneumatic controls, the transmitters use a sensor and a special relay that transmits a pneumatic air signal proportional

Figure 2.2 Typical RTD sensor. (*Courtesy of Johnson Controls Inc.*)

to the medium being sensed. An example is a transmitter under a room thermostat cover that transmits an air signal based upon the temperature being sensed in the room.

The transmitter may look like a thermostat, but it does no controlling by itself; it depends upon a receiver controller in a different location to take the action on the controlled device. The dials of these devices are only used to calibrate and are not moved once they are set. These transmitters come in standard ranges and send out an air signal based upon the medium being sensed. An example is a room transmitter with a range of 30 to 80°F that sends out a signal of 9 psi when the temperature being sensed is 55°F. In this case, the transmitter produces a signal of 3–15 psi as its sensed temperature goes from a low of 30°F to 80°F. Transmitters are not strictly sensors but can be classed in that arena since they are devices that do not do the controlling themselves.

The types of covers deserve some consideration in the discussion of room thermostats and humidistats, since too often the selection leaves a lot to be desired. Take the time to analyze the occupants of the building when deciding whether or not, for example, the room thermostats should have thermometers in the cover or the thermostat should be adjustable by the occupant. Consider, for example, times where designers use adjustable room thermostats in a mental institution when only staff should be able to adjust the temperature. The reason is sometimes not enough effort is spent in the selection of covers. A check with most manufacturers would reveal that they offer various arrangements and are willing to give good advice on the proper type of covers and adjustment models to be used in each application. Suffice it to say that in most public buildings, the thermostats should have minimal control capabilities by the occupants.

The subject of guards is also worth discussing. Remember that the thermostat sensing device under the cover must sense the temperature that surrounds it, and if the mass of the metal near the sensor is higher or lower than the sensor, the sensor will radiate to that higher or lower mass. What all this means is, for example, if there is a heavy cast iron guard around a thermostat, the entire guard and cover must reach the temperature of the air surrounding it *before* the thermostat can sense the temperature in the space. The bottom line, therefore, is to use guards as little as possible and to use ones low in mass (Fig. 2.3). Cast iron guards are archaic and should not be used. They tend to slow the action time down as far as the control system is concerned.

Some manufacturers make recessed thermostats available for special occasions. Like guards, their use should be limited to special cases when their use is absolutely necessary. In those cases, the system must be the type that moves room air across the sensor at all times so

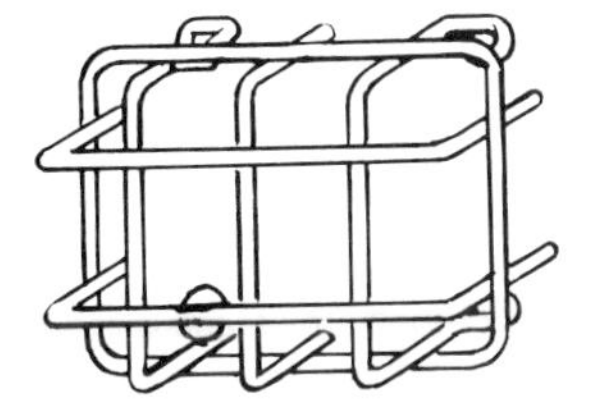

Figure 2.3 Room thermostat guard. (*Courtesy of Johnson Controls Inc.*)

the controller can know what is going on in the space to be controlled. The only type that is available is the one that uses pneumatic control air to induce room air to pass over the thermostat sensor.

Some room thermostats can control at two different temperatures depending upon change in status of the power supplied to the thermostat. In the case of a pneumatic thermostat, changing the supply air (main air) pressure to the thermostat changes it from controlling at one temperature to controlling at a different temperature. In the case of an electric thermostat, wires and/or electronics are switched to make the thermostat control at a lower or a higher temperature. These are called day–night type thermostats and are used in commercial as well as residential systems.

In other situations, thermostats are used that change their action from a remote switching system and allow the summer-winter thermostats to control the same equipment differently in the heating season as opposed to the cooling season. The reason is that on the same rise in temperature the action of the thermostat must reverse itself in the heating versus cooling season. These types of thermostats usually have two dials that can be set independently, but in some cases they have a fixed differential between the high and the low setting.

Care must be used in designing the systems and selecting the thermostat. Studies have shown that night set back of room temperatures can save on energy, provided the night set-back temperature is not too low. After the initial surge of the energy crunch of the 1970s, some designers decided that in the interest of saving energy on commercial systems, a dead band in the set point of a thermostat was acceptable to the average occupant of a commercial building. Thermostats were designed that would allow the temperature in the space to drift up or down a few degrees at the set point without using either heating or cooling energy. It must be remembered that when these types of thermostats are used, the occupants may complain and research involving the type of system is in order. Some systems, such as certain types of multizone systems, will not work with these types of stats.

Another thermostat that needs to be mentioned is the *one-pipe thermostat.* This type of thermostat does not have a relay as a part of its

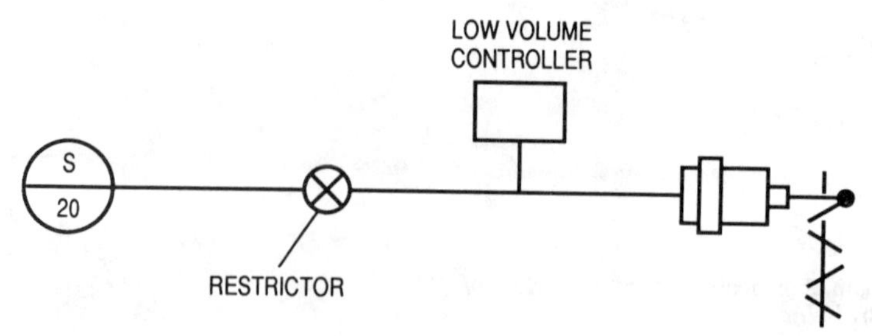

Figure 2.4 Piping of a one-pipe thermostat. (*Courtesy of Johnson Controls Inc.*)

makeup, and although a study of Fig. 2.4 with the restrictor required shows it can work about the same as thermostats with a built-in relay, there are major differences. First, since there is no relay in the thermostat, the *capacity* (as well as speed) of the thermostat to pass or exhaust air to the device it controls is limited. This means it should never be used in applications where there are long air lines to the control devices. Also it should never be used where there is a multiplicity of devices being controlled (maximum of three) with a lot of lines to fill up or if the device being controlled is large and has a large diaphragm to fill up. The control engineer should never use it if the required control actions dictate fast action. It works best if it is used in systems where the primary fan system does most of the work and the room thermostat merely has to knock the rough edges off the system. An example is a fan-coil system where the mediums are controlled by indoor-outdoor systems and the room thermostats only have to adjust the valves a small amount in the spaces where the loads do not change drastically.

Let us look at some typical problems and solutions with room thermostats. One of the first ones that come to mind is room thermostat sensitivity, or throttling range (TR). *Sensitivity* is defined as the amount of the output of the controller per unit change in the item being sensed. Throttling range is the change in sensed item to make a full change in output. As an example, with a pneumatic thermostat the sensitivity would be the pounds per degree output of the stat. If the sensitivity is too high (or the TR is too low), cycling can occur like a two-position controller. If the sensitivity is too low (or the TR is too high), the control becomes sluggish and drifts. The sensitivity that must be set is the one that gives the closest control without cycling. Some thermostats come from the factory with a fixed sensitivity, which is adequate for some applications. About 2½ lb/deg (or a TR of 4°) is a well-recognized standard and will work with most simple control jobs, but if there is any doubt about the need for adjustable sensitivity the more modern approach is to use thermostats that can be adjusted in the field, specially if the system being controlled is complicated by sequencing required on multiple devices.

Room thermostats as well as other controllers are available with what is called feedback systems. Feedback systems are sometimes compared to proportional with integral (PI) controllers, and although not strictly the same, both systems allow much closer control than straight proportional controls. Somctimes you will hear people use the term PID, which means proportional with integral and derivative control. The use of derivative in ordinary commercial control systems is not necessary and/or recommended. Thermostats with PI only are common today and give better control in 90% of the systems used in modern buildings. If the room thermostat is not able to control properly and it does not have PI built into it, a switch to the type that uses PI might be in order. The simpler the system, the less the need for PI control; the more complicated the system, the more the system might need PI control. An example of a simple system that would *not* need PI control is a room thermostat controlling a steam radiator in one room.

Room thermostats and other controllers that use air as the power to operate the system, as in the case of a pneumatic system, do not operate well if the air supplied to the thermostat is not clean and free of oil or water. The subject of water and oil in the air lines is covered in other chapters, but here we will look at what affect water and/or oil has on the thermostats and what to do if that situation occurs.

Most thermostats have built-in filters, and if the water and oil is discovered early in the maintenance program, there is the possibility that the thermostats can be disassembled by a qualified technician, repaired so the conditions that caused the thermostat to become contaminated with oil or water no longer exist, and returned to the system. It is not always necessary to throw away thermostats that have been contaminated with oil and/or water.

Pneumatic thermostats often have pivots and lid assemblies that can get out of alignment and cause the thermostat to not control at all or to stick at one position or the other. The lids that operate against the leakports that bleed air must line up with the bleed ports perfectly for the thermostat to function. These lids, and so on, can get out of alignment if they are tampered with by someone not familiar with their design. Lid alignment tools are available from some manufacturers but should only be used by qualified personnel.

Two-position and modulating electric thermostats can also have problems with pivots and lid assemblies, but their problems usually involve such things as low voltage being supplied to the system or wiring that is too small for the loads being controlled. They also have a tendency to arc at the contact points if the loads being controlled are too large for the thermostat being used. This applies to two-position as well as modulating thermostats where the arcing can occur at the wiper arms of the modulating devices. When there is arcing, there will be pitting of the contacts and eventual burn-out. The cause of the arcing must be determined and the situation corrected for the thermostats to provide troublefree long life. Electronic controllers, on the other hand, tend to be more trouble free, since they have little or no moving parts and the problems with electronic devices usually surface in the first few months of operation. Loose connections as well as overheated ambient temperatures at the thermostat location, along with stray excessive magnetic currents, are about the only conditions that can damage electronic controllers. Remember, usually the controllers themselves are not in the room (only a wire wound resistance or thermistor) with the controller in an equipment room, probably in a panel.

Another issue that needs to be discussed concerning room thermostats and sensors in the room is location of the thermostats and sensors. The location of *any* heat-producing device must be taken into account when trying to diagnose problem situations. An example is a lamp directly under a room thermostat. Excessive air motion, which can change fast, blowing on the thermostat will cause all kinds of problems. Furniture, desks, and so on, that block air movement at the thermostat can be a problem. Room stratification with different layers of air can give false information to the thermostat trying to control the space. If, for example, the room was designed for full laminar flow below the 7-ft level with the placement of the registers at that level and the room thermostat at the ceiling, there is no way the thermostat can control. Common sense will give the average engineer the solutions. Locations on outside walls or cold spots are to be avoided because thermal coupling to the wall will give errors.

If a room thermostat is used in a zone control situation, where one thermostat is controlling the air supplied to a *number* of rooms, it is impossible for that thermostat to know what is going on in the adjacent rooms. Therefore, to blame the controls for a problem of one room being hotter or colder that the other in a zone control situation is to read the system completely incorrectly. The problem is usually not the controls, but the *balance* of the system from one room to another, and the only way to correct the situation is to *balance* the delivery of the air to the zone.

One way to solve some of the problems with room controllers is through the use of recorders to see what is going on in the space 24 h a day, 7 days a week. Using a recorder, you can produce a chart of accurate records of what goes on in the space at various times, allowing you to analyze the problems and come up with solutions. Be sure and place the recorder *at the thermostat.* It may also be necessary to record other items simultaneously, such as the outside air temperature and the cubic feet per minute (cfm) of the airflow to the space. Some chart recorders that record humidity and temperature at the same time on the same chart, and some record the output air pressure from the controller and/or the voltage output of the controller, while recording the temperature. A little imagination will help you design your own system of recorders to check out control system problems.

Just about all of the problems and solutions discussed concerning thermostats also apply to room humidistats. The types of sensors used in room humidistats have been the subject of much discussion in other control books and will not be approached here.

The only additional item we will emphasize with room humidistats is the requirement that time be taken to be sure the humidistat is in a truly representative location, not where it will be affected by false humidity factors such as something that generates humidity in the atmosphere. As an example, a humidistat near plants that are watered frequently can give a false reading. Also, good maintenance procedures must be followed to keep the humidistat in good working order. The elements of a room humidistat tend to drift over time due to dirt buildup on the element and chemicals in the controlled space. Also, the response time is slower than a thermostat because a humidity sensing element must give up its moisture on its surface. The moisture in the space may be lower, but it is still on the element. Some materials used to sense humidity can take a set or stretch. This means that recalibration is in order more often than with room thermostats.

Unit Thermostats

The following discussion will apply to unit *thermostats and unit transmitters* (Fig. 2.5). They are interchangeable since almost all of the fac-

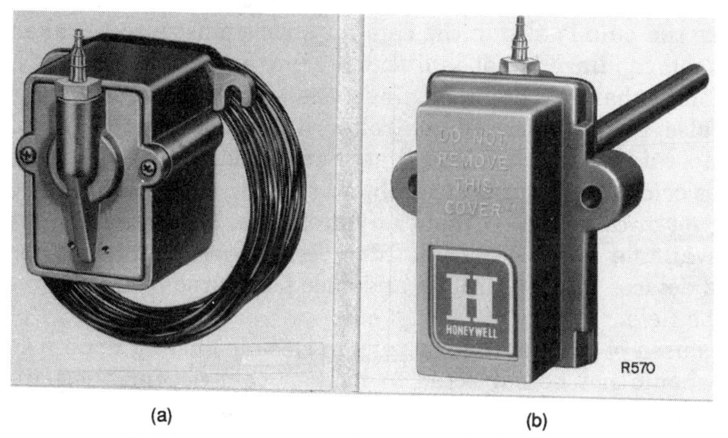

(a) (b)

Figure 2.5 Unit thermostats (rigid stem and capillary). (*Courtesy of Honeywell Inc.*)

tors that affect one, affect the other. For an explanation of the operation of transmitters, refer to the discussion on room thermostats.

Unit thermostats are the devices used in terminal units, mixing boxes, fan-coil units, and induction units. They typically measure return air and have set points and TRs like room thermostats. Remote bulb controllers are used in duct work, water lines, tanks, vessels, and so on, to control the air, water, steam, and so on, used to maintain the comfort conditions in the building. Their set points are matched to the hot or cold supply they control, and they have wider ranges of TR adjustment.

Transmitters are available in two basic configurations—capillary types and insertion types. Capillary types, as can be seen, enable the control engineer to place the operating head at one location and the sensing part at another location. Care must always be used in the mounting of the controller itself, since most pivot style controllers are sensitive to vibration and erratic control can result. This is particularly true if the controller is mounted on the duct work of a fan system. The vibration can be intermittent and not always apparent. The controller or transmitter should always be mounted where the dials and the face can be easily read and the adjustments made without difficulty. Gauges should be used where appropriate since it is difficult to adjust, set, and operate the controllers and/or transmitters without them. If the devices are hidden, they will be forgotten.

The types of capillaries available vary from company to company, but generally they come in different lengths and in different styles. The common ones have a sensitive bulb at the end that is placed in the medium stream, with a length of capillary between the bulb and the transmitter. The capillary is liquid filled, and extreme care must be exercised to be sure it is not bent too sharply, which can cause it to kink and prevent the system from working. Since the fluid that expands inside the bulb is also in the capillary, care must also be taken to see that the capillary is not laid across some areas that are much hotter or colder than the ambient air in the equipment rooms. This will give false readings at the controller if the mediums involved change temperatures from time to time. Some older capillaries were furnished as compensated or cross-ambient types that were able to be laid across hot or cold pipes without any problems. Today those types would be available only on special order. Remember, capillaries are liquid-filled devices that are fragile and once punctured cannot be repaired in the field.

The bulbs used on the capillaries vary in size, depending upon use. The bulbs should not be subjected to heat or cold beyond their intended design, or the bulbs and capillaries may rupture. Besides the standard bulb at the end of the capillary, there are what is usually referred to as averaging bulb capillaries. These capillaries are of a special nature that are designed to sense the average temperature along the path of the capillary. There is usually no bulb at the end of the capillary, and the design is such that the liquid inside the capillary responds to the average temperature along its path.

Another capillary that needs to be mentioned is one used with low limit stats that are designed to protect coils and other devices from freezing in excessively cold weather. These capillaries are strung out in the airstream in such a manner as to sense the coldest possible air entering a coil or other device and shut down the fan when a temperature is sensed that might damage the coil or other device. The freeze stat will operate if any point along its capillary senses a temperature below the set point.

There are many suggestions from control manufacturers and others as to the way to mount the capillaries of the freeze stats. Suffice it to say, the important thing is to mount the capillary where it will sense the critical areas of the airstream and protect the equipment from freezing. The expert control installer checks the stratification in the duct work before installing the capillary of the freeze stat to protect the equipment. The bottom line is that capillaries must be treated with respect; if they are, they will give years of service. Capillaries seldom lose their charge after they are installed and work properly unless disturbed.

PI and PID controllers have been discussed in the section on room thermostats, and the same principles apply to unit controllers discussed in this section. Controllers involving the equipment rooms of the capillary type and insertion type are, however, furnished with PI and PID control more than the room controllers. The wider TRs for stable control of fast discharge loops benefit more from PID. Remember that the closer the control is to the equipment room involving the air-handling unit and the water mediums, the less the room thermostats have to do for the final actions. Therefore, good PI and PID control is seen more often in the equipment rooms and less in the occupied rooms.

Like the room thermostats and humidistats, the unit controllers in most cases have the capability of sensitivity adjustment. Care must be taken not to readjust the sensitivity set up at the factory unless there is a need. Usually the factory set sensitivity is adequate for most applications.

The use of real one-pipe thermostats as unit thermostats is limited to situations where only a small actuator is being controlled. When a one-pipe device is encountered on a job, chances are it is a transmitter not a thermostat controlling anything. Transmitters (Fig. 2.6) are sometimes of the one-pipe variety for the simple reason that the only item that must be filled up by the branch line of the transmitter is the bellows of the receiver–controller. Thus, since the mediums being

Figure 2.6 One-pipe capillary transmitter. (*Courtesy of Johnson Controls Inc.*)

sensed usually change slowly and there are no long, large lines involved, the one-pipe transmitter makes sense. Remember, one-pipe devices are *slow* and sluggish.

Insertion thermostats and humidistats are nothing more than the unit thermostats and humidistats discussed above that do not have capillaries attached to them. Capillary thermostats can be used with wells in pipes as well, but generally speaking, in most cases it is more practical to use insertion thermostats and humidistats. There are no capillary humidistats available in today's market, so all humidistats used with duct work are of the insertion type. Insertion thermostats and humidistats have some limitations as to the size of the sensitive bulb that sticks into the duct work. Thus, if the area being sensed is large and the spot that is picked to sense is not near the side of the duct, a thermostat with a capillary is in order. With a humidistat there is no other choice than the insertion type.

When using immersion thermostats to sense fluid in a pipe, it is advisable to use a well for the thermostat bulb since the system will not have to be drained if the bulb has to be removed for any reason. Most manufacturers suggest that the well be filled with a special grease or oil to help transmit the temperature of the well in the fluid to the bulb of the thermostat. Wells are made of brass, stainless steel, as well as other metals. It is important to check and see that the material of the well is compatible with the fluids in the pipe so that the well will not be eaten away by the fluids. Wells should be placed in the pipe so as to sense the fluid being controlled, and they cannot do this if they are in an area of a pipe where the fluid may be stratified. The best place for a well is at a tee in the pipe so the fluid tends to ram against the well. Sometimes placing a well at an angle will help the well sense the temperature properly.

Some manufacturers have at times made dual bulb thermostats for special applications (Fig. 2.7). These have been used in indoor–outdoor control setup to sense the temperature of the medium indoors and have its set point adjusted by the outdoor temperature. So there is no one fixed control point on the same controller, it adjusts its set point continuously as the outdoor temperature changes. This is also done electronically by having resistance bulbs in series in the same circuit so the effect of the resistance of the circuit is the sum of the two resistances.

Duct humidistats are always of the insertion type as stated above, and the elements used are generally the types used in room humidistats. Since there is more space for sensing elements, however, there is some flexibility in that area. In this case, there are also some systems that can be used that could not be used in room controllers; in particular, wet bulb thermostats and Dewcell™ controllers, which will be discussed later in this chapter.

Capillary thermostats lend themselves nicely to the concept of master–submaster control, indoor–outdoor control. It is important to know how this concept works, as it is used extensively in all types of control schemes on various systems.

The basic idea is to vary the control point of one controller from another device that is sensing. For an example, outdoor temperature. The

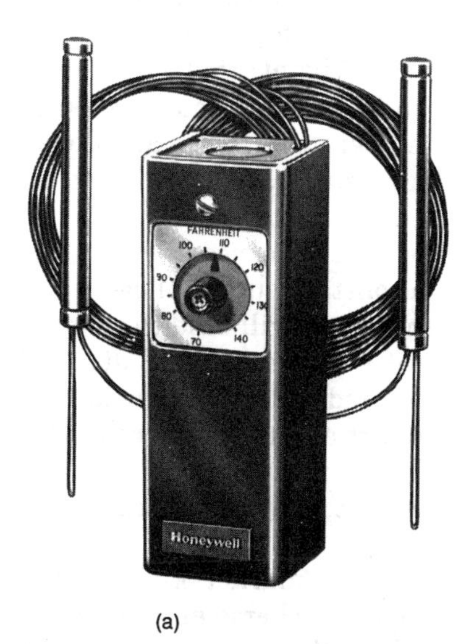

(a)

Figure 2.7 Dual bulb capillary thermostat. (*Courtesy of Honeywell Inc.*)

theory behind this is that the building system does not need to supply the same temperature of air when it is 60°F outside as when it is −20°F outside. So the controller that controls the temperature of the air going to the space in the duct work is constantly being varied in accordance with the outside temperature. It is as if a controller sensing outdoor temperature would have a long arm changing the dial of the controller doing the controlling in the duct work as the outdoor temperature changes. In this case the controller sensing the outdoor temperature is referred to as an open-loop system since it is *not* controlling the temperature it is sensing. The setting up of the master and submaster controllers can be difficult if a chart is not used but is simple if a chart is used and the instructions are followed. The dials on the controller only show what the setting was at the time of calibration.

The electronic equivalent of the master–submaster systems is different since the operator is able to change the authority of one controller over the other. The charts will show that the principles are the same, but the setup is different and involves setting the ranges and the authority the sensing elements have over the controller. A check with the instructions provided by the manufacturer will usually give all the data needed to make the adjustments.

Since humidity elements are not always stable and sometimes require constant recalibration, it may be necessary to control on the basis of wet bulb instead of humidity. *Wet bulb* is a measure of the humidity in the air stream, and humidity is a percentage of the amount of moisture in the air as compared to the amount it can hold without condensing out of the air. Wet-bulb control is accomplished by using a standard capillary-type thermostat (not humidistat) that has a wick or sock of clothlike material around the bulb of the thermostat to keep it wet at times and sense the wet-bulb temperature in the airstream. This temperature along with the dry-bulb temperature will give an accurate indication of the humidity in the airstream.

There are, like all things, problems with this concept. First, the wick must be kept wet at all times with a constant supply of clean water. This means that any impurities in the water will tend to clog the wick and give a false reading. Therefore it is suggested that distilled water be used. If that is not possible, deionized water is the next best solution. The wicks tend to pick dust and dirt and therefore require a lot of maintenance. The systems that do work are costly to install, and when maintenance problems are added to them, they do not seem to be worth it.

There is another proprietary device on the market known as a Dewcell™ or Dew Probe™ (Fig. 2.8), which uses a chemical sensitive to humidity that changes its resistance as the cell absorbs more or less moisture and that regulates the amount of heat applied to the cell.

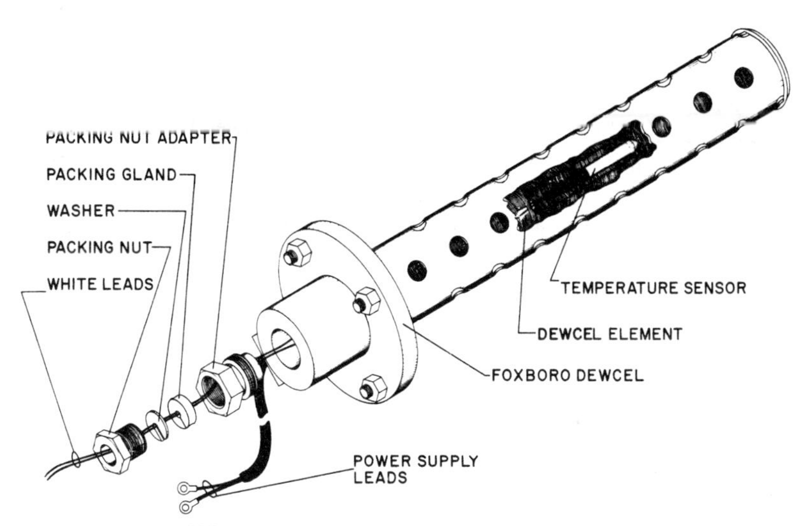

Figure 2.8 Dewcell™ for capillary thermostat. (*Courtesy of Johnson Controls Inc.*)

The temperature in the cell is indicative of the dew point of the air being sensed. Here again the first cost is higher than an ordinary humidistat and the usual HVAC system does not merit that degree of sophistication.

The location of the thermostat that is the master in the scheme of master–submaster controllers is important and should not be overlooked in the design of a good system. Examples of poor locations are in an outdoor air duct work, which will not do the sensing job when the fan is off, or where the sun will have an adverse affect on the bulb. Another one is in a pocket where true temperatures are never read. Some designers have found that locating the bulb above a window can give a false reading if the occupants open the window from time to time and let the warm air of the building affect the bulb. If a sunshield is used, consideration must be given to the type of sunshield and what is to be accomplished with it. There is no one answer to the problem, except to say that all factors need to be considered when selecting the location of the bulb of the master outdoor stat.

Sometimes the concept of master–submaster is used in an application that does not require one of the temperatures being sensed to be the outdoor temperature. This principle also works well for systems where the item being controlled (a chilled water coil as an example) is a great distance from the controller so there is a large time lag from controller to automatic valve and from coil to room sensor. Using the room controller as the master and a controller right at the coil as the submaster, the long time lag can be reduced and more precise control be affected. The time lag in some situations can be a real problem, and the above solution has been used effectively in many instances.

Electronic thermostats have some of the problems described above. In addition there are the problems of the correct type of wire to use between the sensor and the controller and the distances involved. Modern technology has just about eliminated most of these problems with such things as shielded wire and an additional wire that cancels out the resistance or capacitance problems that can develop, but the control engineers need to pay attention to these issues. Otherwise, problems can develop that are not always apparent to the untrained eye.

Special Thermostats

Most HVAC systems have some thermostats and humidistats for special applications, such as fire stats, smoke detectors, and freeze stats, as well as humidistats used for high-limit devices. Fire-stats, as they are called, are used to protect life and limb, and most codes require them in the system; but often the wrong kind is used. They must be the type that require a manual reset lever, so the reason for them taking action is determined *before* the fan is restarted. They must be sensing the temperature in the correct location, which is usually the return air to the fan. They must not be the type that has an adjustable set point so tampering will defeat their purpose. The standard temperature is usually 125°F, although there may be some cases where 135°F is acceptable. Smoke detectors are also special controllers used in some cases as required by code. In some cases the codes require that the fan shut down and alarms sound to alert the operators, but in all cases they should be of the types accepted by the local code authorities.

In some applications the manufacturers provide special thermostats mounted as an integral part of a valve or damper operator (Fig. 2.9). The competitive pressures of the market may have driven them to do this in areas where many of these devices are used, such as on fan-coil units where the control is specified to be from the return air to the unit. These controller–valve combinations may also be of the summer–winter type, requiring a main air changeover system that further complicates the controllers. The above items can also be furnished as controller–damper operator combinations.

There are also thermostats that sense surface temperatures that are used as a Strap-on in an attempt to sense the temperature of a fluid inside a pipe from outside the pipe. Care must be taken to see to it that the contact between the sensor of this type of thermostat and the pipe is clean and the contact is firm. In some cases it is advisable to wrap the strap-on thermostat so as the prevent ambient temperatures from affecting the sensing of the element.

Figure 2.9 Valve top capillary thermostat. (*Courtesy of Johnson Controls Inc.*)

If the requirement of the system is that devices controlling the temperatures and humidities be explosionproof, there are some special considerations that need to be mentioned. First, pneumatic controls are by their very nature explosionproof, since there is no electricity connected with most of the devices. It must be remembered, however, that the pneumatic-electric items such as pressure switches do have the potential of causing an explosion in a volatile atmosphere. They must be furnished in the explosionproof cases that meet the codes of the type of severe atmosphere involved. *Not all explosionproof devices meet all explosionproof requirements.* The explosionproof case of the device does not prevent the spark from causing an explosion; it merely *contains* the explosion in a confined area *if* it occurs. Some atmospheres are more volatile than others, so care must be taken when specifying and/or using explosionproof devices.

In summary, this chapter has provided a picture of the types of thermostats and humidistats available on the market, as well as a description of the uses of those devices and a glimpse into the problems and solutions involving thermostats and humidistats. Other chapters will expand on some of the items from this chapter as they apply to the systems being controlled, so that combining what has been learned in this chapter with information in others will give you a better feel for the total picture as it applies to controllers.

Chapter

3

Dampers and Damper Motors

This chapter will discuss dampers, and damper motors, including the problems involved in sizing dampers and some misconceptions about dampers. It will discuss different types of dampers, including pneumatic, electric, and so-called electronic damper motors, and describe the differences among automatic dampers, smoke dampers, fire dampers, and balancing dampers. The chapter also covers the manufacture of dampers, along with coatings that are sometimes used. A portion of the chapter is devoted to dampers manufactured by the air-handling unit manufacturer.

The chapter has information on sizing and mounting of damper motors and the use of pilot positioners. It covers problems in the mounting of motors on air-handling unit dampers.

The chapter will give you some direction on the selections of dampers and damper motors and clear up misconceptions in the area of sizing dampers.

Dampers

Automatic dampers have various classifications; the two classifications used in the HVAC field are the parallel blade (Fig. 3.1) and the opposed blade (Fig. 3.2). The parallel blade types were the first ones used in the industry, and their use could be considered the beginning of industry's entry into damper blade science.

We begin this discussion by describing the process of controlling air with a damper. In other words, we will describe what happens in the duct work when a damper begins to close or open. To start, keep in mind that the fluid (air that is) being controlled, or what we are trying to control, can be an incompressible fluid at pressures below 12 in. of water. Above that, compressibility should be considered. Furthermore, all fan laws and damper characteristic curves are based upon an

Figure 3.1 Parallel bladed damper. (*Courtesy of Johnson Controls Inc.*)

incompressible fluid flow. Gases (air) can *bend* so that the volume will not be affected, and, as such, it may not be controlled at all. Air can easily stratify in the duct work and usually does when we do not want it to. Therefore, in effect a damper can be considered a poor control device at best. At the same time, dampers *can be* as good at controlling as valves, provided they are sized properly.

As far as stratification is concerned, water can stratify with valve control, but the results are not as dramatic. We must generally use every engineering technique at our disposal to accomplish our ends when it comes to controlling the air streams with a damper. Having said this, a discussion of parallel blade dampers versus opposed blade dampers is in order.

Parallel Blade versus Opposed Blade Dampers

Parallel blade dampers tend to bend the air during the first few degrees of rotation as they go from full open to closed, and they do very little controlling in the first 20%–30% of movement. They bend the air streams, rather than modulate them. There are many instances, how-

Figure 3.2 Opposed bladed damper. (*Courtesy of Johnson Controls Inc.*)

ever, when that bending is of value, as can be seen from the Fig. 3.3. The bending is appropriate, for example, when we are trying to mix the airstreams.

Opposed blade dampers are usually used where better control of the airstreams are desired and we want to prevent large amounts of stratification in the duct work. This is especially true when face and bypass dampers are involved and we are trying to wipe the face of the coil with the same volume of air. A careful study of the actions of an opposed blade damper will reveal, however, that the linkage of some of the older style dampers, where the seals had to overlap, must be special since the closure rate cannot be constant as the center shaft of every other blade turns. Newer styles of dampers that use opposed blades have seals that butt together, making the above kind of rotation unnecessary.

The control engineer and those who use automatic parallel and opposed blade dampers need to study characteristic curves so as to understand flow characteristics (Fig. 3.4) and to see why certain dampers are used in certain applications. The bottom line is that the control

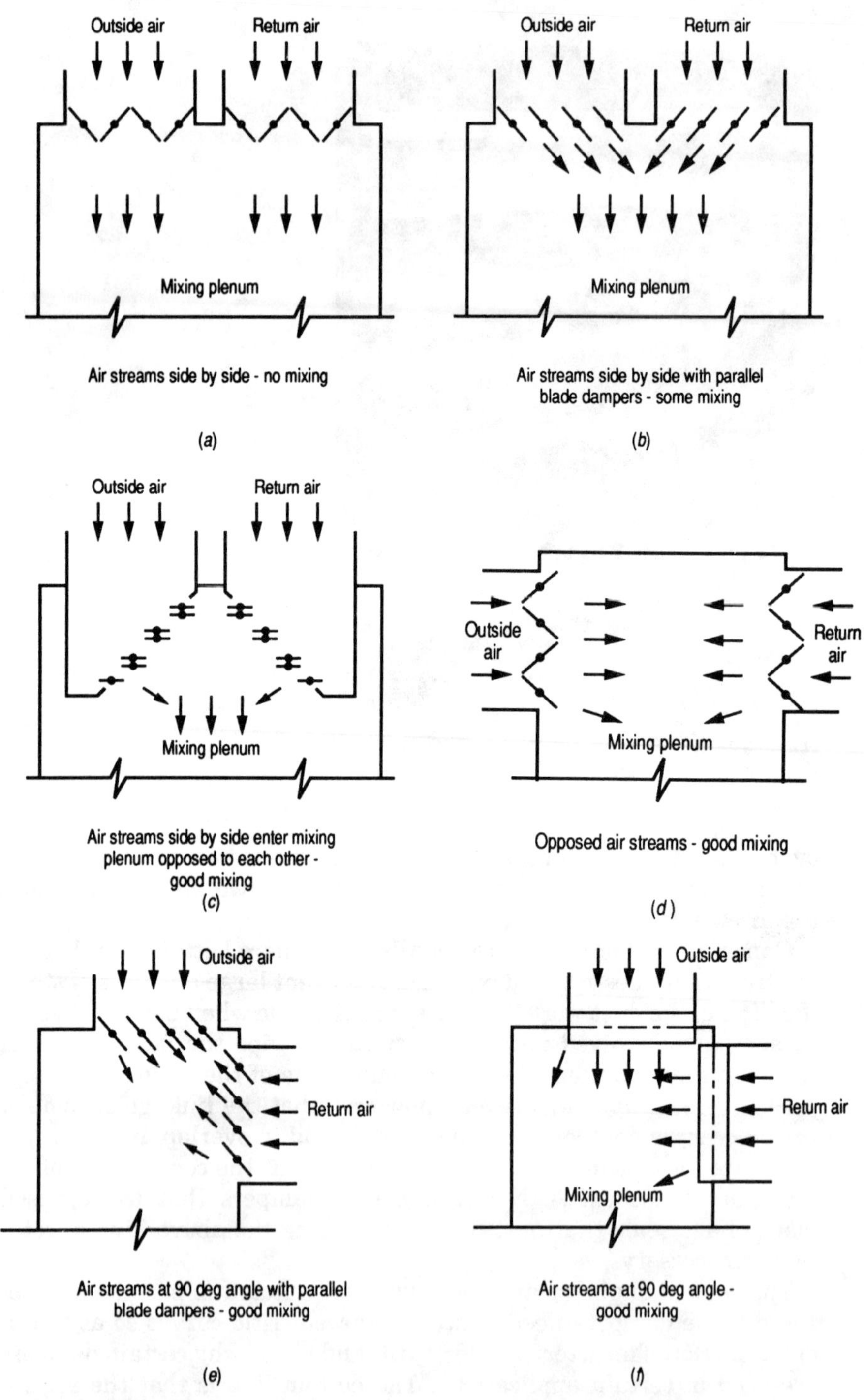

Figure 3.3 Typical various mixing box arrangements. (*Courtesy of ASHRAE.*)

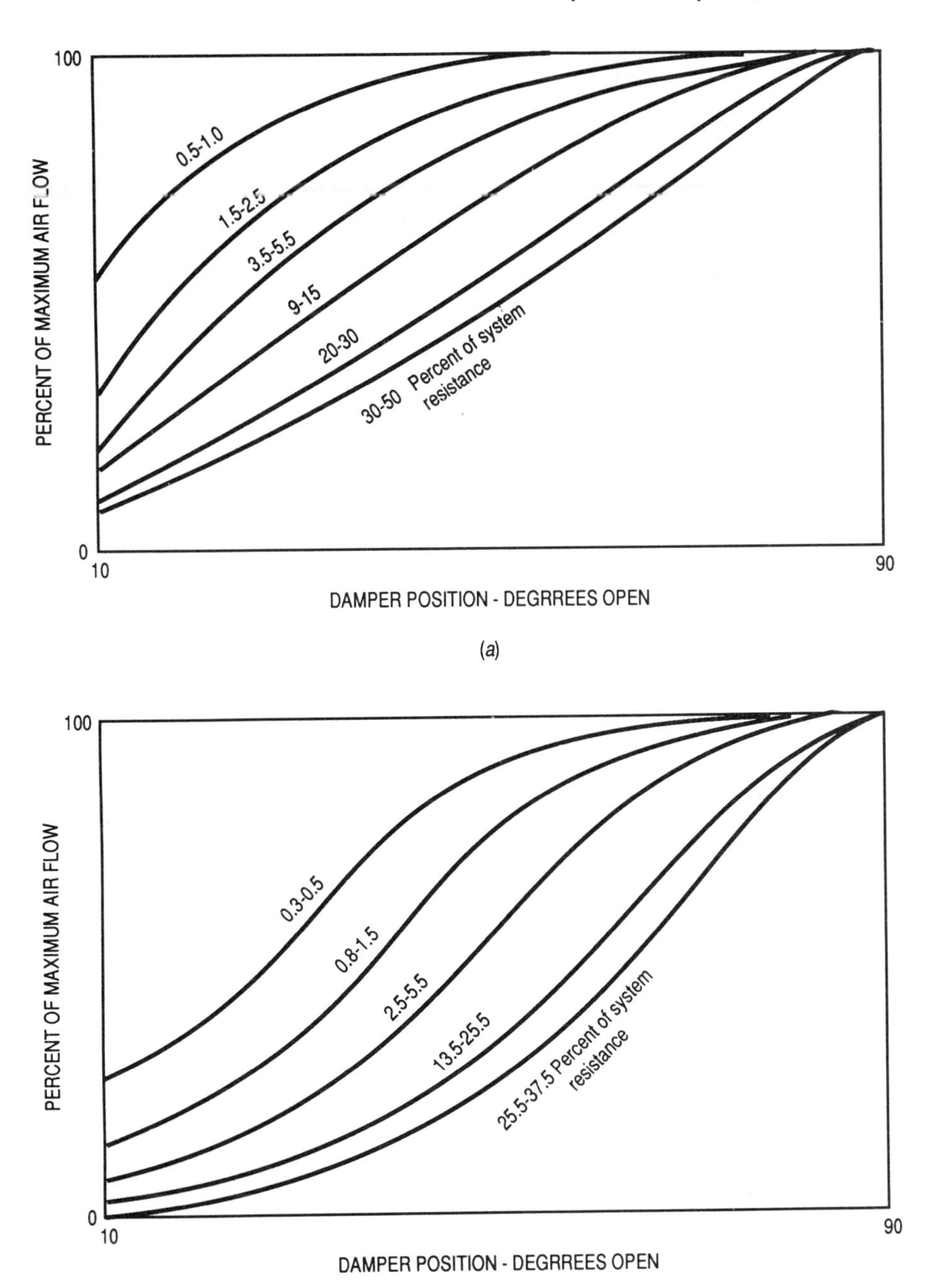

Figure 3.4 Damper flow characteristics; *(a)* parallel blades, *(b)* opposed blades. *(Courtesy of ASHRAE.)*

system would like to have a linear relationship between the movement of the dampers and the flow rate of the airstream. This is an ideal situation that has never been accomplished, and the control industry is trying to meet that goal.

Some dampers are not used for control or to maintain comfort but are used for safety. These are the fire and smoke dampers used in HVAC systems.

Fire and smoke dampers

Fire dampers are put in the duct work to stop the spread of fire and confine the fires to one area of the system. As such, they need to be of a rugged material that can withstand the heat generated in a fire. They are seldom if ever pivoted in the middle like an automatic damper, although they can be similar. They are almost always held in the open position by a linkage system that can be fused and is designed to melt and close when the temperature reaches about 165°F. The closure is accomplished by springs or weights, and the dampers must never be of the type that can be easily opened after they close during a fire situation. In all cases, the dampers must meet the requirements of such organizations as the National Fire Protection Association (NFPA) and Underwriters Laboratories (UL). The locations of the dampers is clearly spelled out in most codes that apply to a particular type of system.

Smoke dampers (Fig. 3.5), on the other hand, are not required by all codes. They are manufactured by vendors that supply automatic dampers and in some cases are used as both control dampers and smoke dampers. With these dampers, as with fire dampers, there are codes that apply, but they usually are not as stringent as those used with fire dampers. From their name, we see that smoke dampers are used to stop the propagation of smoke and the resulting panic in the event of a fire. Generally, they are involved with the central control and monitoring systems.

The material requirements of smoke dampers are also not as stringent as with fire dampers, but the tightness of the seals is of prime importance. Smoke dampers can be operated by automatic damper motors or can be spring loaded to close as a result of a signal from a remote controller. The key to smoke dampers is the tightness that is available from the manufacturer. Tightness is also important on ordinary control dampers, since the tighter the seals of the dampers, the less the leakage and the better the ability to control air.

Some manufacturers rate their dampers on percent leakage with tests that are certified to back their claims. The problem with the percentage method of rating a damper is that a 1% leakage factor does

Figure 3.5 Smoke damper. (*Courtesy of Ruskin Corp.*)

not say what the actual cfm of leakage will be. It could be as low as 10 cfm or as high as 10,000 cfm. It has been suggested that the leakage be specified not only in terms of percent but also of the static pressure drop through the damper, the size of the damper, and the total cfm flow in the wide open position. Air Movement and Control Association (AMCA) Publication 511 presents the damper rating based upon volume per square feet through a damper's face (Fig. 3.6) versus measured pressure differential across the damper. Remember, however, that the costs for low-leakage dampers must be weighed against the need for the more expensive dampers, and the applications must be checked so that we do not, as an example, use an expensive damper for the hood exhaust of a restaurant.

Manual dampers

Some mention needs to be made on manual dampers, even though they are not technically used for control. Manual dampers are an important part of the overall system, in particular the ability to balance the system and assist the control system to work better.

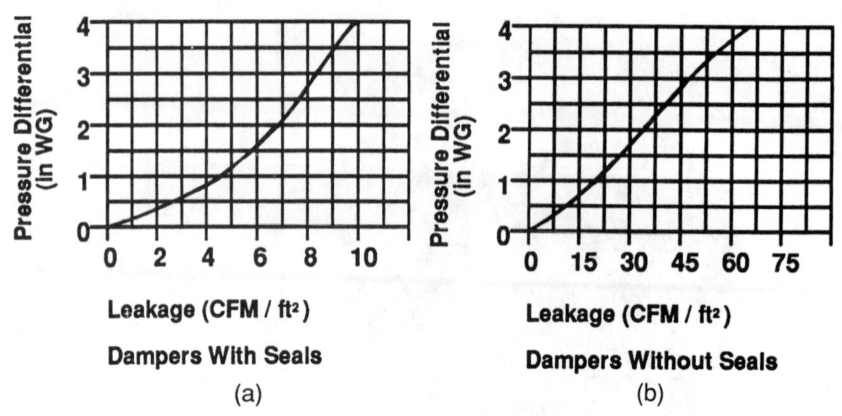

Figure 3.6 Typical damper leakage charts. (*Courtesy of Johnson Controls Inc.*)

They are used primarily to balance constant volume systems so the correct amount of air is distributed to the various places at the proper rate. They can be of the splitter type or the usual closure type. Splitter dampers try to split the air and redirect it to various sections of the duct work. Balancing dampers try to apportion the air to the various sections of duct work so the correct amount gets to the correct place at all times.

Static control dampers

Static control dampers are different from the usual control dampers in that they maintain the static pressure in the duct work based upon the action from a static pressure controller. Since static pressure is usually difficult to control in a system, static control dampers need to be the best available for their applications. First, theoretically they never have to close tightly; thus there is no concern for the seals of a static control damper. Second, there is no need to have overlapping lips on the blades since the damper never closes, and the lips might create noise when the blades are near the closure point in a high static pressure situation. A static pressure control damper should have blades as small as possible to give precise and accurate control. The use of parallel bladed or opposed bladed dampers is not a consideration with static dampers, as they are usually in a area where stratification is of no consequence.

Sizing of automatic dampers

The sizing and selection of automatic dampers is probably the most important part of this chapter, since more problems have arisen because of improper damper sizing and type selection than anything in the control industry (except perhaps the improper sizing of valves, which will be treated later). In order for a damper to control the amount of air going through it, there must be a *pressure drop* and that pressure drop must be large enough to allow the damper *to control, not just bend, the air.*

The biggest mistake made by engineers is sizing the damper on the basis of the size of the duct work and/or the opening in the wall of a building. The size of the duct work should have nothing to do with the size of the damper. The *cfm and the pressure drop allowed* are the important factors in sizing dampers for full flow.

The objective of modulating dampers is normally to control air temperature or pressure in a feedback control loop. Problems to be avoided are hunting control and excess restriction to airflow. Both problems are avoided when the volume of air throttled is proportional to the control signal driving the actuator throughout the full stroke. This gives a constant gain as seen by the control loop for maximum stability. It also gives a constant total resistance in parallel paths such as face and bypass dampers or outside air–return air, OA–RA, air mixing dampers. In these applications nonlinear operating characteristics would starve a fan suction.

The resistance in series with a damper is the largest factor in selecting the type. Less resistance uses parallel bladed, and more resistance uses opposed bladed to give the most linear operating characteristics as the dampers reduce flow. Sizing down is necessary when there are parallel paths and the path with the least resistance needs a smaller damper to match the resistance in both paths and to avoid a short circuit in the bypass path. Face and bypass dampers and recirculating versus OA exhaust air paths are examples of this. Typically, the bypass dampers and the recirculating dampers are the low-resistance paths that need to have the dampers sized down to balance alternate paths.

In some cases sizing down could affect the horsepower of the fans and other equipment, but if good control is important, proper sizing of dampers is a must. Basically, the static pressure drop across a damper in the wide open position should be known in order to give the best alpha ratio (which is the pressure drop of the system divided by the pressure drop of the damper). An alpha ratio in the vicinity of 4 will usually give the most linear curve results for the opposed blade damper, whereas an alpha ratio of 10 will give the most linear results for a parallel bladed damper.

Another way to look at damper sizing is to use charts to calculate the best ratio of damper to duct size for a particular set of pressure drops and face velocities. An understanding of the relationship of damper position and flow of mediums through the duct work it is important to use the chart method of sizing dampers.

The flow versus position characteristic of a damper in a system is dependent upon the damper design with the inherent characteristics of the damper, the amount of flow resistance in series with the damper, and the percentage opening of the damper. The damper blade design and other design features will give the damper the inherent characteristics mentioned above, but the shape of the operating characteristic curves will also depend upon the flow resistance in series with the damper. Examples are outside air stationary louvers and heating or cooling coils in series with the damper. In almost all applications when we use a damper, with the exception of the case when we try to control static pressures by the movements of dampers, we are looking for a straight-line curve. That is, we want a curve where the flow is exactly proportional to the damper position (50% open equals 50% flow as an example). Since the control of static pressure is a square root function and not in itself linear, the curve for proper control is not a linear one. For this discussion, however, we will assume that the straight-line curve is what we want. The flow resistance in series with a damper can greatly modify the characteristic curves of both the parallel blade and opposed blade dampers. The effect of a series resistance is to modify the flow characteristics of the damper, with the greater the resistance, the greater the modification being the rule.

The application of dampers in a system is best analyzed by the use charts available from the major control manufacturers. First, the system needs to be sketched out with all points of constant flow and variable flow identified. Next, we need to determine the lineup with the greatest resistance to flow. After that, using the information as to the cfm of the system, the square feet of the duct work, and the fixed resistances such as the outdoor louvers, the approach velocity of the air at the louver can be obtained. Then, given a fixed resistance of the outdoor louver determined from the manufacturer, a damper can be selected from the appropriate charts, depending upon whether a parallel blade or an opposed bladed damper is desired.

The charts and this method of sizing will let you select the best damper for the job whether it be a parallel blade or an opposed blade damper. The technique used in the sizing of dampers shows that the sizes used for *duct work* in the modern HVAC system should have nothing to do with the sizing of dampers for control of the system. Generally, the use of the various systems to size dampers will almost always present a damper that is *smaller* than the duct where it is to be installed.

The sizing of some dampers is limited by the applications such as in the case of face and bypass dampers. Here the coil manufacturer limits the velocity of the air going through the coil, and the face and bypass dampers when placed at the coil usually have to be sized to cover the face of the coil. They can, however, be moved up or down the duct work and their size changed to give a reasonable pressure drop if there are no space limitations involved. If the face and bypass dampers need to be next to the coil, the best dampers available and ones with small blades should be used to get good control. The bypass, remember, does not need to be as big as the face damper and should be sized to have the same full flow drop as the face damper.

Fresh air and return air dampers need to be sized as suggested above and to be installed to try to prevent stratification as much as possible as shown in some of the figures. The object is to get the airstreams to mix well before they arrive at the fan, since even after going through the fan [if it is a double inlet double width (DIDW)] the air can and will remain stratified. Other tricks can be tried to be sure the airstreams mix, and a little thought will suggest some. An air balance diagram will help you understand what is going on in the system and enable the engineer to size the dampers with the proper pressure drops and make the dampers control.

Dampers can be made from various materials, but in general they are made from steel with parts of the frames and bracing made of aluminum. The steel is usually coated with a galvanized material to help prevent corrosion from foreign items in the airstream. All types of coatings are available for special circumstances, including for salt spray air and to preserve the dampers when the airstream is acidic. For example, Heresite™ can be used on the blades when there is a highly corrosive atmosphere.

The types of bearings are also an important factor when looking at the quality of a damper. Each manufacturer has a preference and in the past most used oilless types made of oil-impregnated sintered bronze and similar materials. Today, the bearings are generally made of synthetic materials such as neoprene nylon and similar compositional materials. There is seldom if ever a need to oil the damper bearings; as a matter of fact the oil will attract dirt to affect the action of the damper blades. One caution that needs to be mentioned when dealing with damper bearings: When the blades are *vertical,* the bearings on the bottom must be the type that will take thrust forces. Also, special side seals may be required in these cases. Therefore, if the blades are mounted in a vertical position, a damper cannot be moved from one orientation to another.

The interlinking of dampers, such as OA and RA dampers as well as face and bypass dampers, requires some discussion since many mistakes are made by installers and others. Dampers can be interlinked either by using the damper shafts on the outside of the duct work or by interlinking the blades internally. When the blades are interlinked, care in observing the damper actions is needed, since if they are improperly interlinked, binding can take place and affect the control cycles. As an example, note that some older opposed blade type dampers cannot be interlinked with parallel blade dampers since the linkage of the opposed blade dampers is such that the blades do not rotate at a constant speed to enable them to open and close. The installer and engineer should follow the instructions provided by the manufacturer in interlinking dampers.

When external interlinking is used, the biggest problems come from crank arms that are too long or too short and crank arm pins that slip on the damper shaft. In some cases, the shafts should be drilled and a pin used to replace the set screw. Most manufacturers place a slot on the end of the damper shaft to indicate the position of the damper blade inside the duct work. If there is no slot on the end of the shaft, the installer should make one before the damper is installed in the duct work.

The use of long rods to tie dampers together from crank arm to crank arm on the external shafts is discouraged and additional damper motors should be used instead. There is always the possibility that in the case of long rods play in the linkage can affect the controls.

Whenever the dampers do not operate smoothly, there is usually a good reason, and most often it is because the dampers were not installed properly. They must not be forced into the duct work, and, after being installed, if they cannot be operated by hand when the fan is off, they will never operate smoothly when the fan is on and the damper motors are attached.

Damper manufacturers are constantly trying to come up with the perfect damper, and the emphasis lately has been on the tightness of the dampers so leakage is minimized. Various types of seals are being used on modern dampers, and one of the problems involves the fact that the tighter the seal, the more power required to open the damper once it is closed. The seals being used are of synthetic plastics and other polymers that will resist cracking and drying out and at the same time do the job of sealing effectively.

Sometimes the tighter the seal, the more difficult the operation of the damper. The problem is particularly evident with opposed type dampers since they require the operation of one blade of a pair as the damper begins to open. Leakage (or the lack of it) is the most important subject for damper experts, and dampers are rated on percentage leakage with the pressure drops to allow the manufacturer to present curves to the engineers and end users.

Since the air-handling unit manufacturers can provide mixing boxes and units with face and bypass dampers, they also provide air-handling units with some of the dampers for those units. The unit manufacturers sometimes try to cut costs in the area of automatic dampers, and the results can generate a control problem. In most cases, the unit manufacturer is not a damper expert, and the dampers show it in the way they are installed in the unit and the way they are manufactured. This has been particularly true of the zone dampers in a multizone unit, where leakage causes all kinds of problems. Hot air is leaking into the zone when the hot portion of the zone damper is supposed to be closed. In the past, the unit manufacturer has resisted the idea of the control contractor furnishing the zone dampers on a multizone unit. Not enough emphasis is placed on the types of bearings, types of seals, size of blades, and all of the other features the damper manufacturer normally tries to incorporate into the damper line.

Damper Motors

Damper motors are not motors in the true sense of the word, but the term has been used and has stuck with the industry. *Damper motors* are the devices (both pneumatic and electric) that operate the dampers in a system from the control signal of a device.

Pneumatic damper motors

Pneumatic damper motors are (Fig. 3.7) in almost all cases operated by air pressure pushing against a spring with an airtight diaphragm

Figure 3.7 Pneumatic piston damper motor. (*Courtesy of Johnson Controls Inc.*)

to seal off the chamber and allow the motor to work. Damper motors of this type are usually rated in pounds of force they can do on the device (damper) being operated. The length of the crank arm on the damper shaft changes the force being presented to torque, so the sizing of the damper motor depends upon the force that can be derived from the motor as well as the length of the crank arm on the damper. All control manufacturers provide charts that tell the novice how to size damper motors for a particular damper. The problem comes when the force required to operate the damper is not known because there is insufficient information about the static pressure, cfm, and pressure drop across the damper. The torque required to operate the damper *must be known* before the damper motor can be sized. There are some published standards that can be used most of the time, but remember that dampers that are not maintained can, over time, require more and more torque to operate them. Care must also be exercised when using more than one damper motor on a single damper. It is possible to get two or more damper motors fighting each other on a single damper if care is not exercised in the adjustment of the damper motors.

Pneumatic damper motors have what is known as nominal spring ranges over which they operate. Some examples are 5–10 psi, 8–13 psi, and 3–13 psi. Notice the word *nominal* is used to describe these spring ranges. This means that with no external forces involved (e.g., on the bench in a lab), the damper motor will begin to operate at 5 psi and finish its stroke at 10 psi.

The spring used for this 5–10-psi damper motor is a precision spring manufactured to close tolerances, but it will only operate within that range when no other forces are on the damper motor. Thus, to say that a damper motor is not operating correctly when it does not operate within its standard spring range is to misunderstand the facts. Any damper load to be overcome requires pressure outside the operating range. Remember that in the power stroke (with the air being supplied to the motor) it is possible to increase the capability of the motor by increasing the available air pressure, but on the return stroke, the only thing available is the strength of the spring in the damper motor. Therefore, care must be used when specifying which portion of the stroke (power or return) has to do the most work.

Most of the damper motors have stop screws built into the motor that allow minimal adjustment on the spring range by changing the span of the spring slightly. If the system is a simple one that involves a controller controlling one damper motor with no sequencing required, it is not necessary for the motor to operate at exactly the spring range specified on the motor itself, since calibration adjustments can compensate for the shift in the spring range caused by the external forces on the damper motor. The problem arises when there is complicated sequencing involved in the control cycle. An example is when the cycle requires that a damper operate before a valve opens. In this case a shift in the spring range of either the damper or the valve can cause the control system to malfunction and not give the desired results in the space. The only way we can determine if the external forces (friction, pressure drop, etc.) are affecting the sequencing of the control cycle is an on-the-job evaluation of the results. If the sequencing is not that critical and the end results in the cycle seem to show that the temperatures, and so on, are being maintained, then nothing further need be done. If the sequencing is critical, however, an alternative is the addition of pilot positioners to the damper motors and valves.

The operation of pilot positioners is described in other books and are mentioned here to warn you that they are not a cure all and should not be used unless the system warrants their use. They are used only *when complicated sequencing is required and when the external forces (friction, etc.) can be excessive and variable.* There are some pitfalls to using pilot positioners in that they require adjustments when the system is operating and they can get out of adjustment even after the system has been operating for some time. Thus, periodic maintenance is required. There is also a significant problem when there is more than one damper motor on a single damper with pilot positioners involved. That is the problem of getting the motors out of sync with each other. There is a way to tie the dampers together and use one pilot positioner that feeds all motors.

Pneumatic damper motors operate well under almost all ambient conditions, but there are times when engineers have used them under circumstances that cause them to fail. Some damper motors have composition diaphragms and those diaphragms do have some temperature limits. Therefore, it is wise to check the manufacturer's literature to determine the limits. Pneumatic damper motors have been known to operate outdoors under varying conditions of cold and heat, but they are not recommended for all areas of the country. Corrosive atmospheres such as near the seashore can be a problem with some metals used in damper motors.

Piston damper motors, as they are sometimes called, exert their force with a push–pull motion, and that movement must in most cases be turned into a rotating motion to operate a damper. This is done with crank arms and linkages. One of the best places to look for problems with a control system is in the linkages and crank arms that are out of adjustment and may be slipping on a damper shaft. These linkages and crank arms can also be binding if they are not operating in the same plane. The dampers should operate freely even without the damper motor hooked up, as we cannot expect the motor to make up for poorly adjusted linkage.

Two kinds of brackets are used for pneumatic damper motors on automatic dampers: those used where the damper motor is mounted outside the duct work and those used where the damper motor is within the confines of the duct work. Sometimes, for example, the blade mounted motors must be used since there is no way the damper shaft can be extended outside the opening. This would be the case when a damper is installed in a masonry opening in the building. It is wise, however, always to try and use the system where the damper motors are installed outside the airstream, as the dust and contaminants in

Figure 3.8 Frame mounted piston damper motor. (*Courtesy of Johnson Controls Inc.*)

the airstream can affect the operation of the damper motor in time and the out of sight, out of mind principle applies here as far as maintenance of the damper motors is concerned.

The brackets and linkages used for the two types of mounting arrangements are altogether different and not *interchangeable*. One of the biggest problems with the brackets and linkages mounted inside the duct work is the possibility of bending the blades if the motors (where more than one is used) are out of sync, since when this type of arrangement is used the motors operate by pushing and pulling on the blades.

Damper motors are generally made from a die-cast material that has proven over the years to be the best all-around material for that application. Some modern damper motors are made of plastic or extruded aluminum. Each of these materials has its place, but some are more adaptable to certain ambient conditions and the specifications of the motor should be adhered to.

The diaphragms of the damper motors are composed of a material designed to last years, but even then they can fail due to unusual ambient temperatures or foreign matter in the control air being fed to the motor. Other than that, piston pneumatic damper motors require little servicing except to check and see that shaft bearing is free of dust and corrosion and to add an occasional lubricant such as a graphite to the shaft to ensure a smooth operation of the piston. Do not use liquid lubricants, as they tend to attract dust and cause more problems.

There are some special damper motors of the pneumatic type that need to be mentioned here. These are the "two-spring" damper motors used in special applications such as with unit ventilators in schools. The two-spring types are used to allow for a "hesitation" in the stroke of the damper motor as the air pressure increases or decreases. The reason for this is found in an examination of the cycle involved with the sequencing that is required.

Two other special dampers require some discussion: the inlet vanes of a centrifugal chiller and the inlet vanes of a large fan (Fig. 3.9). The brackets and crank arms as well as the linkages used with these two applications are special and always almost require the use of large damper motors and pilot positioners on the damper motors. The torque can be excessive for the average damper motors, so care must be used in specifying or using damper motors under these circumstances.

Electric damper motors

Electric damper motors (Fig. 3.10) are closer to being motors in the true sense of the word, since they do have special induction motors that rotate to provide the motive power to operate the dampers and valves. In this case the rotational motion is converted to a push–pull

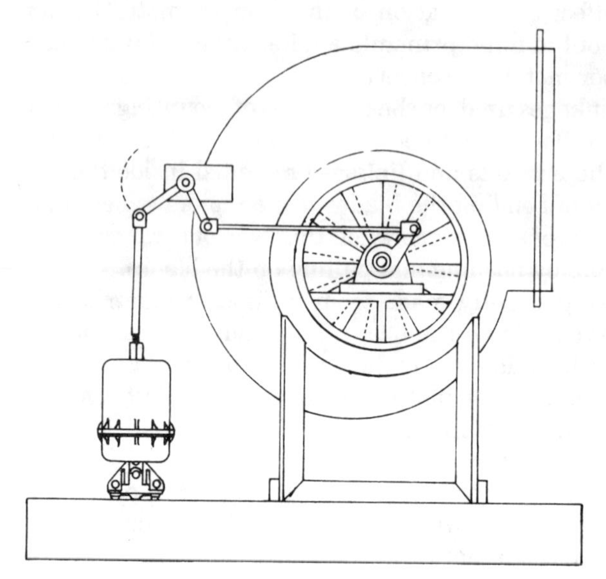

Figure 3.9 Damper motor on fan inlet vanes. (*Courtesy of Johnson Controls Inc.*)

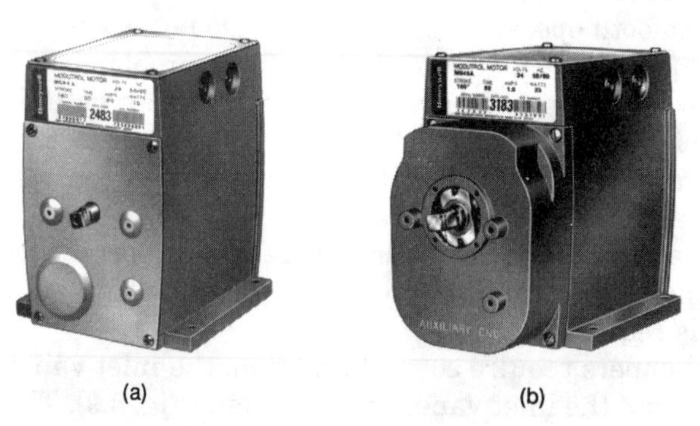

(a) (b)

Figure 3.10 Electric damper motors. (*Courtesy of Honeywell Inc.*)

motion by the use of crank arms and universal joints as well as rack and pinion gears to allow for this conversion. Some newer electric and electronic motors are designed to mount directly to the damper shaft and valve stem requiring little or no linkage connections. Some original damper motors were bidirectional alternating current (ac) motors that rotated either direction depending upon the way the ac voltage was impressed on the windings of the motor. All of the original and most of the damper motors today are of the low-voltage (24 volts, alternating current, VAC) style used in modern low-voltage control systems.

Before the advent of the spring return styles, the first low-voltage damper motors required that the signal be reversed in order to get the damper motor to change directions. The first innovation that was developed was called the Series 90(TM) control system. Essentially, it was a three-wire system that allowed the motors to modulate through the use of a feedback potentiometer and balancing relay. The result was as close to a modulating system as possible with electric controls. The system allowed for all of the additional auxiliary devices such as heating–cooling switches, low-limit devices, and reversing switches. The addition of large coiled springs operating against the normal movement of the motors added the spring-return concept and allowed the initial motors to compete for motors with the specifications for spring–return capabilities, which was a natural with pneumatic actuators.

Some newer actuators incorporate a battery for return to a minimum position. The life of the battery, however, is limited and needs to be checked periodically. An examination of Fig. 3.11 will show that the controllers that were used required a wiper potentiometer and the motors had a comparable potentiometer to balance the system and allow that actuator or motor to stop at a predetermined position.

Some control companies developed a less expensive style of motor that does not require a potentiometer but uses a unidirectional motor that is stopped when the current is interrupted. With that type of motor, a single-pole–double-throw type of controller that has a center position when there is no contact with either side of the switch is used. When the temperature moves either up or down, the controller makes contact with one side or another and drives the motor in the direction that compensates and either raises or lowers the medium tempera-

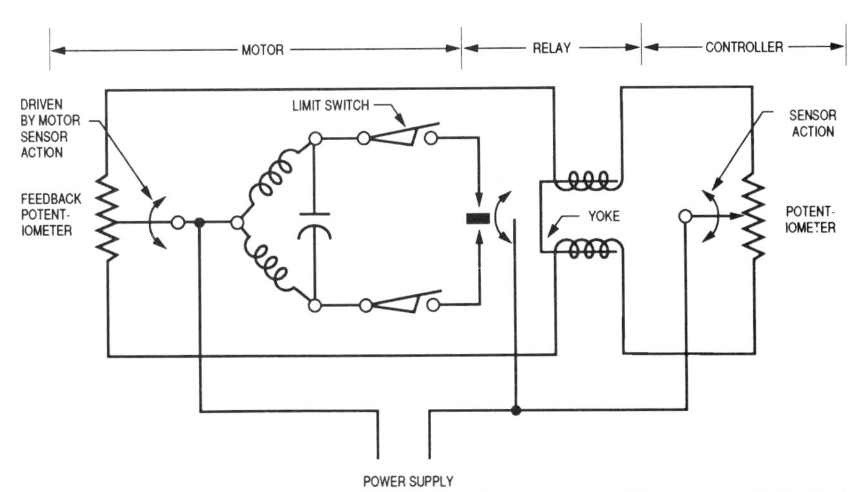

Figure 3.11 Typical wiring diagram of electric damper motor. (*Courtesy of ASHRAE.*)

ture. When this happens, the controller senses it and brings the switch to the center off position and stops the motor movement. When the temperature changes again, the process is repeated. This is sometimes called *floating control.* Since the motors do not require a potentiometer for the feedback circuit, they are less expensive and the controllers are also less expensive to manufacture.

Two-position electric motors for valves and dampers can be the motorized type as well as the solenoid type. Spring-return types, both two position and modulating, require a stronger motor than the ones without spring return. This is because the motor must wind up the spring on the power stroke so when the power is off the spring can return the actuator to the normal position. This makes the spring-return type more expensive for the same application. Sometimes relays are added that can operate a nonspring-return type and supply current to the motor when a fan shuts down and return the actuator to a normal position. This requires that power be available from a nonfail source, and will not work if power to the whole system goes down.

Electric damper motors are always of the type where the windings on the motor are immersed in oil for cooling purposes. Since the oil can become very hot when the motor is operating, it can be dangerous to open the seals of an electric damper motor. That maintenance needs to be left to the experts. Some newer types use self-lubricating materials in their gear trains to eliminate the need for oil-immersed gears. These types are used primarily on light duty actuators.

Electric damper motors are rated on the basis of the speed at which they operate in going from one position to another and the gear boxes involved with that speed. Modern electronic damper motors are an offshoot of the electric damper motors. The big difference is that the solid-state control boards that are a part of the electric damper motors determine the speed, the action, and many other functions that were previously fixed at the time of manufacturer. In other words, the basic body of the damper motor is the same and the electronic solid-state boards can be changed to meet the requirements of the control cycle. The changes can even be made in the field just by changing the solid-state electronic board in the damper motor. The modern electronic damper motors are more flexible and more reliable.

Some problems with electric and electronic damper and valve motors (they are the same device) involved applications in which the torque and forces required to operate the devices were too large for the motors, causing the motors to fail by stripping the gears. These gears were generally made of a composition material that gave long life if not overstressed. When those gears failed, the motors were so much junk. Using the motors in an atmosphere that was higher in ambient temperature than specified also caused many of the failures attributed to electric and electronic damper motors. Crank arms that slip off the damper motor shafts can be a problem if care is not exercised in the tightening of the set screws against the proper place on the damper motor shaft. Improperly sized transformers that supply the 24-V power to the control system can also cause premature failures of the motors. Like the pneumatic motors, the improper use of crank arms and connecting rods that cause binding can cause failures in the motor bearings and shafts. The brackets used to mount the electric and electronic damper motors are special and require a rugged mounting system and sufficient bolts and screws since these types of motors are heavier than pneumatic motors and will literally tear a bracket off the wall or the duct when the motor operates if not properly secured.

All types of damper motors need to be checked and at times serviced to maintain their integrity. The biggest mistake made is mounting the damper motors out of sight. This is particularly true when the damper motors are mounted on damper shafts provided with the unit dampers. Time and time again HVAC jobs wind up with the damper motors completely inaccessible and the maintenance people not even aware they exist. In the case of pneumatic motors, the least that should be done is to pipe an air gauge from each motor to an area that can be viewed so the personnel responsible can check on the operation of motors that are out of sight. Even though the gauge does not indicate the actual operation of the motor, since it is merely the branch line pressure from the controller, it can be a help in the operation of the system. Sometimes the motors are installed before the duct work is insulated and the insulator, who is not as concerned with the operation of the controls as the control installer, winds up preventing the motors from operating by the way he or she installs the insulation on the duct work next to the motor.

Electric and electronic damper motors are also vulnerable to damage from dirty air if they are installed in the airstream, as is the case with pneumatic motors. Electric and electronic damper motors are sensitive to position mounting and should not be mounted, for example, upside down. This is particularly true in the case of valves since a leaking packing nut can cause fluid to leak and ruin the motor if the valve is above the motor. The best positions are from vertical to a 90° position.

Electric and electronic damper motors and valve motors have, like anything else, limitations on the amount of square footage of damper that they will operate. They can be grouped together so that more than one motor can be used on a single damper. Here again the wiring diagrams used under the application must be followed to the letter since troubles can develop if the slave motor is not wired correctly. Usually, the slave motor is different than the master motor to ensure that it tracks the master motor properly. In the case of electronic motors, this is accomplished by changing the solid-state boards in the motor. Some motor actuators have a travel adjustment pot built into the solid-state board. There are also end switches available on some models of electric and electronic actuators to facilitate special control cycles and actuate two-position devices during the operation of the damper actuator.

In this chapter we discussed the types of dampers and damper motors, with emphasis on maintenance and proper installation. Dampers and damper motors will give years of service when care is used in their installation and maintenance.

Chapter

4

Automatic Valves

This chapter will discuss all types of automatic valves and all applications of them that are appropriate for the HVAC industry. It will consider body styles, the connections to those body styles, and how they apply to the HVAC industry. The chapter will present applications of automatic valves and will consider the body materials and inner valves and how they play an important part in the valve itself.

The discussions will include packings, as well as the applications of packless valves. They will also include the various types of operators, both pneumatic and electric and electronic.

Valve sizing will be an important part of the chapter, and the problems involved with improper sizing will be emphasized. Self-contained valves will be treated, along with industrial types of automatic valves, butterfly valves and their uses in the HVAC industry, and treatment of maintenance procedures in the area of automatic valves.

Types of Automatic Valves

The body styles of automatic valves can be divided into two categories:

1. Two way
   A. Normally closed
   B. Normally Open
2. Three way
   A. Mixing
   B. Diverting

These styles of bodies can further be divided by the type of connections to the valves, as well as the type of unions used with the automatic valves.

Two-Way Automatic Valves

For two-way valves, it can be seen from Figs. 4.1 and 4.2 that in one case the valve is closed when there is no air in the actuator and in the other case the valve is wide open. The reasons involve the control cycles that are used and the need to have a normal position when the power or the air is off in the control system. A typical example is the necessity to open a valve on the heating cycle so as not to freeze important parts of the system on a power failure. The implication is that it is better to overheat in the dead of winter than to freeze up the pipes of a system. There are many other instances where normally closed valves are needed to provide the correct cycle of control including sequencing of the valve with other devices on the fan system. Usually, the valve on a hot water convertor being supplied with steam is closed because of safety. From the figures, you can see that the normal position of the inner valve depends upon the construction of the body and the action of the spring. All automatic valves must be piped where the

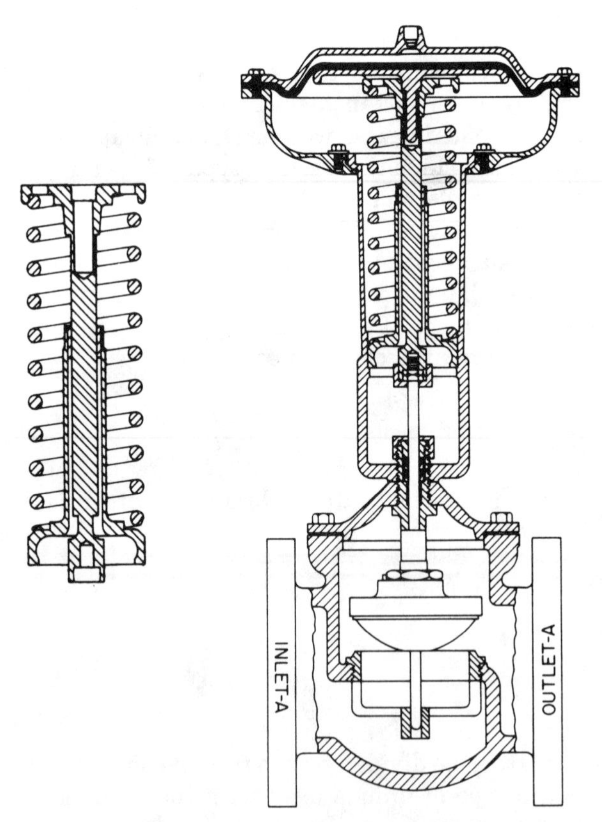

Figure 4.1 Cross section of typical normally open pneumatic valve. (*Courtesy of Johnson Controls Inc.*)

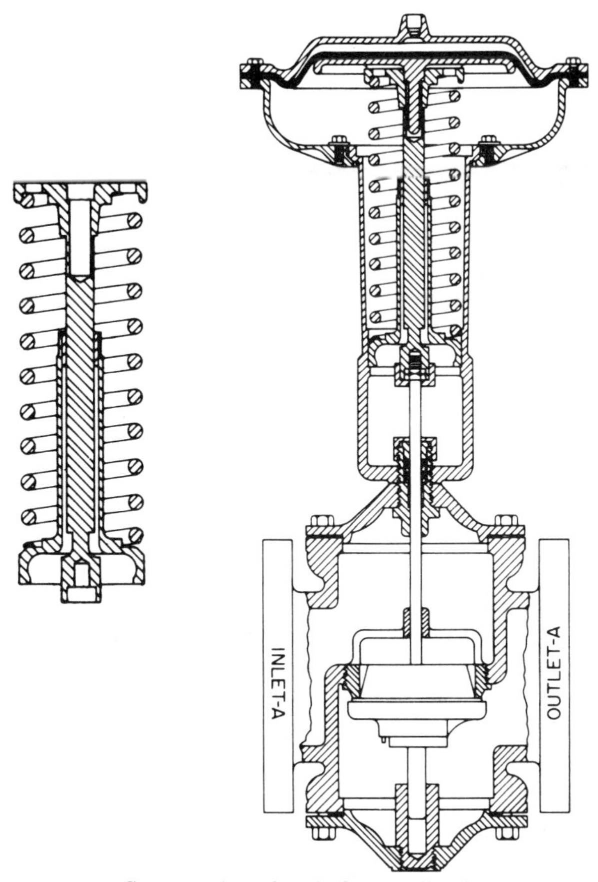

Figure 4.2 Cross section of typical normally closed pneumatic valve. (*Courtesy of Johnson Controls Inc.*)

inner valve operates *against* the pressure of the system. If piped backwards they can cause a hammering sound in the pipe work. Some industrial pneumatic valves have springs on the opposite side of the diaphragm so the normal position of the valve is quite different from the run-of-the-mill HVAC controls valve, this means that a valve can be normally open or normally closed even if the configuration of the body indicates that the valve is only normally open or normally closed.

The connections to the body of a valve can be screwed or flanged, as well as provided with flare and/or solder connections. The standard pipe threads used in the United States are those used in almost all cases, but European metric systems do not use U.S. standards for screwed and flanged valve bodies. Even in the United States, there are different forms of flanges with different bolt hole dimensions, so it is important to specify the American Standards Institute, ASA, standard flanges used with the

Figure 4.3 Example of ground joint union on a valve. (*Courtesy of Honeywell Inc.*)

control valves from the industry. Generally, valves are available through 2 in. with screwed connections; 2½ in. and above generally have flanged connections. Some screwed valves come with ground joint unions as part of the valve body so they can easily be replaced without cutting out a whole section of pipe (Fig. 4.3).

With flanged valves the replacement problem is not as acute, but too often thought is not given to how to service or replace a valve if it is defective. This presents a real problem when it comes to three-way valves which are not easily serviced in the line. It is usually expected that hand valves are placed so that complete shutdown of a system is not required to service an automatic valve. Without the hand valves mentioned, the complete shutdown of the system may be required for servicing.

Three-way automatic valves

Three-way valves come in basically two configurations—three-way mixing and three-way diverting (Fig. 4.4). The illustrations will show how the two styles work and the use of those styles. In almost all cases the three-way mixing can be used if properly piped for all applications. Only rarely will you have to use the three-way diverting valve. The problem comes when someone tries to pipe the three-way mixing valve in a diverting application. A close analysis of the inner valve of the two will show that there is a difference and why one will not work in place of the other.

The diverting valve has *two* inner seats and discs, and the mixing valve has *two* inner seats and *one* set of discs on a stem. If the mixing valve is used as a diverting valve, the valve can hammer and cause

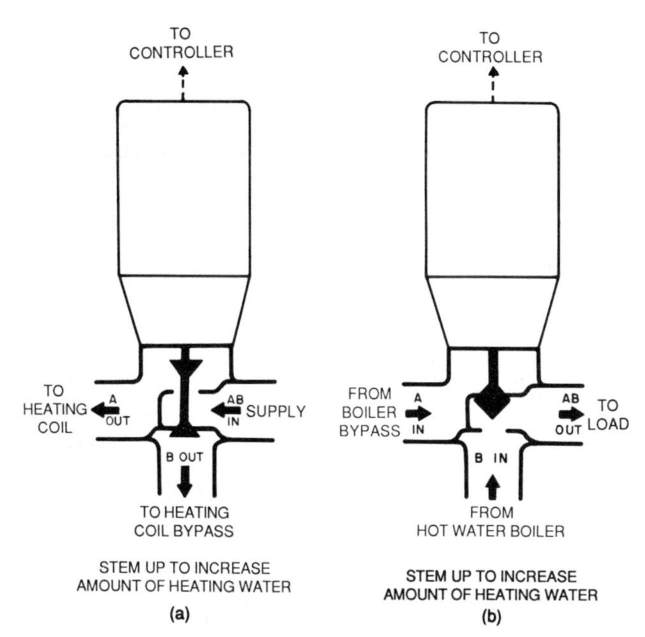

Figure 4.4 Typical operations of diverting and mixing valves. (*Courtesy of Honeywell Inc.*)

problems. One application that comes to mind where a diverting valve must be used is the case of a cooling tower bypass control cycle where the system is open and there is no pressure with which the mixing valve can operate. Another application is when the diverting valve is the valve on the return piping used to send the return water either to the chiller or the boiler in a hot water and chilled water system.

It is advisable to pipe the valve so that the actuator of the valve is in the vertical or upright position, at least in the horizontal position. The reason is that if the packing starts to leak, the steam, water, and so on, will not flow by gravity to the actuator of the valve and cause corrosion problems.

The inner valve parts consist of a seat of some type and a disc that closes against the seat. Remember, however, that in this case as opposed to a manual valve the seat and the disc do not rotate as they close against each other, and the only thing that keeps them closed is the force created by the actuator. The discs themselves are made of various types of compositions designed for the service being used. Steam discs are made from the same material used for discs on hot water control valves. The seats in some of the valves are cast and ground into the valve bodies at the factory. The bodies of automatic valves are made of various materials from cast iron to brass, as well as such materials as carbon steel, carbon moly, chrome moly, 304 stainless steel, and 316 stainless steel. The bodies are rated based upon standards set up by American National Standards Institute (ANSI), and are based not only on the pressure involved but also the temperature of the fluid in the system. One standard used is the ANSI B16.15.

A rule of thumb sometimes used is to consider low-pressure valves as those rated at 125 psi, medium-pressure valves as those rated at 250 psi, and high-pressure valves as those rated at 400 psi and above. The higher pressure valves will sometimes have special metal seats screwed into the body of the valve that are replaceable. Most flanged valves have screwed-in seat rings. Some of them are made of stainless steel or other special materials for special applications. In some valves (usually high-pressure valves) the seating is metal to metal, and there is bound to be some leakage when the valve is closed. The percentage of leakage is small, but metal-to-metal seating valves cannot be designed to close off bubble tight.

A lot of research and effort has gone into the design and manufacturer of the inner plugs used in automatic valves; the illustrations will show a few of the types (Fig. 4.5). Suffice it to say that the effort has been to try and create flow characteristics suitable for the application. When the flow of the fluids is directly proportional to the movement of the valve stem and inner valve system, the characteristic is called *linear*. There are also equal percentage characteristics that give linear heat output from variable flow through a hot water coil. In this case, smaller changes in flow when a valve is barely open are offset by larger coil output because the water temperature drops so much at low flow through the coil. There are also quick-opening valves that can be used with two-position applications where control of the medium is not important other than being off or on. And there are V-port valves modulating plug valves along with other configurations.

The characteristic flow graphs in Fig. 4.6 illustrate how the various types of inner valves control the flow based up the style of the plugs in the valve. The other important item in the inner valve of an automatic valve is the stem connected to the disc. The stem can be exposed to the mediums, and if they are corrosive, problems can result. Sometimes stems are of special materials such as stainless steel, and in situations in which corrosion can result after minimal usage a stainless-steel stem might be in order.

Some manufacturers have designed different inner valves called *cage-trim* type valves (Fig. 4.7) intended to make the servicing of inner valves easier and to provide better guiding of the internal parts. Some industrial valves use the cage-trim concept. In this case, the inner valve characteristics are machined into the valve bodies.

(a)

Figure 4.5 Typical inner valve configurations of industrial valves. (*Courtesy of Fisher Controls Co.*)

All automatic control valves are rated on the basis of pressure and temperature and are designed to withstand the rated pressure at a specific temperature as well as an overpressure as a safety factor. This rating is a *body rating,* not the rating the valve will operate against. That is a matter of the size of the operator of the valve, its power, and other factors related to the construction of the inner valve. The higher the pressure rating of the valve, the thicker the metal used in the body construction. In the case of very high-pressure valves where high temperature is also a factor, the bodies are not screwed since screw connections are seldom used in high-pressure piping. In this case, flanged valves will be used in the smaller sizes. High-pressure valves (400 psi and up) will almost always have inner valves constructed of stainless steel or other exotic materials. They usually will incorporate metal-to-metal seats without composition discs.

The subject of packings for valves has received a lot of attention over the years, and there have been many improvements from the initial as-

(b)

Figure 4.5 (*Continued*)

bestos and graphite packings that were first used. Again, it must be remembered that the stem must slide up and down but still keep the mediums from escaping into the atmosphere. If a packing is too tight, it may keep the mediums from escaping but also keep the stem from sliding up and down. Therefore, the stem must be lubricated, must be free to move, and must prevent the mediums from escaping.

Some of the first packings consisted of asbestos ropes impregnated with graphite. Later, Teflon rope impregnated with graphite was used on some larger valves (the bigger the valve, the larger the stem, and the more the tendency to leak). U-Cup™ packings were introduced in the late sixties and in some places have become the standard of the industry. Some earlier ones were constructed of BUNA-N material; later a form of silicone-type U-Cup packings were used. Some manufacturers provide spring-loaded Teflon packings that are also quite successful.

In an attempt to solve some of the leaking packing problems in the 1960s, some companies began to provide and specify packless valves. *Packless valves* used a bellows that moved with the stem and that was

(c)

Figure 4.5 *(Continued)*

sealed around the stem to prevent leaking of the steam or water. The packless automatic valves also had a secondary packing like the ordinary valves of the nonpackless type. This was to prevent the mediums from escaping into the bellows area and preventing the stem from moving since the bellows was not able to compress the water (or condensed steam) back through the leaking packing into the valve body. The problem with packless valves is twofold: First, when the secondary packing starts to leak, as all packing will do sooner or later, there is no indication on the outside of the valve and the bellows will try and compress the fluid inside the bellows as the stem moves up and down, causing the valve to lose control and not operate. Second, sometimes the leaking of the secondary packing into the bellows cavity will cause corrosion from inside the bellows that is not apparent from the outside, causing the bellows to fail when least expected. Packless valves can be successfully used if they are maintained and watched closely.

Some mention needs to made of industrial valves used in the process industry, such as those made by Fisher Co., Taylor Instrument Co., Foxboro Corporation, Keilley Mueller, and Brown Instrument Co. These types of automatic valves are a breed apart and should not be

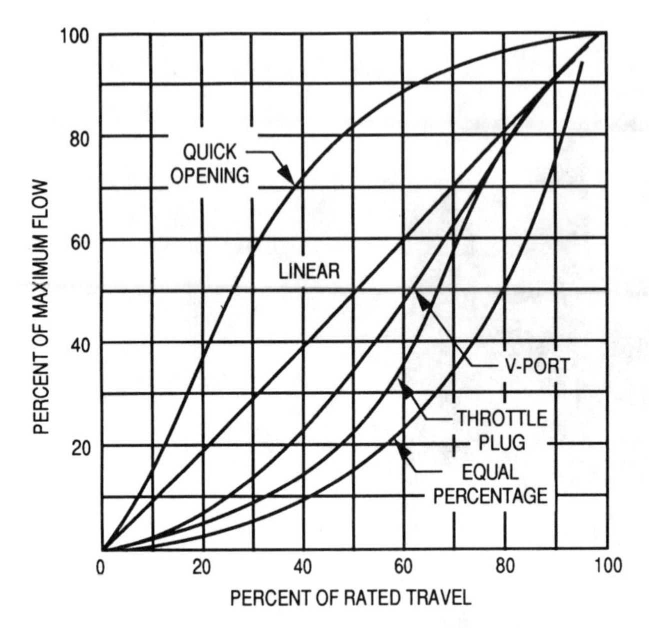

Figure 4.6 Flow characteristics of various valve types. (*Courtesy of ASHRAE.*)

confused with the HVAC industry-type automatic valves. They are used in all types of processes with special inner valves for the exotic fluids involved. In particular, they are used with high-pressure situations that require the special inner valves available from those companies. Figure 4.8, although not complete, gives some idea as to the style and configurations of these types of valves. Of particular note are the different types of inner valves available from these companies.

Valve Actuators

The motive power to operate an automatic valve is called the *actuator*. The actuator can be pneumatic, electric or electronic, or in some cases hydraulic using oil or another fluid. Pneumatic operators that use air as the power to move the valve stem require an airtight cavity and a spring to oppose the movement in order to function properly (Fig. 4.9). Rack and pinion double-acting actuators are also available from the industry. These airtight chambers can use a bellows, diaphragm, or piston arrangement to accomplish the end results. The bellows can be made of a brasslike material, with the diaphragms and pistons made

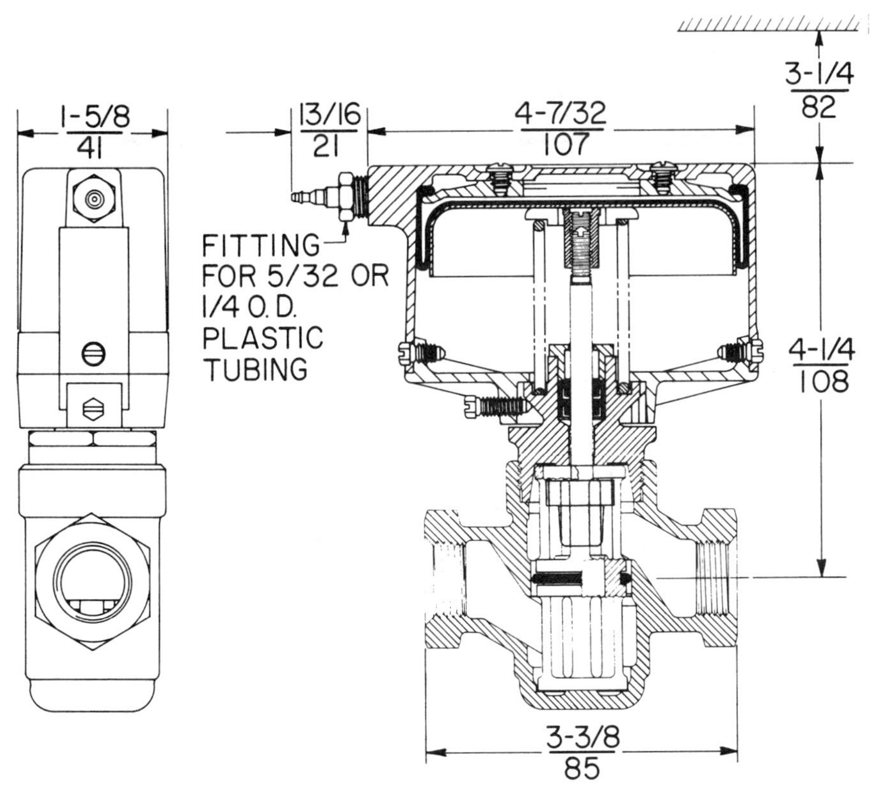

Figure 4.7 Cross section of a cage-trim valve. (*Courtesy of Johnson Controls Inc.*)

of a rubber composition. There are ambient temperature limits that need to be adhered to in the case of the diaphragms used in these operators. In some cases where the pneumatic system has been contaminated with water and/or oil, damage can result with the diaphragms of the valve operators. The area of the diaphragms or bellows is important if the valve is to operate against the dynamic pressure being used in the system. This means that care must be used in selecting the valve, with the temperature and pressure rating of the valve being taken into account. It is possible to specify a larger operator for the same size valve to compensate for the higher pressures encountered.

Like the damper operators, valves come with different springs calibrated to a tolerance that permit the valve to operate within a spring range through the full stroke of the valve. Like the damper actuator, that spring range is *nominal* and does not take into account the external and internal forces acting on the stem, seat, and disc of the valve. Those forces can change the operating points of the valve so that the nominal range no longer applies. Some pressure outside the nominal

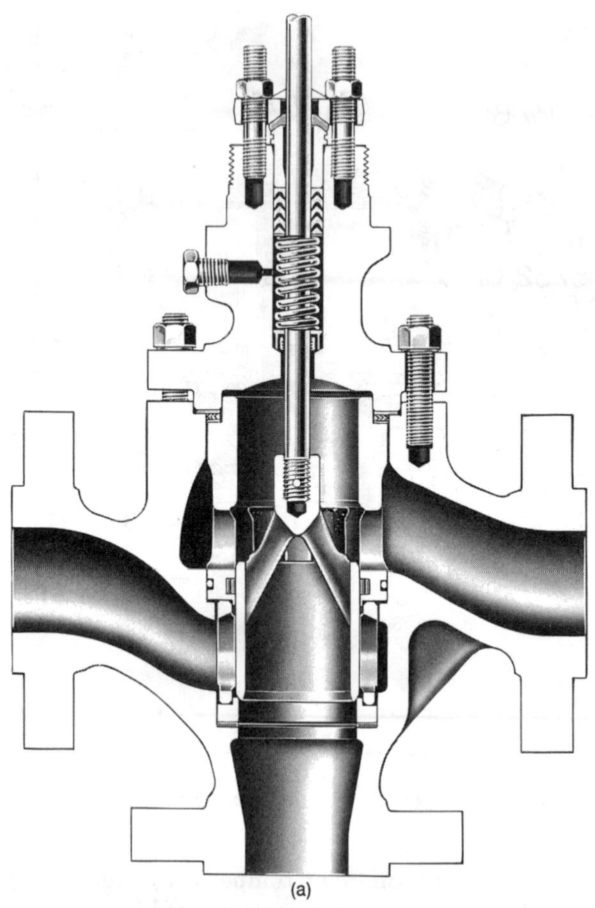

(a)

Figure 4.8 Cross section of industrial valves. (*Courtesy of Fisher Controls Co.*)

range is needed for close-off force. If the control system is simple, such as one controller controlling one valve, the balance point will shift and the controller will control at a different point, which will allow for adjustments to the desired point. Some adjustment, even though small, is permissible with some valves. Whenever the available spring ranges are not satisfactory, particularly in the case of complicated sequencing of valves and damper actuators, the best thing to do is use positioners (Fig. 4.10) on the automatic valves.

The valve positioner operates with the motion of the stem of the valve being the action that either adds additional air to the diaphragm or exhausts air from the diaphragm until the motion of the stem is comparable to the signal coming to the valve from the control-

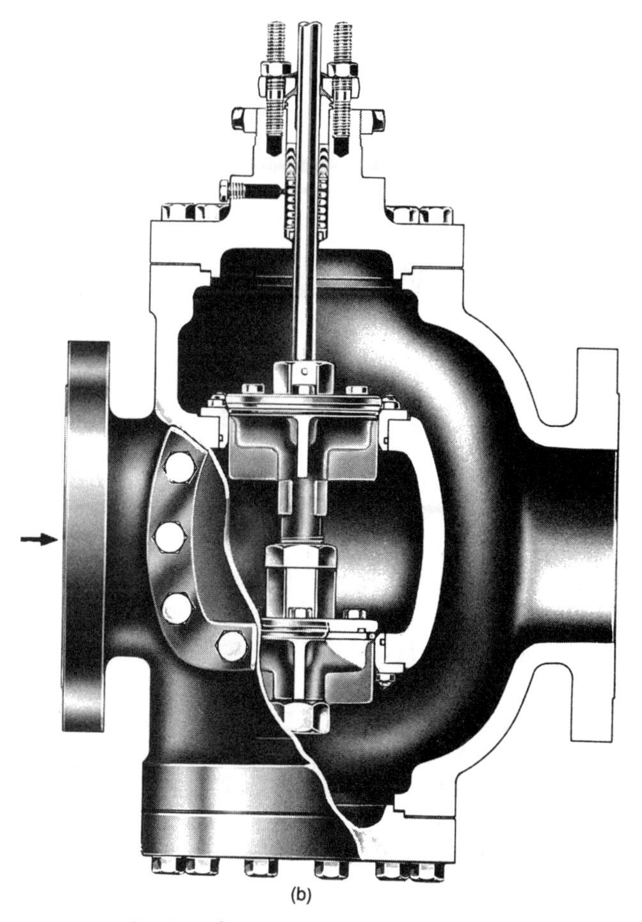

Figure 4.8 *(Continued)*

ler. To put it another way, the stem must move and that motion be sensed by the positioner to stop the positioner from adding more and more air to the diaphragm. The advantage of a positioner is that both the span of movement and the starting point of the movement of the valve stem are adjustable, enabling the system operator to set the sequencing of the valves and damper actuators with greater accuracy. Positioners should not, however, be used to correct deficiencies in system design such as the use of valves and dampers not sized properly for the pressures and temperatures encountered.

As far as maintenance is concerned, the operators and service people need to set a regular schedule for maintenance checks wherein the valves in the system are checked for packing leaks, worn packing, packing that is too tight, corrosion on the valve stems, dry packing

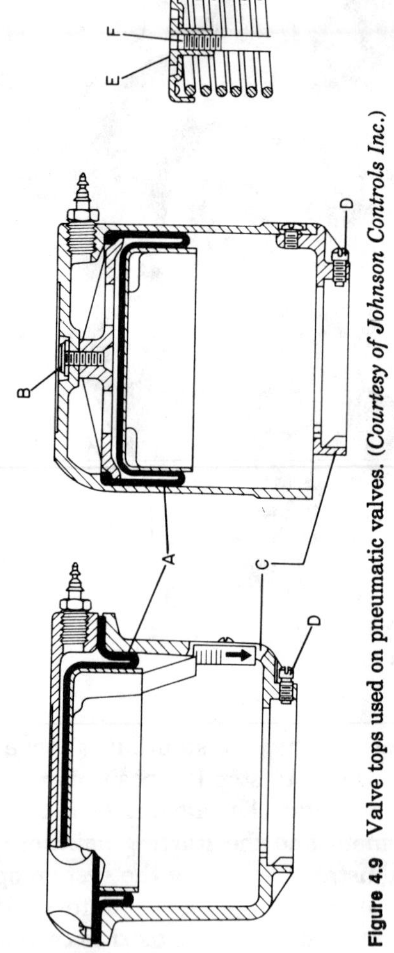

Figure 4.9 Valve tops used on pneumatic valves. (*Courtesy of Johnson Controls Inc.*)

Figure 4.10 Valve with pilot positioner. (*Courtesy of Johnson Controls Inc.*)

that causes the stem to stick or operate with jerky action, as well as worn discs and wire drawn seats due to too great a velocity across the seats. The diaphragms also need to be checked to see that they are still airtight and have no leaks. This can be done with a handy tool called a squeeze bulb obtained from most control companies. This device allows the operators to stroke the valve through its normal range without having to have the controller connected to the valve. With an air gauge connected to the squeeze bulb, the operation of the valve can be watched in conjunction with the span and end points of the spring of the valve. This will also help to check the positioners if there are any and reset or readjust them as necessary.

Electric actuated valves (Fig. 4.11) can, as with damper actuators, be either solenoid or motorized. Further, the motorized type can be two position or modulating. Solenoid types are always two position and are usually used only in the sizes below 1 in., since the larger sizes would require too much power to operate with the magnetic coil of a solenoid. Motorized valves can be modulating or two position, depending upon the equipment supplied with the motor. The big differ-

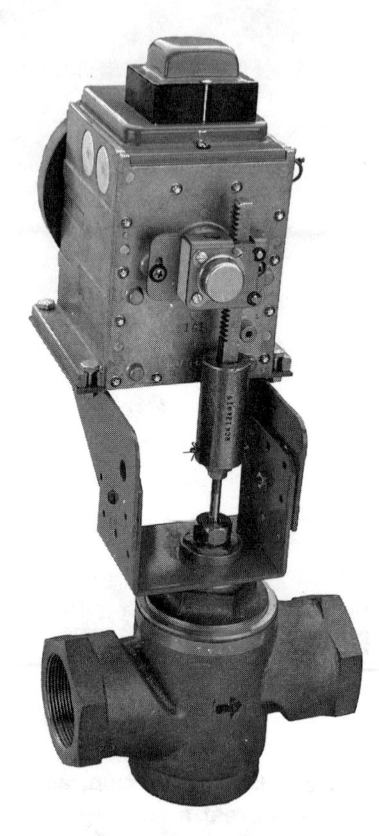

Figure 4.11 Valve with electric damper motor. (*Courtesy of Johnson Controls Inc.*)

ence between damper motors and electric valve operators is the stroke of the motor, which must be much smaller than with a damper actuator. The action of the motor must first be changed from a rotary action to a push–pull action (except for butterfly valves as will be seen later) and must match the stem travel. This is done in most cases with a rack and pinion gear set or a set of crank arms that is adjusted to suit the small stroke of the valve stem. The brackets used to mount the electric motors on the valve body are very special and must be designed to fit the particular valve body. They are generally not interchangeable between valves of the same company or of a competitive company. Again, electronic operators are basically the same as electric operators as far as the valve bodies are concerned.

Electric and electronic valves should never be installed with the operator below the horizontal position, as damage can certainly take place if the packings leak into the motorized operator.

Valve Sizing

One of the least understood control principles is that of valve and damper sizing. In the case of dampers, it was shown that too small a pressure drop can cause the damper not to control but merely to bend the air. In the case of valves with a compressible medium such as steam, the problem is even worse as far as controllability is concerned. *When a valve is used to control the mediums to a coil, as an example, the pressure drop through the valve should be at least 50% of the individual system total pressure drop.* This principle is hard to swallow for some engineers, but it is the only way the valve will control under all circumstances.

The large pressure drops required with steam valves can cause other problems in the case of steam coils situated where freezing can occur, since the pressure drop can also cause a lower pressure in the coil and a premature condensing of the steam, which can allow for freezing of the condensate in the coil. This can be engineered out of the system with the proper use of coils and steam traps used with the coils. If the application is a mixing valve where there is constant flow to the load, sizing is not critical. Too often the comment is made that pressure drop is not important with three-way valves. If the application is a coil bypass, however, nothing could be further from the truth, since part of the time the fluids are going through the coil where there is a pressure drop and part of the time the fluids are going through the bypass where there is little or no pressure drop. To do the job properly, a balancing valve must be placed in the bypass so the pressure drop is the same for both lines. Then the pressure drop must again be figured for 50% of the total system pressure drop for both ports of the three-way valve. Most experienced control engineers will agree that the occasion of an undersized valve is so rare as to be almost nonexistent. Line- size valves are seldom used by the knowledgeable control specialist or engineer.

There is a group of valves that although used only for special applications needs to be mentioned here. It is the self-contained (Fig. 4.12) valves sometimes used on radiators and in equipment rooms. These valves as can be seen from the illustrations are operated by self-contained closed hydraulic bulbs that sense the temperature changes and operate the stems without the use of separate controllers. They may also be used where great accuracy is not a factor and where power (air or electricity) is not available. They are not noted for their accuracy, and the bulbs that do the sensing are sometimes not very attractive when used in occupied spaces. Since they require large expansion of the sensitive mediums in the bulb to operate, they are usually of the vapor tension type and the position of the bulb relative to

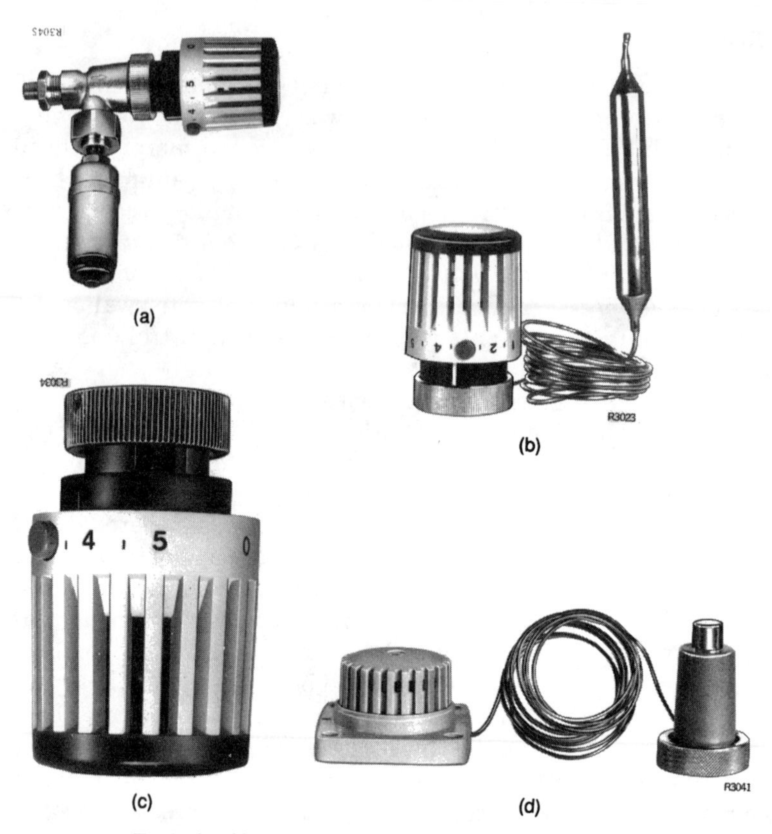

Figure 4.12 Typical self-contained automatic valves. (*Courtesy of Honeywell Inc.*)

the operating head of the valve is critical, thus limiting their use in certain cases.

Butterfly Valves

Butterfly valves although used almost exclusively as manual valves in the past, have come into their own in the last few years as automatic control valves (Fig. 4.13). Some reasons people are using them more often as automatic valves can be seen in the following advantages:

1. They are more compact than globe valves of the same size.
2. They are less expensive.
3. They have fewer parts.
4. They are available in larger sizes than globe valves.

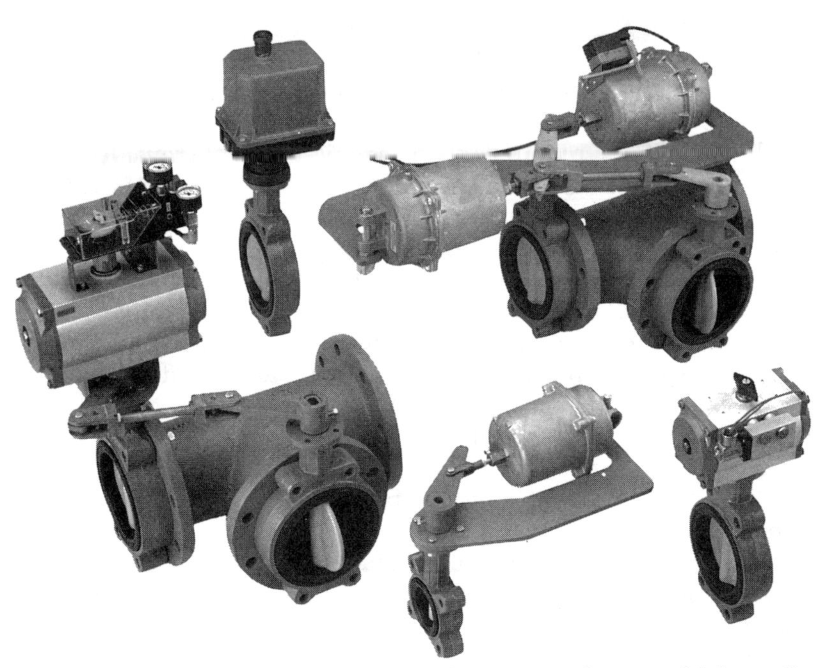

Figure 4.13 Examples of butterfly valves with operators. (*Courtesy of Johnson Controls Inc.*)

5. Actuator sizing is not dependent upon differential pressure across the system.
6. They are bubble tight.
7. When sized properly, they are capable of accurate control.

There are also a few disadvantages to butterfly valves:

1. The cavitation potential is greater than with globe valves.
2. They have a greater potential for water hammer than globe valves.
3. They are not well understood in the HVAC industry.
4. They have an equal percentage-type characteristic not suitable for all applications.

The construction of and materials used for butterfly valves can be varied depending upon the application. The stems and all parts of the valve body, the disc, and so on, are available in a variety of materials such as stainless steel, phosphate-coated carbon steel, and cast iron. The seat materials are also varied, but usually the seats are made of ethylene propylene diene monomer (EDPM). This material is good for all fluids used in the HVAC industry. The discs themselves may be made of ductile iron that is nylon coated.

Care must be used in the sizing of the operators used with butterfly valves, since the seats are tight closing and can take a set after a time. A very high torque is required during the first 20% of travel from the closed position. There is also a bearing friction component of the total torque required to open the valve that must be taken into account. The torque required to operate the valve when it is open beyond the 20% stage is the torque required to overcome the pressure drop in the valve. This is usually much less than the torque required to open the valve. An examination of a curve (available from the manufacturer) showing the various torques required to operate a butterfly valve shows that expert help is needed in sizing the linkage and the operators on a butterfly valve system. The flow characteristics of a typical butterfly valve will give you an idea of what can be expected from a typical butterfly valve under normal pressure drops and flow.

Butterfly valves can be used as three-way mixing valves as well as three-way bypass or diverting valves, since it is just a matter of using two regular butterfly valves operated together from a common set of linkage with one pneumatic or electric actuator. The valves themselves are bolted to a flanged tee in the proper configuration to give either three-way mixing or three-way diverting, as can be seen from the illustrations. The secret is in making sure that the linkage is correctly applied since the breakaway torque can be excessive and there may be an advantage to adjusting the linkage to operate one disc before the other to minimize the excess torque requirements. Also, the equal percentage characteristics gives high-pressure drops to the system where both valves are 50% open. This kind of control engineering requires an expert. Butterfly valves can be used for all applications where globe-type valves are used, but generally speaking they are reserved for the larger sizes where globe valves are cost prohibitive.

In summary, this chapter presented information on valve sizing, valve body styles, inner valves, valve materials, pressure ratings of valves, packing types, packless valves, valve operators, and pneumatic, electric and electronic, and butterfly valves.

Chapter

5

Pneumatic Transmitters, Indicating Receivers, and Receiver Controllers

This chapter will discuss the concept of pneumatic transmitters, indicating receivers, and receiver controllers, also called sensor controllers, and all of the special nonfluidic transmitters, receivers, and receiver controllers. The discussions will include all types of pneumatic transmitters, receivers, and receiver controllers, including temperature, humidity, and pressure control devices.

Receivers come as all types, from those that are as simple as what appears to be an air gauge to those used in the industrial field with charts and chart drives. The receiver controllers discussed in this chapter include all of the models available in the marketplace.

Transmitters

To understand why control companies use transmission principles we must go back to the 1950s when the concept of using more and more control panels was instituted. The market demanded that control companies use panels in control rooms that could do many things, including the ability to read items at a control panel, in an equipment room. The operators also wanted to be able to change set points at the panels, and operate heating–cooling switches at central panels. It also wanted them to start and stop fans, pumps, and so on, from a central panel. The electric control companies were better equipped to provide some of these items since they often used wire to operate items from a remote location. Even they, however, had no solutions for the needs to read temperature, and so on, remotely. There were and still are limitations on the lengths of capillaries. When long capillaries were used and the heads of the thermometers or capillary controllers were placed on panels, the results were far from adequate. The reaction time was slow and the potential damage to the capillaries was pronounced, since they had to be mounted from the sensing bulb to the panel in an area that was subject to damage and high cost.

Submaster resetable controllers were available, so that a rotary pneumatic switch could be mounted on a remote panel to reset the set point of a controller, but that did not solve the problem of how to read the temperature from a remote point. The concept of simple inexpensive resistance temperature devices was available at that time, *but* the amplifiers needed at the panel end were very expensive whether there was 1 or 100 temperatures to be read, not to mention the problem of reading humidities remotely. Thus, some control companies began to use the innovative approach to the problem, called transmission of temperatures, humidities, and pressures. With this approach at a central location both indication and control set point could be accomplished.

The control companies looked around and noticed that industrial instrumentation companies, such as Foxboro, Taylor, Brown, Leeds Northrup, and Bristol had been using the transmission principle for many years in areas like refineries and other facilities that required operations from a remote location using a large panel with all of the data read at the panel. The HVAC control companies, therefore, decided that the concepts were worth pursuing and transmission was born for those companies.

We will discuss principles of the temperature transmitter (Fig. 5.1), and most of the concepts can be transferred to the humidity and pressure transmitters.

Often, temperature transmitters look just like a thermostat, and some concepts used with transmitters are the ones used with thermostats. As an example, the sensing portions of the transmitters and thermostats are the same in almost all cases. The same capillaries (Fig. 5.2) used with transmitters are also used with thermostats. In the case of bimetallic sensors, here too, the same ones are used in both cases.

The physical size of the transmitters are also about the same as that of the thermostats. The covers look the same in both cases, but here is where the similarity ends. To start with, controllers do not always have feedback features in their design, and transmitters have feedback built into the relay portion or nozzle. The reason for feedback in transmitters is to ensure accuracy. The major difference between a thermostat and a transmitter is the rate of output of the transmitter versus the thermostat. Thermostats take the supply air (usually about 20 psi) and send it through the relay system of the controller to the device being controlled at a rate usually greater than a transmitter. As an example, a normal controller has a sen-

Figure 5.1 Room transmitter. (*Courtesy of Johnson Controls Inc.*)

Figure 5.2 Capillary transmitter. (*Courtesy of Johnson Controls Inc.*)

sitivity of about 2½ lb/deg, which means that on a change in temperature of less than 10°, the controller will run its output from 0 to 20 psi. The transmitter, on the other hand, which may have a span of 100° will pass less pressure change for a 10°, change in temperature sensed. The normal operating range of a transmitter is 3–15 psi. That means that if a transmitter has a span of 100°, the output of the transmitter will be 12 psi over a span or range of 100° or 0.12 psi/deg. This is opposed to the 2½ psi/deg of a controller. As might be expected this requires that the transmitter be much more accurate than the controller in terms of the hysteresis and lack of lag built into it. That is the reason transmitters will always incorporate feedback in the design.

Some of the first transmitters had dials on them, but they were not used in the same way we use dials on thermostats. Transmitters come with a fixed range that cannot be changed in the field, but the output of the transmitter can be changed with the dial. In almost all cases, however, they are factory calibrated and should never need to be changed, unlike controllers. Generally, transmitters are available in enough ranges to satisfy all the needs of a HVAC system. Some companies will, however, manufacture transmitters with special ranges for unique applications.

Transmitters are available as two-pipe devices (Fig. 5.3) with pneumatic relays just like thermostats, but they are also available as one-pipe devices similar to one-pipe thermostats. The reason they work well as one-pipe devices is that since their output is dead ended to a low-capacity chamber, the slow action of a one-pipe device is not a problem. Unlike a thermostat that can change its output fast by building up the control line to valve and damper actuators, as well as exhausting the same line, the transmitter's action is very slow so the requirement of large volumes of air to controlled devices is not there. There are, however, some limitations as to the number of receivers and/or receiver–controllers that can be connected to a one-pipe transmitter. The length of the line from the transmitter also needs to be considered in the design of the system.

Temperature transmitters come as unit or remote bulb transmitters for installation in duct work, as well as room transmitters for sensing and transmitting the temperatures in the space. The unit transmitters used at the air-handling unit are available as capillary as well as insertion types, and these are also used with wells for installation in pipes that sense the temperature of the fluids in the pipe.

Transmitters are also used to sense humidity (Fig. 5.4) in the space as well as in the duct work and to transmit an air signal proportional to the humidity in the duct work as well as the space. In the case of humidity transmitters, capillary types are not available. The sensors used with humidity transmitters are the same as the ones used with

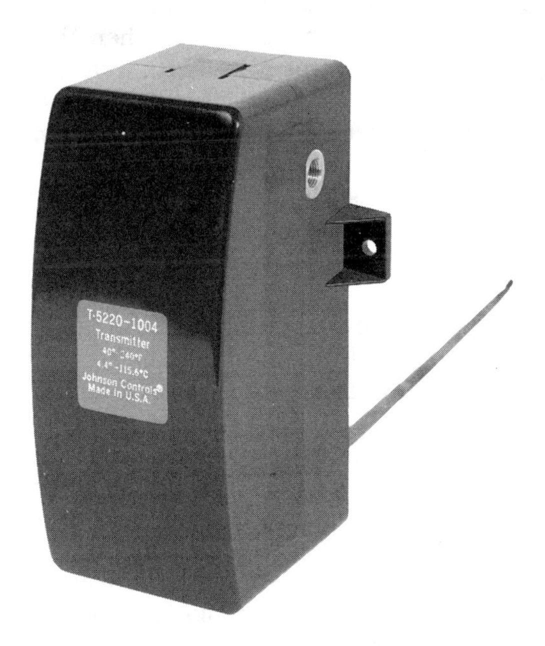

Figure 5.3 Rigid stem duct transmitter. (*Courtesy of Johnson Controls Inc.*)

Figure 5.4 Room humidity transmitter. (*Courtesy of Honeywell Inc.*)

humidistats. Here again they come with various fixed ranges and are factory set so usually there is nothing to adjust when they are installed. As an example, a humidity transmitter with a range from 30% to 80% will have an output of 0.24 lb/percent relative humidity (% RH). Other ranges will have different output values per percent relative humidity.

Transmitters are also used to measure high and static type pressure. Again, the ranges are fixed and not field changeable. The outputs can be calculated just as was done with temperature and humidity transmitters.

With temperature transmitters, the bulbs used with the capillaries may be the same as the ones used with thermostats. That is to say, if thermostats use averaging bulbs, the transmitters can also be the type that use averaging bulbs.

There are also room transmitters that are mass transmitters, which measure the temperature of the wall on which they are mounted.

To field calibrate a transmitter (either room type or unit type), an accurate transmission gauge (Fig. 5.5) must be used and a chart or graph must be checked to find out what the output pressure should be at a particular measured temperature. An example is a room transmitter with a span of 50°F (50–100°F) with its output varying from 3 to 15 psi over that span, producing an output of 9 psi at a room tem-

Figure 5.5 Calibration gauge. (*Courtesy of Johnson Controls Inc.*)

perature of 75°F. If the output were 11 psi instead, the room temperature should be 83.3°F, not 75°F.

Unit-type transmitters can be used with a Dewcell™, in which case they will transmit dew point temperature.

There are also various types of pressure transmitters (Fig. 5.6) used in systems, allowing a system to be basically all transmission instead of conventional pneumatic. Some sample ranges of pressure transmitters are

1.	− 30 in.Hg	to	+ 30 psig
2.	0	to	50 psig
3.	0	to	15 psig
4.	0	to	100 psig
5.	0	to	200 psig

Pressure transmitters are also available as differential pressure transmitters with appropriate ranges to accomplish cycles used in some HVAC systems. Typical examples of needs are the ability to maintain a differential between a clean room inside the clean room

Figure 5.6 Pressure transmitter. (*Courtesy of Johnson Controls Inc.*)

and a reference room outside the clean room and the maintenance of a differential pressure on the supply and return fan systems on a VAV system. Static pressure transmitters are also used in the case of VAV box control. There are even velocity pressure transmitters that use total pressure taps along with static pressure taps, do the required square root calculations, and transmit a velocity reading with a 3–15 psi signal, which can be converted directly to cfm or volume. The available velocity ranges usually match the fan systems used in modern HVAC. Some examples are

1. 200–2000 fpm
2. 300–3000 fpm
3. 400–4000 fpm
4. 550–5500 fpm

For just about every application in the control field, there are transmitters that can be used.

Indicating Receivers

Receivers (Fig. 5.7) used in the transmission systems are usually nothing more than very accurate air gauges that have dials calibrated to the transmitters. Note, therefore, that the receivers come with various spans and ranges engraved on the face of the air gauge dial. That is to say, if the transmitter has a span of 100°F and the output of the transmitter is calibrated to go from 50 to 150°F, the receiver must have a dial that reads a temperature of 50–150°F in the span of the 3–15 psi within which the transmission system operates. Some receivers have multiple ranges engraved on the dial; that requires that the person reading the receiver knows what the range of the transmitter connected to the receiver is. These kinds of receivers are generally only used by the control company technicians who set up and check the systems, as it would be impractical for them to carry a number of receiver gauges to match all ranges used.

The ranges control companies manufacture are usually satisfactory for most applications. But just as it is possible to furnish a transmitter with a special span or range, it is also possible to get a receiver with a special range to match a special transmitter. Most receivers on the market have both the Fahrenheit and the Celsius readings engraved on the face, just as most speedometers have the mph and kmph. Usually, the companies that make receivers will furnish them for flush mounting on a panel, with dials in various sizes. There are specialty companies that will manufacturer a receiver gauge to just about any specification. There are even receiver gauges that have a removable

Figure 5.7 Typical receiver indicators. (*Courtesy of Johnson Controls Inc.*)

front face and dial that can be changed in the field. All of these enhancements cost more than the standard receiver gauges, so some thought needs to be given to the system at the inception, not later when changes may be costly.

Some typical ranges of one manufacturer are

1. 40–65°F
2. 60–85°F
3. 50–100°F
4. −20–80°F
5. 0–100°F
6. 20–120°F
7. 50–150°F
8. −40–160°F
9. −40–240°F
10. −5–64°F (Dewcell operation)

The ranges used with the humidity and pressure transmitters and receivers are comparable to the temperature ranges mentioned above, whether the transmitter is used with a static pressure transmitter or with the velocity transmitter. A check of the manufacturer's literature will show the ranges available for the various transmitters. The important item to remember is that there is no mix and match with transmitters and receivers.

In the industrial field there are receivers that are also incorporated with recorders so that a record can be made of the particular temperature or humidity being observed. These recorders are available as locally mounted recorders as well as panel mounted on a central control panel. They can be the circular type or strip chart type, with either 24-h or 7-day charts. The portable circular chart receivers can even be furnished with wind up clocks so that no power is required at the recorder with just the air line from the transmitter to the recorder.

Receiver–Controllers

Sometimes it helps to think of the transmitter *and* the air tubing that goes from it to the receiver or to the receiver–controller as the *capillary* of a normal temperature controller, which brings us to the discussion of the receiver–controllers. Even though transmitters come in all styles and types, the receiver–controllers are basically one type used for all transmitters. The dials on the receiver–controllers (Fig. 5.8) are

Figure 5.8 Receiver controller. (*Courtesy of Johnson Controls Inc.*)

not marked for temperature, humidity, or pressure; they are marked increase–decrease. Unlike the transmitters, the receiver–controllers are large volume devices, so they have pneumatic relays for the capacity that is needed to control the devices involved. They are what is sometimes called two-pipe devices with a main supply air line and a branch line going to the device or devices being operated. Some of them incorporate feed back control circuitry. Others have the capability of switching from proportional only control to PI control automatically or through a field adjustment. Most of the receiver–controllers on the market are designed to be changed in the field from direct to reverse acting and vice versa.

Models are available that are used with two transmitters for a dual input system (typical for master–submaster cycles). The ones on the market at the present time have dials used for all inputs and therefore are just marked increase–decrease. They have the capability of being reset from a remote switch, and the gain or sensitivity can also be changed at the receiver–controller. The combination of the transmitter and the receiver–controller determines what the span or range will be and whether or not the system is being used to control temperature, humidity, pressure, velocity, or level.

Like the receivers, receiver–controllers can also have either circular or strip charts receiver controllers. These types are often panel mounted and have built-in capabilities of changing the action from direct to reverse and vice versa. The charts are available in 24-h as well as 7-day styles and with all of the other features of receiver–controllers mentioned.

Receiver–controllers can take the place of any thermostat, humidistat, pressure controller, and so on, that has been used in the past. The big advantage to the use of transmissions systems is the ability to have the sensor remote from the controller and put the controller on a panel.

About the only problem with room transmitters is that the occupants sometimes think they are adjusting the thermostat when in fact they are upsetting the whole system by turning the dial of the room transmitter. This problem can also be transferred to the unit transmitters if the maintenance engineers do not understand the system.

Some receiver controllers presently installed in systems are of the fluidic type without the stacked diaphragms of most modern receiver controllers. The fluidic types used a principle known for many years of involving the concept of a jet of air that can be bent by shooting another jet of air perpendicular to the first jet. That action was used by harnessing the power of the impingement of the first stream of air. Unfortunately, those receiver–controllers require air that is always perfectly clean because of the size of the ports and jets that are involved. Thus, the slightest problem with the control air in so far as purity is concerned caused the receiver–controllers to malfunction. As a result, most of the older fluidic receiver–controllers have been or are being replaced by the more modern stacked diaphragm types of receiver–controllers.

This chapter has presented information about most of the types of transmitters, receivers, and receiver–controllers available on the market today. The discussions involved the reasons transmission of temperatures, and so on, are desirable when operators can use panels they can operate from a remote location, as well as visualize what is going on in the system by seeing the temperatures, humidities, pressures, and so on. Modern electronics also has systems available that permit the designers to use other concepts that do just about the same thing pneumatic transmission does. Those systems will be discussed in subsequent chapters.

Chapter

6

Auxiliary Devices

In order for control systems to function properly, they need more than just thermostats, humidistats, control valves, and so on. This chapter will be concerned with all the other devices needed to make the control system a complete integrated system.

Today's modern systems need many auxiliary devices to condition the signals sent from controller to controller and from the controllers to the controlled devices. The chapter will discuss the technology of the auxiliary devices that condition the signals. It will include information about complicated procedures, such as BTU calculations, using some of the relays mentioned. The discussions will involve both pneumatic and electric devices as well as electronic devices that multiply, add, subtract, and divide the signals from the various controllers.

Where to use and not to use the devices, as well as problems that can be encountered with the improper use of the devices will be discussed.

The proper use of pneumatic as well as electric switches with applications in control systems will be covered. Discussions will include the set points, span width, and so on, of PE switches and information on explosion proof switches and devices.

The subject of transducers will be discussed, since they are playing an increasingly larger role in the integration of the systems given the modern control and monitoring systems as well as the use of computers in central control systems transducers.

All types of switches (both pneumatic and electric) as well as controllers will be covered. The unusual and different items used on occasion for special applications will also be part of this chapter.

Pneumatic Relays and Cumulators

There is a series of devices used to condition the signals from pneumatic controllers to each other and to the devices being controlled

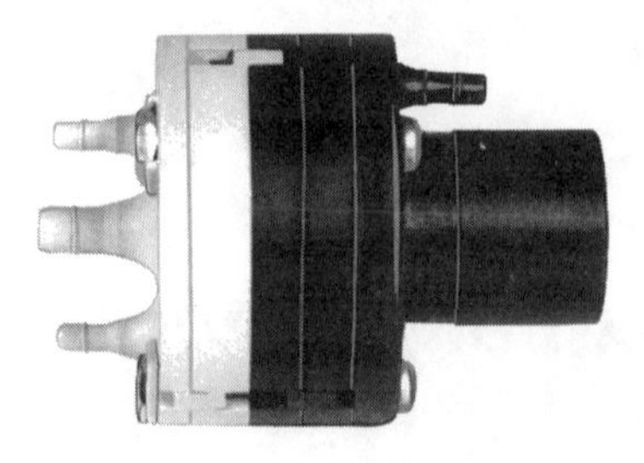

Figure 6.1 Direct acting to reverse acting relay. (*Courtesy of Honeywell Inc.*)

such as the valves or damper motors in an HVAC system. These devices are often called *relays* or *cumulators* and come in a large variety of configurations. The simplest types are relays that change the output signal of a controller from direct to reverse acting and vice versa (Fig. 6.1). They are often used when the signal from a controller has to control more than one item from a single branch line, with some devices requiring reverse action and others requiring direct action from the same controller. With such a relay or cumulator in the branch line from the controller, one device can be controlled with direct action and another device in the same branch line can be controlled with reverse action. The relays discussed are generally used in the branch lines from the controllers to change or accumulate information from other signals and thus present the correct information to the device being controlled.

Some relays take one or more signals from controllers and pass on the highest multiple signal (Fig. 6.2) received or the lowest signal received. Sometimes these are relays that accept two signals and pass the lowest or the highest of the two; other times they are relays that pass the lowest or the highest of many signals.

Other relays average the signals (Fig. 6.3) from up to four controllers and pass the average of those inputs. Another relay type device is used to change the sequencing action of two devices controlled by the same controller. This is used in the case of a valve and damper motor controlled by the same controller that have gotten out of sequence due to external

Figure 6.2 Highest to lowest pressure relay. (*Courtesy of Honeywell Inc.*)

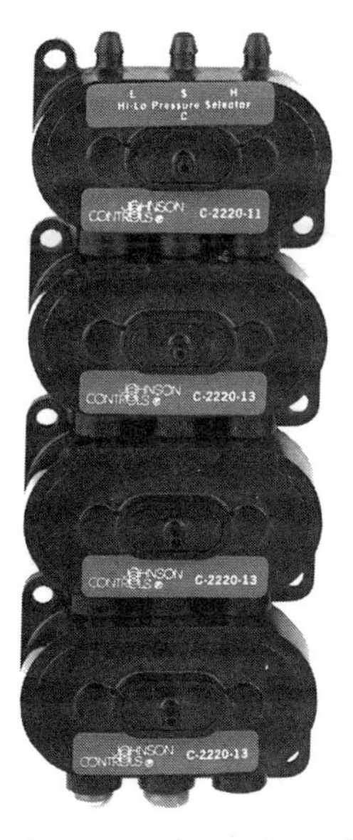

Figure 6.3 Averaging relay. (*Courtesy of Johnson Controls Inc.*)

forces on the valve or the damper motor. This device allows the system to be adjusted and the two controlled items to be sequenced again. There are also relays that take a proportional signal from a controller and at particular adjustable start–stop points raise the output from 0 to 20 psi or whatever the main air pressure happens to be. Also available are relays that boost the signal from a controller to a higher *volume* to operate many devices that could not have been operated directly from the controller itself due to the limited air output volume of the built-in relay of the controller. These types are generally one-to-one relays that boost the output signal on that type of ratio.

One of the most interesting types of relays is the biasing relay (Fig. 6.4) used for special applications such as allowing a manual switch to change the pitch of the curve in a master–submaster control sequence. It is a pressure actuated device that applies its output pressure to pilot a pneumatic controller or operate a controlled device in accordance with pressure signals received from *two* other sources. The illustrations clarify the application much better than a description can.

There are relays or cumulators that can add and subtract and are sometimes called *repeaters* (Fig. 6.5). These are often used in pneumatic transmission systems to increase the sensitivity of a transmitter so that it, in effect, can use its signal as a controller signal. There are even relays that can extract the square root of a signal or divide two signals and get a quotient. The square root extraction relays are used in determining flow of fluids in pipes in gallons per minute (gal/

Figure 6.4 Biasing relay. (*Courtesy of Johnson Controls Inc.*)

min). To read flow in a pipe in gallons per minute, we must first read the pressure drop across an orifice or other artificial device designed to create a pressure drop in the pipe, take those two readings, subtract the lower one from the higher, extract the square root of the answer, and multiply the result by a constant for the pipe size and other factors. The square root extractor relay is also used in VAV applications. The relay converts the input signal from a velocity transmitter to an output signal that is linear to the controlled volume.

Pneumatic signal limiters are another type of relay used in pneumatic systems. They allow an adjustable high or low limit to be put on the output of a controller so that a controller operating one device can operate another one with a limited signal.

Care should be used in the use of the relays mentioned since the improper application of some can cause the control cycles to act just the opposite of the desired sequence. Some of the relays and pneumatic cumulators have limits as to the amount of air they can pass, and trying to operate too many devices will make the control action sluggish and unresponsive. Some of the relays have ratios that must be adhered to or they will not operate as expected. Also remember that some relays and cumulators require a main air supply and will not

Figure 6.5 Repeater relays. (*Courtesy of Johnson Controls Inc.*)

function without it. This means that when planning an installation in a panel or at a remote location, plans must include a 20-psi or similar main air.

Some relays have fixed switching points and cannot be field adjusted; others can be changed in the field. Therefore, it is important to be sure the correct one is chosen for the application. It is frustrating to realize that the switchover point, for example, of a two-position relay is not adjustable in the field after it is installed, and must be replaced by an adjustable one.

Relays and cumulators are versatile devices, but even they have limitations so that sometimes specifiers will call for cycles that all the relays in the world cannot accomplish. Also, sometimes one control company will manufacturer a relay that performs a certain function differently from the relays manufactured by other control companies, and specifying that relay cuts out the competition. It is necessary to specify only the sequence and allow the control company to use its relays to accomplish the sequence desired. In general, however, what one company can provide is usually available from all or most of the control companies.

The specifier should also realize that there are limitations to items that can be accomplished through the use of relays. As an example, just because the literature says a relay can be used to sequence different devices does not mean that the number of devices to sequence is unlimited. A study of the literature will usually prevent the misuse of relays and cumulators if the information is followed.

Electric Relays

The above discussions involved the use of pneumatic relays and cumulators to condition the air signals from controllers and other pneumatic devices. There is a family of devices used to accomplish just about the same thing with electric signals. These are called *electric relays,* and the term *relay* is more commonly used with these devices than with the pneumatic devices mentioned previously.

One reason for using relays is that they allow the switching of wires from one circuit to another, where high voltage on the wires is switched *without* passing the high voltage through a switch. The danger involved in switching high voltage is far more acute than in switching air signals in a pneumatic system. The danger is low from switching an air signal but it can be high when switching, for example, a 440-V power signal from one place to another. A relay properly suited to the high voltage involved can do that job nicely. A remote switch can operate the lower voltage relay, and the people operating the switch would be in little danger from shock.

Relays also allow the use of smaller wires to control high-voltage situations that are remote from, for example, a central control panel. Thus, the use of relays usually implies that the voltage operating the coil of the relay is lower than the voltage being switched. In a few cases this is not true because the switching requirements are complicated, and it is not practical to use switches alone.

Most relays use a magnetic coil, which, when energized with electricity, creates a magnetic field that pulls on a plunger, thus making a movement that controls the action of some contacts to do the switching. This arrangement is the basic concept behind almost all relays. The variable parts of the relay types involve the configurations of the contacts.

The contact points of a relay are rated on their ability to transfer power from one source to another. The higher the voltage and amperage involved, the greater the potential for arcing or sparking at the relay contacts, and the greater the requirements for heavier contacts that can withstand the arcing. The speed at which the contacts are make and broken is also an important factor in the design of a relay since that too plays an important part in the ability to resist arcing and damage to the contacts of a relay.

Relays are available in an almost unlimited variety of contact configurations. They are single pole single throw (SPST), single pole double throw (SPDT) double pole double throw (DPDT), three pole single throw (3PST) and so on.

The magnetic coils of the relays are available in just about all the normal ac and dc voltages that can be encountered in the HVAC and/or control industry. There are relays made for all special foreign voltages, as well as to operate at adjustable set voltages. Some relays are sensitive to certain amperages instead of voltages.

A large percentage of electric temperature control systems use a 24-V source of power since low-voltage wiring is easier to install in buildings. It is less dangerous to personnel working with it, and anything less than 30-V does not always come under the codes that apply to wiring higher than 30-V. As a result, control cycles that require a lot of switching and conditioning of signals use 24-V relays extensively. In those cases, the relay coils will be the same voltage as the system voltage. Those types of control systems require a step-down transformer to provide the 24 V for the system.

The use of relays in an electric control system is common today because of their versatility and almost unlimited variety of configurations available.

The transformers used in low-voltage control systems have volt ampere, VA, or watts ratings and should not be used in systems with relays that require larger capacities than are available with the transformers being used. Thus, the ratings of the coils of the relays need to be checked to be sure the transformers are large enough to do the job.

Relay coils are also used in part of the magnetic starters used to start and stop large three-phase motors. Here again, the reason is that the motors cannot be started directly across the line without arcing problems and danger to those starting the motors, not to mention the fact that starting from a remote location would be very difficult without a starter to close all three-phase wires simultaneously. Thus, there is a need for a relay in the starter. Also, there are safety controls normally of lower voltage to stop a motor by opening all three phases simultaneously.

The main problem with mechanical contact relays involves the arcing that takes place with the contacts of the relay when the relays are too small for the voltage and amperage in the system. The arcing causes pitting of the contacts and eventual failure of the relay. The manufacturers of relays have made great strides in research and development of theses contacts, and some contacts are even made with rare metals to try and make them last longer even when they are overloaded. The bottom line, however, is that the contacts be properly selected for the service intended.

Relays are available with their own enclosures as well as in open styles that can be mounted in other enclosures, such as central control panels.

Some relays use hermetically sealed mercury contacts to cut down on the arcing problem and make the relay last longer under very heavy duty conditions. The theory is that if the contacts are in mercury and in an oxygenfree atmosphere, the arcing problem will be reduced and pitting of the contacts will be eliminated. Since mercury in itself will conduct electricity, it makes an ideal medium for the transfer from one contact to another. These types of relays are more expensive in general than the mechanical contact types.

Some relays are used to delay an action and are called appropriately, *time delay relays*. They are available with a delayed start or with a delayed stop. These relays are also called dashpot relays because there is a small dashpot that when used leaks air at an adjustable slow rate to retard the action of the contacts even after the coils are energized or deenergized. The dashpots are connected to the contacts mechanically, and the delay is a restrictor used to adjust the speed of the air leaving the dashpot. Dashpot time delay relays are considered old technology; today, with the advent of electronics, the time delay relays are all electronic with much more flexibility.

There are so many special-purpose relays that an entire chapter could be devoted to them. As an example, mechanical latch relays require a momentary pulse of current to close the contacts and no power to keep the relay contacts closed, with another pulse of current to open the contacts. This is unlike the usual relay, which requires power on the coil at all times to keep the contacts closed and the removal of power to the coil to open the contacts. There are also relay contacts that are closed when the relay is supplied with power as well as when the relay is without power to the coil. These are called normally open and normally closed relay contacts. Some relays can be plugged into special bases that allow more flexibility in certain cases by changing relays easily. The bases have multiple contacts so that just about all the configurations can be used and changed at will. Other relays are plug-in relays, which are of the reed type that are encapsulated in a glass capsule to seal out the air and protect the contacts from corroding. Reed types usually have a small capacity but they are very flexible and available in many configurations, such as five-pole double throw.

As stated above, and important enough to be reemphasized, the main problem with electric relays is the overloading of the contacts in some applications. Sometimes the coil of a relay will burn out after years of use, but that is the exception rather than the rule.

Pneumatic Switches

The use of pneumatic switches in pneumatic control systems has been around since the advent of pneumatic control systems many years ago. The first switches were simply small air valves that either put air into the system or exhausted it from the system when the handle of the valve was turned in one direction or another.

Today there are many types of pneumatic switches used in the industry. Almost all of them are mounted on a panel, even if it is only a small panel for just one switch. The more sophisticated systems have large panels with many switches and other devices mounted on them. Large industrial control panels in places like refineries always have a variety of pneumatic switches installed on their panels.

Lever-type pneumatic switches are available as two-position, three-position, four-position, five-position and, six-position switches (Fig. 6.6). These types of pneumatic switches require an inspection of the inside paths of the air to determine how they work, but basically they are nothing more than machined tubes and passages allowing the air to get to the correct passage when the center shaft is turned a certain amount. Note that passing air to the system is only one requirement in the switches. The switch must also have the capability of exhausting the air from the air line after the lever is turned to a different position. A typical example is a switch that takes main air of 20 psi and sends it to any one of three or four different places.

Figure 6.6 Pneumatic switch. (*Courtesy of Johnson Controls Inc.*)

Push-button switches mount in a room or on a panel. Although they look like electric push-button switches, they are pneumatic switches and pass or exhaust the air from their branch line.

One of the most common of pneumatic switches used in pneumatic systems is the gradual switch (Fig. 6.7). This switch takes the main air pressure (usually 20 psi) and passes it on gradually as the dial turns from 0 to the 20 psi of the main air. It can be set to any position and is used to operate any pneumatic device with gradual action. A gradual switch can be used, for example, on a panel to operate an exhaust damper in a remote location that is not automatically controlled. It can also be used remotely to reset the set point of a thermostat that has the capability of being reset from a remote position. The gradual switch is really nothing more than an adjustable pressure reducing valve that can take the 20 psi main air and pass it on to the branch in small increments. Toggle pneumatic switches are available for specifications that call for the pneumatic and electric switches on a panel to match.

Figure 6.7 Pneumatic gradual switch. (*Courtesy of Johnson Controls Inc.*)

Generally, there is little that can go wrong with pneumatic switches. The problems arise when pneumatic switches are connected incorrectly and the results are not what was specified. All of these switches have markings on the bodies that correspond with the numbering system on the literature supplied with the switch. You must be sure the switches are hooked up in accordance with their instructions. If drawings are not available or lost, blowing through the switch with your mouth will tell what port is connected with what port.

Electric Switches

Just as there are switches for pneumatic control systems, there are switches for electric and electronic control systems. Electric switches are also an integral part of the control wiring of motors for fans pumps and other HVAC items. They, like relays, are rated on the ability of the contact surfaces to withstand arcing and the resultant pitting from arcing. Care must be used when selecting a switch to check the capacity rating and not use the switch for service above the published rating.

Electric switches come in many styles; the main ones are toggle, lever, rotary, push button, and slide, which span the requirements of the HVAC industry. Like relays, electric switches can be single pole and many pole, with the action being single or double, such as a single-pole single throw, single-pole double throw, up to more complicated types.

Toggle switches can be as simple as the toggle switch in a room that turns on the lights or as complicated as large disconnect switches that interrupt the power to a very large motor starter. Small toggle switches are sometimes called bat handle switches to distinguish them from the usual toggle switches used for house wiring. Rotary switches allow the transfer of power from a source to any one of a number of circuits. These are usually used in low-voltage systems.

Push-button switches are momentary or maintain contact type, which when pushed maintain their contact. These switches, like relays, can have a normal position; that is, they can be normally open as well as normally closed so that when pushed the contacts are reversed. The most common use of momentary push-button switches is for starting large motors like fans and pumps through an across-the-line starter. Momentary push-button switches can also be used in systems where the button is used to make a momentary connection between an RTD and a meter to read a temperature on a panel. There are many uses of maintain contact push-button switches. Push-button switches can be more than one pole, almost in unlimited fashion.

In some cases, especially in the area of low-voltage controls, slide switches are used. Their application is limited, however, since they are usually very low in terms of capacity. You may see them in residential control systems on the thermostats that control the household furnace and residential air- conditioning.

Other switches are explosionproof and weatherproof switches as well as some specialty switches. Any device that transfers power from one circuit to another and requires the human hand to operate it is a switch. To repeat what was said earlier, the main problem with switches is using one that is not adequate for the service. Occasionally, switches will wear out mechanically, but that is the exception instead of the rule. This depends upon the type and style of switch and the way it is constructed.

Transducers

Transducers is a term that has come into use in the last 35 years or so with the advent of electronic controls and energy management systems. The term is used to describe the device that changes a signal from one medium to another in a *proportional* way. That is, it coverts a pneumatic signal to an electric signal and vice versa. The key here is that the transfer is proportional during the transfer of signals. If, for example, the transducer passes 0–20 psi as the input goes from 0 to 18 V dc, when the input is 9 V dc, the output is 10 psi, and so on, proportionately. The same thing is true in the case of a transducer that changes 0 to 20 psi to either 4–20 mA or 40–200 mV of output amperage or voltage.

Some transducers can convert from pneumatic to electric and/or electronic signals (Fig. 6.8) to match all of the controllers on the market today. The most common are those that convert the electronic or electric signals from the modern electric and electronic controls to a pneumatic signal to operate a pneumatic valve or damper motor. These types are more predominant in today's control systems for the simple reason that although there is more and more emphasis on electric and electronic controls, the actuator end of the business has not kept pace with the controllers and other items. The bottom line is that a pneumatic actuator, whether it be a valve or a damper motor, is still the simplest, most troublefree, most economical device that can be manufactured and used in a control system. None of the electric and electronic actuators on the market today can come close to a pneumatic actuator in terms of cost or easy operation. Even the latest DDC control systems used in energy management systems usually use pneumatic actuators with transducers.

Most research in the last few years has gone into the development of transducers that change the signals from one type of signal to another *accurately* with the ability to repeat the information continuously over

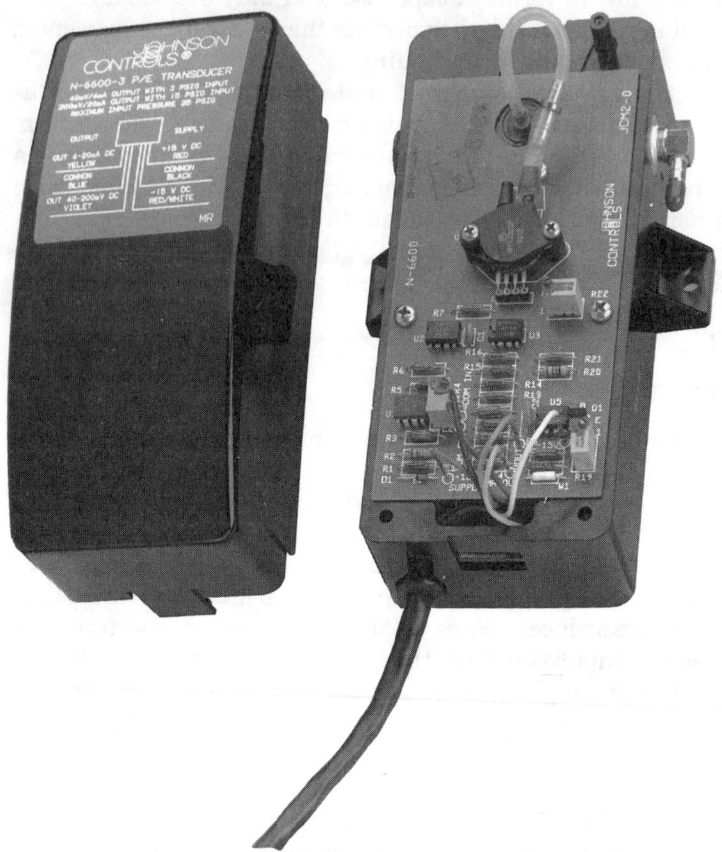

Figure 6.8 Transducer. (*Courtesy of Johnson Controls Inc.*)

a long period of time. There are transducers that use ac voltage, dc voltage, ac amperage, as well as dc amperage. The pneumatic outputs are only limited by the pneumatic inputs to the transducers.

Some modern electronic energy management systems use plug-in transducers in panels. This allows flexibility for modifications to energy management and control systems at the panel itself. These plug-in transducers are subminiaturized, and are usually nonrepairable and what is sometimes called throwaway.

The problems that occur with transducers involve the repeatabilty of the signals from the input medium to the output medium. The transducers can require repeat calibration and checking to be sure they are accurately transferring the information from one medium to another. It has been suggested that in critical situations gauges and meters be used on both sides of the transducer to ensure the repeatabilty mentioned above. Transducers should never be used if vibration is a problem since they are highly sensitive to vibrations. Voltage drop to the coils of a transducer can be a problem, so adequate wiring is important and skimping on wire size and other factors can cause problems when using transducers.

There is a class of devices sometimes considered under the classification of electric-to-pneumatic and pneumatic-to-electric devices that need to be mentioned here. They are the electric-pressure EP, and pressure-electric PE, switches, and they are very different. The fact is, EP switches are *not* switches; they are valves (Fig. 6.9). And PE switches are devices that switch electric current based upon a signal from a pneumatic air line.

An EP switch is a device that takes an electrical current and sends an air signal to its branch line from main air and exhausts its branch line when the voltage is discontinued. The device is nothing more than a three-way air valve operated by a solenoid coil that allows for the passage of control air to control devices when energized and exhausts that same branch line when the coil is deenergized. The most common use of this device is when the coil of the EP switch is connected to a fan starter to supply the control system with air when the fan is started and to shut down the control system when the fan is stopped.

A PE switch (Fig. 6.10), on the other hand, is truly a switch that is activated by an air signal from a pneumatic control system so that something electrical can take place. An example is a pneumatic signal that starts a pump when it reaches a certain level. The PE switches available today are adjustable so that at any point along the 0–20-psi signal from a controller, a fan, pump, and so on, can be started or stopped. These switches come in a variety of configurations, such as SPST, SPDT, and DPDT. They are available in many pressure ranges and differential settings, and some are explosionproof. Some of the most popular PE switches use a mercury switch to make the electrical contact from a bourdon tube arrangement just like an air gauge. The one PE switch that almost everyone will recognize is the one used on

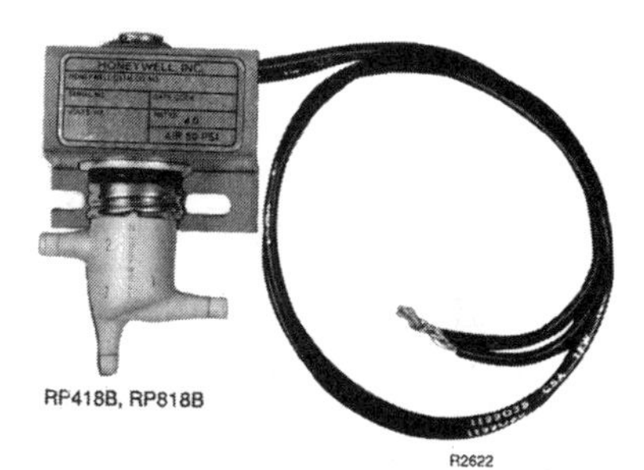

Figure 6.9 EP switch. (*Courtesy of Honeywell Inc.*)

Figure 6.10 PE switch. (*Courtesy of Mercoid Corp.*)

an air compressor to shut off the motor when the pressure in the tank gets high enough to prevent the tank from blowing up.

Pressure Controllers

Although pressure controllers are not strictly auxiliary devices, their position in the spectrum of controls indicates they are not common and therefore might be called auxiliary equipment. There are static pressure, differential pressure, low-pressure, high-pressure, and velocity pressure controllers.

Static pressure controllers (Fig. 6.11) are probably the most common since modern VAV systems require their use static pressure controllers at the VAV box in the space or just above the space. It is important to note that one of the most difficult control schemes used in the industry is *static pressure control.* This is because with any static pressure controller there must be a *reference point* as well as the static point being measured. The problem develops when the designer, operator, or control engineer tries to pick the correct reference point. Since static pressure is in the range of inches of water pressure, very slight changes in the reference static pressure can affect the ability of the controller to control the system. For example, if the reference point is outside atmospheric pressure, slight changes in wind direction and velocity will upset the static pressure controller inside the building. Other chapters in this book will explore this subject further. Those

Figure 6.11 Static pressure controller. (*Courtesy of Honeywell Inc.*)

sections should be studied to understand static pressure controls thoroughly.

Low-, high-, and differential pressure controllers are available as pneumatic controllers as well as electric and electronic controllers. The ranges of these devices can go from −30 in.Hg to hundreds of psi positive pressure. Pressure transmitters are also available with matching receivers and receiver–controllers. Differential pressure controllers and transmitters have arrived on the scene since the popularity of VAV systems increased with the energy crunch a few years ago. They are used extensively to maintain a differential pressure between the supply fan and the return fan in a VAV system. The controller tries to maintain the *difference* between two sensing tips in the duct work that are connected to the controller. The actual value of the pressure at either sensing tip is not important to the controller, only the difference between the two pressures.

Clocks and Miscellaneous Devices

The final items that need to be mentioned under the category of auxiliary devices include clocks, programmers, air gauges, stepping switches, solar compensators, sun shields, thermometers, and restrictors, to mention a few of the more important devices.

Clocks come in two basic styles—24 h and 7 day. They used to be mechanical devices with clock motors and large wheels with adjustable trippers to activate a switch and start or stop something. Today, they are generally electronic and do not have any visible moving parts. Clocks often have multiple switch mechanisms for operating more than one device with either SPST or SPDT action. In some cases the action is DPDT on two switches in the clock. Usually the contact ratings of the clocks are such that they are used for pilot duty and require relays or starter coils to operate large motors, and so on.

Astrological clocks are available that are programmed to take care of the changes in the time of sunset as well as holidays and can be programmed for many years in the future. They have override systems that allow for last minute changes. These clocks are very expensive, however, and should only be considered if warranted.

Warm-up and cool-down programmers can save energy by delaying the start of a large system in the morning based upon the mass temperatures in the building. These devices can also be used to shut down the ventilation during the warm-up and/or cool-down period so as not to waste energy when the building is unoccupied.

Air gauges are simple but important parts of any pneumatic control system. They are available as surface or flush mounted and in a variety of diameter sizes. They come in an almost unlimited number of configurations as far as the dials are concerned, but the most common ones are usually 0–30 psi. Some gauges change color when they sense high pressure and reverse the color when they sense low or no pressure. Other gauges read temperature, humidity, static pressure, and so on, when the dials are properly installed and the gauges are connected to transmitters. Gauges are sometimes forgotten in the rush to specify and complete a control system installation. They are, however, about the most important auxiliary device that can be installed.

Stepping switches (Fig. 6.12) are used progressively to close or open small pilot duty switches in sequence as the air signal from a controller increases or decreases. The cut-in and cutout points as well as the differentials between the switches are adjustable. Step controllers are sometimes used to operate the capacity reduction solenoids of a piston-type refrigeration compressor. They can come with as few as 4 switches and as many as 12 switches. If more than 12 switches are required, two units can be tied together.

Solar compensators and sun shields are used to protect the bulb of a thermostat or transmitter from the elements as it tries to sense true ambient temperature. A manufacturer's literature will reveal what is available and the proper use of those devices.

Lastly, there is an almost unlimited number of types and styles of thermometers that can be used with any modern HVAC system and too often the systems are cheapened by the lack of sufficient thermom-

Figure 6.12 Stepping switch. (*Courtesy of Johnson Controls Inc.*)

eters in the system. Here is a case where a preponderance of thermometers in the system will never be too much.

In summary, this chapter described the other devices that go along with the controllers, transmitters, dampers, damper motors, and actuators. Although the devices are considered auxiliary items, they are no less important than the major items in a complete control system. Without them no control system would be able to function properly and accomplish the cycles required by today's technology.

Chapter

7

Construction Systems and Devices

All control systems, whether pneumatic, electric, or electronic, require a source of power. For pneumatic controls, the source of power is clean compressed air. For electric or electronic systems, the source of power is electricity. A few systems use expanding fluid, which does the sensing to operate the control devices themselves. Examples are the power of the refrigerant to operate expansion valves used on a direct expansion, DX, coil. A few systems operate radiator valves directly with the power of the expanding fluid in the sensor. Some experimentation has been done with systems that operate with steam as a motive power, but those systems were never used to any extent with HVAC systems.

Air Compressors

Most commercial control systems began years ago as pneumatic systems, and the science of air compressors has come a long way with years of experience behind systems that use control air. Commercial pneumatic control systems that modulate came long before comparable electric or electronic systems.

The concept of using air as a control power is even older than the concept of an electric-motor-driven air compressor (Fig. 7.1). Some earlier models of air compressors were water driven. That is, they were served by air compressors powered by the domestic water systems in a building. The air compressor consisted of a water-powered piston directly connected to another piston that compressed the air for the control system. Some of these air compressors can be seen in the museums of the control companies.

Figure 7.1 Air compressor. (*Courtesy of Johnson Controls Inc.*)

The first models of electric-motor-driven air compressors had large flywheels with flat belts. Generally, they were slower than the air compressors of today, and the pistons in the compressors were larger than the ones seen today, with the lubrication systems being primitive by today's standards. Not until the development of modern metals and the availability of today's compression rings were designers able to speed up the revolution per minute (r/min) of the compressors and thus use smaller pistons with increased volume capacity in the same space as the older models. The slow speed compressors lasted longer, however, and ran much cooler than the ones in use today. The act of compressing air to a higher pressure creates heat, and that heat can cause wear. The older ones, again, did not run as hot as the high-speed ones of today. By the same token, the older compressors were not able to pump up to the high pressures of today, and as a result the problems with moisture in the systems were more acute than today.

In addition to being classified by size, air compressors can also be classified by whether they are base mounted or tank mounted or whether they are a single or a duplex type (Fig. 7.2). Base mounted means the tank used in the system is in another location and is not part of the compressor. A tank is used to store the air that is compressed, since if there were no tank the compressor would be cycling

Figure 7.2 Duplex air compressor. (*Courtesy of Johnson Controls Inc.*)

on and off all the time. The tank acts as a buffer for the system and is absolutely necessary in any air control system.

Duplex compressors are usually two compressors on a single base or on a single tank that are the same size and sometimes picked to alternate in operation, thus reducing the wear and tear on one unit. Using a single compressor that is twice the size is also an option, but here the operators are putting all their eggs in one basket and gambling that the failure of that one compressor will not cause a major failure in the system. The standard procedure in larger systems is to use a duplex and size each compressor for about 75% of the load so that in an emergency the system's important functions will still be operable.

To understand sizing air compressors, we need to go back a bit to the pressures and use of the air in a pneumatic control system. The original systems settled on 15 psi as a control air pressure, probably because they were able to pump up to that pressure without too much difficulty and there was little damage that air at that pressure could do. Some industrial systems today use anywhere from 13 to 1000 psi. In the HVAC world, 20 psi is the standard for Johnson, and 18 and 22 psi are standards for some of the other control companies.

In order to function, all controllers and devices being controlled must *use air*; therefore, the air compressor is there to replenish the air being used with an even, steady pressure. This is accomplished at the compressor by pumping up to a certain pressure, at which point a pressure switch senses the set pressure and turns off the air compressor. When the pressure in the tank drops to a lower pressure as a result of the systems using air, the pressure switch turns the compressor on again and the process is repeated. This may sound elementary, but it must be understood to visualize the elements of "sizing" of the compressor.

Sizing is important, since the compressor will fail or require excessive service if it is sized too small and money will be wasted if it is sized too large. The sizing is based on the usage of air in the system so that the *standard cubic inches* rating of each instrument (SCIMS) is the figure used to determine the air compressor size. The compressor should be sized to run about 33% of the time, even though some of the specifications that are written today call for 50% run time. Any time that it runs more than that, there is the possibility of shortening the life of the compressor. The formula for sizing an air compressor properly is total SCIM of all devices ÷ percent run time × 1728.

After the compressor is installed the run time should be checked at least every month, with a stop watch or the sweep second hand on a watch to see that it is not running over 33% of the time. Also in most systems the number of starts per hour should not exceed three or four. It is possible to run less that 33% of the time and yet have the number of starts per hour be excessive because the size of the tank is too small for the system. An example is using a 15-gal tank on a system that requires a 20-hp compressor to supply the required cfm of air. Such a system would supply about 81 cfm to the controls and require a 200-gal tank. All of the manufacturers have charts that will give the recommended tank sizes along with the capacities of the compressors at the various pressures being used. Excessive starts per hour can be detrimental to the motor and the starters being used.

After the system is installed and sized to run 33% of the time, if the compressor runs more than the 33% it did when installed, it usually means the system has developed some leaks that need to be pinned down and repaired. It could also mean that the compressor has problems, such as broken piston rings, broken intake valves, or a faulty intake filter. To determine if the latter is true, isolate the compressor from the system and check the time it takes to pump up just the tank while checking the charts provided by the manufacturer. If the time is excessive, the compressor needs repair and the problem is not with the air tubing in the system.

Trying to save money by installing a compressor that is too small will only promote problems later on. It is better to err on the conservative side with the heart of the system, which is what the air compressor is in a pneumatic system. This will also help if any add-ons are necessary because someone forgot to add needed controls.

Duplex air compressors are used in the case of standby systems, where each compressor is sized to handle the whole load or very close to it and the compressors are alternately started with a special set of starters in a panel. This allows more even wear on the compressors and provides the standby feature. These alternating starter panels are available from the controls manufacturers as well as motor control starter companies such as Square D, Cutler Hammer, and Allen Bradley Co.

The smaller compressors (those with motors less than 1½ hp) usually do not have magnetic starters for cycling the compressors, but above that size, in particular those with three-phase motors, often require starters as part of the package. In some cases, the starters can be mounted on the compressor or be remote mounted depending upon the size of the motor. With larger systems it is important that the compressors do not cycle off and on too frequently, since this can damage the starter.

The question is often asked about where is the best place to install an air compressor for a pneumatic control system. The answer depends upon many factors. First, the compressor needs to be in an equipment room that is convenient to the operating personnel in the building. Using the old principle of out of sight out of mind, remember that the compressor is the heart of the system, and if it is not readily available and visible, it will be neglected. Second, since it is affected by heat more than by cold, it needs to be in the coolest spot possible in an equipment room. Third, it should be near a floor drain since modern air compressors have drainage systems for the water that is removed from the control air, and that water needs to be drained some place. Fourth, the location has to take into account that suction to the compressor should be as short as possible and the air taken in by the suction needs to be as clean as possible (usually outside air). The longer the suction line, the more likely it will be that the compressor has to work extra hard and the pistons will pull oil from the crank case, which will be pumped into the control air system. Modern control air compressors are installed with special intake suction filters, and the manufacturers will provide charts that indicate the size of the suction line based upon the distance from the compressor to the suction filter, which is usually outside the building. Thus, the compressor location must also take the suction line into account. Fifth, the location needs to involve the piping to and from the compressor and allow for simple and easy installation of the mains and risers of the control air system. In some cases, it is possible to mount the control air compressor on a wall with brackets and bring the installation to eye level. This again helps ensure that the maintenance individuals will see the compressor and service it when necessary.

Air compressors come with instructions for service. If the manuals are followed, the compressors will give years of service. Some important items are that the oil level in the crank case needs to be checked frequently, and the oil lost through vaporization and heat needs to be replaced. As with an automobile, the oil should be changed when it is warm so that it drains easily and it does not take all day to change the oil in the crank case.

There is a trick that can be used to speed up the process. It is to use a squeeze bulb along with a piece of ¼-in. plastic tubing and a cork with a hole drilled in it. Place the cork and the tubing with the bulb connected to it in the dipstick hole of the compressor and force the oil out of the crank case with air pressure using the squeeze bulb. The oil can also break down and needs to be changed in accordance with the recommendations of the manufacturer. Remember, only the recommended oil should be used, not something that was on sale at a local discount store. Also note that synthetic oil should never be used without checking with the control manufacturer since these types of oils can have a detrimental affect on the controllers and the filters used in the system.

Proper belt tension needs to be maintained, and a quick look at the manual will show how to keep the tension and alignment of the belt proper. Some air compressors have belt guards that hide the rotation arrow on the flywheel; if the motor is installed incorrectly the compressor could be operating in reverse, which can be detrimental to the compressor since in some cases the oil pump only operates when the rotation is as specified by the manufacturer. This usually happens with a three-phase motor and the installing electrician reverses the wires to the motor. The solution is to check that the rotation of the air compressor flywheel is correct on all units with three-phase power.

There are four requirements for control air systems. The air must be oil free, clean, dry, and at a constant pressure. To accomplish the first requirement, the compressor must be serviced and maintained so the rings on the pistons do not blow oil by. The suction line to the compressor must be adequate so the suction stroke does not pull oil from the crank case. The temperature of the compressor needs to be watched so high temperature does not break down the oil and make it easily susceptible to blow by the oil in an aerosol or vapor state. The more the running time, the more there is a chance that an aerosol or vapor state will occur in the oil. When that happens, the system will need the proper type of oil filter to protect the controls from contaminated control air. One of the filters that can be used is a coalescent filter (Fig. 7.3) that is 99.9% efficient at 0.03 μ. These types of filters must be sized properly for the system (too large is just as bad as too small); when that is done they are very efficient. Sometimes it may be

Figure 7.3 Coalescent filter. (*Courtesy of Johnson Controls Inc.*)

necessary to add an activated charcoal filter (Fig. 7.4) to the system to catch the vapor phase oil that is too small for the coalescent filter. Also a 5-μ prefilter can be added before the dryer, reducing station area filters to extend their life. An additional refinement is the addition of oil indicators (Fig. 7.5) in the system to show when the system is beginning to pass oil to the controls and that something is wrong. Catching the problem long before it gets out of hand is a sure way to save the controls. A system that is full of oil not only does not work properly, it is a mess to clean up and more expensive to replace completely.

Recall that one of the four items for ensuring a good supply of control air is dry air. All air contains some moisture, and a psychrometric chart will show that the ability of standard air to hold moisture is determined by two things: its temperature and its pressure. The colder the air, the less moisture it can hold; the higher the pressure of the air, the less moisture it can hold. As far as control air is concerned, we raise the pressure as high as we reasonably can without going into special ASME high-pressure vessels, and after that we lower its temperature as low as we can to wring out the moisture further. These two steps will guarantee the driest air available to the control system.

Refrigerated dryers (Fig. 7.6) are used now but not years ago. The first control systems using air were basically used only for heating control systems that operated in the winter. Air for those systems

Figure 7.4 Charcoal filter. (*Courtesy of Johnson Controls Inc.*)

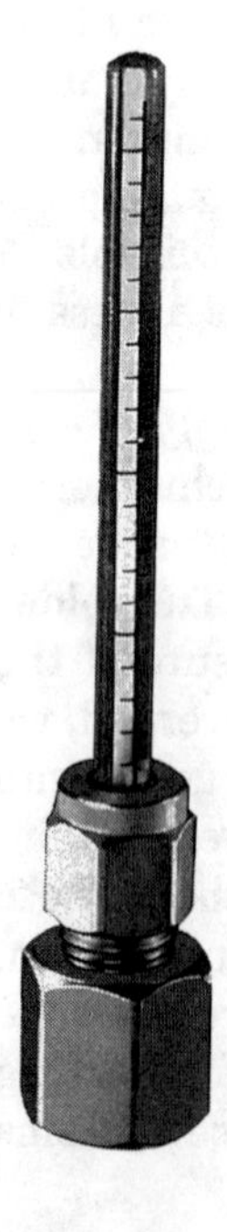

Figure 7.5 Oil indicator. (*Courtesy of Johnson Controls Inc.*)

Figure 7.6 Refrigerated dryer. (*Courtesy of Johnson Controls Inc.*)

came from outdoors, and winter air is the driest. Since there were very few systems installed in air-conditioned buildings that operated in the summer when the outside air is full of moisture, the need to eliminate moisture in the air did not exist. When the technology changed and more air-conditioning systems were being installed, it became apparent that something had to be done about moisture in the control system. One of the first things that was done was to try to take the air from the coldest area available in the summer. This included putting the suction of the compressor in the cold deck of an air-conditioning system to try to get the coldest, driest air available.

That idea worked well except when the air-conditioning system was shut off and the air compressor was still operating and sucking air from the cold plenum that was not cold. When chilled water was available, the air was sometimes cooled by a special heat exchanger using that chilled water. Some systems even used the cold water from a chilled water fountain cooler in an effort to dry the air and squeeze out the moisture. Chemical dryers were and are being used with limited success, but they require maintenance and checking that is not always available. Thus was born the modern refrigerated air dryer that almost all systems today use.

Today, refrigerated dryers come in various sizes and are almost totally automatic and fool proof. They have automatic drains and a control system to maintain the proper temperature of the air leaving the dryer. Desiccant or chemical dryers are also available to dry the control air. The desiccant dryers use silica gel, a chemical that has a great affinity for moisture. Some of these dryers have cartridges that have to be replaced when they are full of moisture; Indicators turn color when that happens, so the operators know when to make the change. Other types use two sets of cartridges (Fig. 7.7) alternately, so that when one is being used the other is being heated to dry out the moisture removed from the control air. The dryers have electric heaters built into their chambers. The electric heaters are turned on and off with a clock. In the twin tower types, the dry tower can be used to dry out the wet tower. If proper oil separation is used, these types can give a long life and eliminate problems with control air moisture. As seen in the following chart, the results from a twin tower dryer with varying operating pressures can be excellent based upon percentage of output cfm used to dry out the alternate tower:

Operating Pressure (psi)	% Purge	Dew Point (°F)
80	23	−40
60	29	−40
50	34	−40

This system requires line voltage power to the set of towers that contain the cartridges of silica gel.

As indicated previously, all modern pneumatic control systems use low-pressure air of about 20 psi. If the pressure switch on the compressor were set to that pressure, the control main air would be up and down like a yoyo and the control system would never settle down. To be sure the control air pressure is constant, the compressor pumps up to a high pressure (90 psi or more) and a series of pressure reducing valves (PRVs) is used to reduce the pressure to that required by the control system. Also remember that the higher the pressure in the

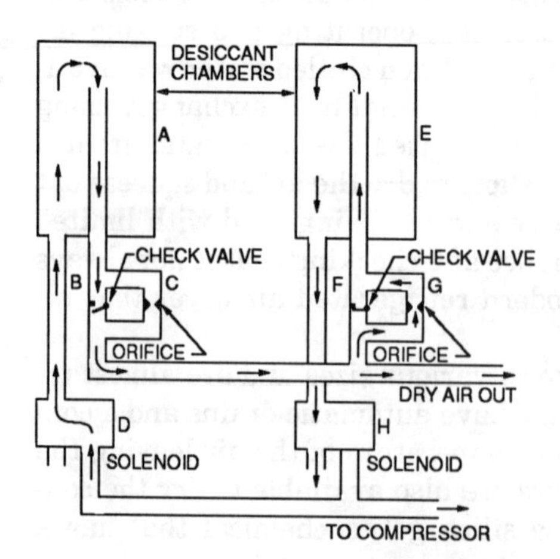

Figure 7.7 Desiccant air dryers. (*Courtesy of Honeywell Inc.*)

tank, the easier it is to squeeze the moisture from the control air at the tank. The reason systems typically use two or more PRVs to reduce the pressure from the high in the tank of 90 psi, to the required 20 psi is that experience has shown that to reduce the pressure from 90 psi in one stage often results is fluctuating main air pressures and the best results are obtained when the reductions are done in at least two stages. On jobs that use pneumatic transmission and plus or minus 1°F is required, it is important to have the types of PRVs that can hold plus or minus 0.05 psi. Sometimes there are more than two PRVs used in a system to reduce it from 90 to 20 psi, then to 15 psi, which is used in a day–night control system or even a heating–cooling system. These systems require two different air pressures to switch the thermostats from heating to cooling or from day to night and vice versa.

Using two air pressure systems is common and allows the thermostats to have different temperatures for heating–cooling and day–night; in some cases separate dials are also used.

Some smaller air compressors do not have automatic drains even though they should. In those cases, care must be exercised to see that the tank drain is opened periodically. Some types of compressors do not have pistons but use diaphragms, with the up and down movement of the diaphragm doing the job of compression. In some of those cases, the air compressors are small and without oil. When those without oil are used, it is important *not* to oil them as damage will result. At times compressors are attached to the fan shaft of a system that do not require a motor. Since they do not operate while the fan is down, if that is a problem, other arrangements must be made. Sometimes control air can be drawn from very large industrial air compressors used in manufacturing or processing plants. Remember, however, that the air used for those systems in those plants is not always as clean as required by a control system. The plant may or may not have requirements for clean, dry, oilfree air like a pneumatic control system.

Control Air Systems for Pneumatic Controls

Pneumatic type of control systems need tubing to be run around a building to the various controllers and the devices being controlled. The types of tubing used for this purpose is the subject for this section. By the way of history, the first systems used a combination of flexible lead pipes and ⅛-in. galvanized threaded pipe with threaded elbows, tees, and couplings. The installation of those systems was labor intensive, since it required a lot of work to cut and thread the pipe every time a connection had to be made.

With the advent of seamless copper tubing it soon became apparent that galvanized pipe was too difficult to use and copper pipe, with its ease of connections through soldering, was the way to go. At first the systems only used the hard copper; but after some experimentation it was discovered that the annealed soft copper also had a place in the control industry.

Soft copper tubing is available in 50- and 100-ft rolls, and hard copper tubing is available in 12-ft sticks. Soft copper tubing is usually available in sizes from 3/16 in. up to 1/2 in. (OD). Hard copper tubing is available in sizes from 1/4 to 2 in. OD.

Tubing used in the plumbing and heating areas has a wall thickness in the range of 0.035–0.050 and more. These types of tubing are referred to as types K, L, and M.

Type K is used for fuel oil, air-conditioning, refrigeration, underground service, and low-pressure hydraulic lines. It is also used in interior plumbing, steam and hot water heating systems, and some industrial process applications. Type L is used in medium pressure interior plumbing, panel heating, domestic steam and hot water heating systems, and some industrial application with processes involved. Type M is recommended for drain and vent lines. As the lettering system goes from K to L to M, the wall thickness of the tubing decreases. K has the thickest and M the thinnest wall.

Copper tubing used by the control industry does not follow the K–L–M wall thickness series. This tubing generally is sized on the basis of outside diameter with a wall thickness of 0.025 in. up to 0.625 in. OD and a wall thickness of 0.030–0.035 in. up to 1⅛ OD. An example of the two types of tubing are as follows: 3/8-in. control tubing is 3/8 OD, and the same tubing in K is 1/4 in. with an OD of 3/8 in. and a wall thickness of 0.035 in. This is not to say that the tubing with the thicker walls cannot be used in the control industry; it is just that the tubing and fitting used are less expensive than the other.

In the past, aluminum tubing was tried in the control industry, only to be dropped because of it being more difficult to make the connections than with copper and soldering. Aluminum tubing is less expensive and lighter in weight than copper, and resists damage from corrosion; however, other workmen such as carpenters and electricians can do damage to aluminum tubing more easily than copper tubing. With the advent of the crimp-type fittings, it is now possible to use aluminum tubing in place of copper tubing in the control industry. Crimp-type and other fittings used for pneumatic tubing will be discussed later. Aluminum tubing is available in hard drawn 12-ft sticks and in soft 100-ft coils. A disadvantage of aluminum tubing is that a protective coating must be used in direct burial applications in the ground and in concrete.

In the last few years, polyethylene, plastic, tubing has made its way into the HVAC industry, in particular, the control industry. In 1965, it was almost never used or specified. Today, it is almost the standard of the pneumatic control industry.

Polyethylene, or plastic tubing as it is referred to most of the time, is available in 3/16 in., 1/4 in., 3/8 in., and 1/2 in. OD sizes. Standard lengths are 100 ft, 250 ft, 500 ft, and 1000 ft. The lengths are usually shipped on rolls; smaller lengths are shipped in cartons. The standard color used in the control industry is black, although the tubing is available in various colors. The tubing is also available in multitubing bundles, where there is a number of color-coded tubes wrapped in a bundle with a protective PVC sheath over the entire bundle. In this case, the use of various colors allows the tracing of control lines from one end of a bundle to another.

Dekabon™ bundles are semirigid and consist of single tubes enclosed in a PVC sheath followed by a metallic barrier and a thermoplastic outer layer. Armored bundles are single tubes enclosed in a PVC sheath and galvanized flexible steel armor.

Plastic tubing can be installed easily and is particularly adaptable to the piping inside panels. There are considerable savings with the installation costs when compared to copper and aluminum tubing, since the number of tools needed is minimal. Changes are easy to make with this type of plastic tubing, and if care is taken in the installation, it functions just as well as copper or aluminum. It is corrosion resistant and does not transmit vibrations as with other solid metal types of tubing.

The disadvantage of plastic tubing is that it softens with heat and, as such, care must be taken not to place it in contact with heat sources that can soften or melt it. It also will sag if the ambient temperatures where it is installed change dramatically from normal to very warm, so extra fastening methods need to be used. Expansion can be as high as 1 ft/100 deg of temperature change. It has been suggested that if there is the possibility of physical damage or radical temperature changes, plastic tubing be encased in protective sheaths such as thin-wall conduit or in troughs of metal. It should be protected against chafing due to vibration and movement in and around mechanical equipment. Also, the bends made with it need to be carefully checked for kinks, since plastic tubing will kink if the bends are too severe.

In some cases, larger sizes of PVC (rigid type) tubing is used for control air lines, and it is available in schedule 10, 40, and 80. The schedule 80 pipe can be used with screwed fittings as well as with PVC glued fittings. The schedule 10 and 40 is always used with the glued fittings. PVC pipe is always rigid and should only be used where that type is satisfactory.

The advantage of PVC plastic pipe is that its less costly than copper, it weighs less than copper or steel pipe, and it can be used in almost the same areas as large copper or steel pipe. In the higher schedule types in some cases it has the same pressure rating as copper or steel pipe. The main disadvantage of PVC plastic pipe is that when glued fittings are used and a mistake is made or the fitting leaks, there is nothing that will dissolve the glue. The fitting must be cut out with a cutter or saw and the section replaced with a new one. Figure 7.8 gives a summary of the data mentioned here.

With any of the types of tubing used for pneumatic control systems, fittings are necessary to make the various connections to devices and so on. Soldered, compression, crimped, flared, barbed, and glued fittings are available for the tubing mentioned here.

Over the past decades, soldered fittings, which are only used with copper tubing, have had the largest usage in the control industry. The key to proper usage of soldered fittings with copper is using proper types of flux and solder and being sure the pipe and tubing at the juncture point are *clean*. The recommended solder to use with these types of fittings is 50% tin, 50% lead or 95% tin, 5% antimony. The latter is used where pressures may exceed 450 psi. The 50/50 solder is most commonly used in copper control tubing applications. A silver solder technique is used where constant vibration and corrosion are involved, but that type requires a very high temperature torch and is almost never used in the control tubing. It is reserved for Freon™ and similar piping systems.

The flux used in this type of work is generally the zinc chloride acid paste type. There are also self-cleaning types in stick form, where supposedly the flux can be used without cleaning the pipe or the fittings. Long-time control installers will dispute those claims, and most of them tend to use the paste types listed above. The key, as mentioned above, is to be sure the pipe where the fitting is to be soldered is clean and the inside of the fitting is clean of any oxides that have formed. After the fitting and tubing are properly wiped with flux, they need to be heated to about 365°F and solder touched to the joint where the fitting and the tubing meet. If the solder sucks into the joint with a capillary action, the fitting and tubing are hot enough. If it does not, the system is still too cold. Good installers never try to build up solder outside the joint to compensate for not getting the fitting hot enough.

Compression fittings are used only at the terminal points of the system, for example, as the final connections to the controllers and/or actuators such as damper motors and valves. That is done so changes can be made and because service is easier than having to unsolder a connection at a thermostat or valve. Compression fittings can be used with copper, aluminum, or plastic pipe if the proper ferrule is used. They are made of brass, plastic, and other special metals and come in all styles, such as elbows, tees, and couplings, and in sizes for 3/16–1/2

BUNDLED LOW DENSITY "P" TUBING

NO. OF TUBES	OUTSIDE DIAMETER OF BUNDLE				POUNDS PER 100 FT. OF BUNDLE				MINIMUM BEND RADIUS OF BUNDLE			
	1/4" Poly-Cor	1/4" Dekabon	3/8" Poly-Cor	1/4" Armor	1/4" Poly-Cor	1/4" Dekabon	3/8" Poly-Cor	1/4" Armor	1/4" Poly-Cor	1/4" Dekabon	3/8" Poly-Cor	1/4" Armor
2	5/8	11/16	7/8	13/16	9	15	14	40	1-1/2	5	2	5
3	5/8	—	7/8	13/16	12	—	18	41	1-1/2	—	2	5
4	13/16	13/16	1-3/16	7/8	13	20	21	45	2	10	2-1/2	6
5	7/8	—	—	15/16	15	—	—	48	2	—	—	6
7	7/8	1	1-1/4	1-1/32	18	29	30	53	2-1/2	12	4	7
8	1-1/32	—	—	1-3/32	20	—	—	63	2-1/2	—	—	7
10	1-1/8	—	1-5/8	1-1/4	23	—	44	72	3	—	5	8
12	1-1/8	1-9/32	1-45/64	1-5/16	26	44	61	76	3-1/2	15	6	9
14	1-1/4	1-11/32	—	1-3/8	29	48	—	81	4	16	—	10
19	1-3/8	1-1/2	2-5/64	1-1/2	37	56	86	92	5	18	10	11
37	1-61/64	—	—	2	75	—	—	131	9	—	—	14

NOTE: Two Tube Dekabon is Elliptical in Shape.

Figure 7.8 Piping data. (*Courtesy of Johnson Controls Inc.*)

PLUMBING AND HEATING COPPER TUBING

(Dimensions in Inches)

OUTSIDE DIAMETER	NOMINAL SIZE	CLASS	WALL THICKNESS	POUNDS PER 100 FT.	INSIDE VOLUME (CU. IN.) PER 100 FT.
3/8	1/4	K	.035	14.5	87
		L	.030	12.6	93
		M	.025	10.6	99
1/2	3/8	K	.049	26.9	152
		L	.035	19.8	174
		M	.025	14.4	191
5/8	1/2	K	.049	34.4	262
		L	.040	28.4	280
		M	.028	20.3	306
3/4	5/8	K	.049	41.8	400
		L	.042	36.2	418
		M	.030	26.3	448
7/8	3/4	K	.065	64.1	524
		L	.045	45.4	580
		M	.032	32.8	619
1-1/8	1	K	.065	83.8	932
		L	.050	65.3	990
		M	.035	46.4	1050

POLYETHYLENE TUBING

(Dimensions in Inches)

Standard Low Density "P" Tubing and "FR" Tubing
Recommended Temperature Limits: −100°F to +175°F – Black
−100°F to +130°F – Other Colors

OUTSIDE DIAMETER	WALL THICKNESS	POUNDS PER 100 FT.	INSIDE VOLUME (CU. IN.) PER 100 FT.	MINIMUM BEND RADIUS
3/16	.030	0.6	15	1
1/4	.040	1.0	27	1
3/8	.062	2.6	60	1-1/4
1/2	.062	3.6	133	2-1/2

NOTE: Type "FR" Flame-Resistant Tubing has a minimum temperature of 0°F.

High Density "HP" Tubing
Recommended Temperature Limits: −60°F to +220°F

OUTSIDE DIAMETER	WALL THICKNESS	POUNDS PER 100 FT.	INSIDE VOLUME (CU. IN.) PER 100 FT.	MINIMUM BEND RADIUS
1/4	.040	1.1	27	1-1/2
3/8	.062	2.7	60	2
1/2	.062	3.7	133	3-1/2

NOTE: Type "HP" Tubing is available in Black only.

Figure 7.8 *(Continued)*

in. OD tubing. The adaptors have a portion that is standard inch pipe size, IPS, threads for use with standard pipe; the nut and sleeve portion of the adaptor has different threads and cannot be screwed into a controller or a controlled device. It is important that the metal ferrule be used with metal pipe (copper or aluminum) and the plastic ferrule be used with plastic pipe.

JOHNSON SERVICE COMPANY ALUMINUM TUBING
(Dimensions in Inches)

OUTSIDE DIAMETER	TYPE	WALL THICKNESS	POUNDS PER 100 FT.	INSIDE VOLUME (CU. IN.) PER 100 FT.
1/4	Hard Soft	.032	2.6	32
3/8	Hard Soft	.035	4.4	87

OUTSIDE DIAMETER	TYPE	WALL THICKNESS	POUNDS PER 100 FT.	INSIDE VOLUME (CU.IN.) PER 100 FT.
3/16	Soft	.025	4.3	20
1/4	Hard Soft	.025	6.9	38
5/16	Hard Soft	.025	8.7	65
3/8	Hard Soft	.025	10.5	100
1/2	Hard Soft	.025	14.4	191
5/8	Hard	.028	20.3	306
3/4	Hard	.030	26.3	448
7/8	Hard	.032	32.8	619
1-1/8	Hard	.035	46.4	1050

Figure 7.8 (*Continued*)

Crimped fittings are available that can be used with copper, aluminum, and plastic pipe. They are available for use with tubing from 3⁄16 to 3⁄8 in. OD. The crimp fittings are made of plastic with an aluminum ring at each point where a joint is to be made. The inside of the fitting is serrated to grip the tubing and prevent twisting or turning. If these fittings are used with soft copper, plastic, or aluminum tubing, a metal insert is used to prevent the tubing from being crushed when the aluminum ring is crimped. A special crimping tool is necessary when using these types of fittings; using that tool makes the system almost foolproof, as the tool will not release until the proper pressure has been applied and the ring around the fitting is crimped tightly. The disadvantages of this fitting are that they are more costly than soldered or barbed fittings and a special tool must be used.

Barbed fittings are the most economical and efficient way to connect plastic (polyethylene) tubing. The installation is simple and requires no special tools. As a matter of fact, except for the final connections at the controllers and controlled devices, an installer only needs a pair of scissors to cut the plastic tubing; the barbed fittings can be installed with hand pressure. Barbed fittings that have threads for the final connections to devices require a thread sealant. The plastic fitting must be torqued tight and given a one-half extra turn. If tightened too much, the plastic product may crack. Samples of typical pneumatic control fittings are shown in Fig. 7.9).

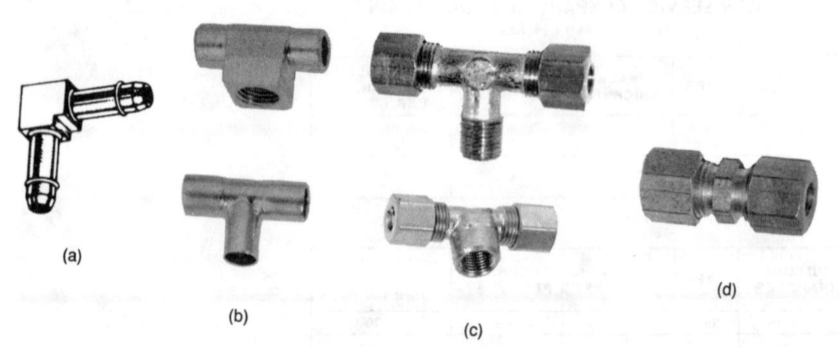

Figure 7.9 Typical pneumatic control fittings. (*Courtesy of Johnson Controls Inc.*)

Flared fittings are almost never used with control tubing systems because of their high cost and the need for special tools. They are used in refrigeration work since they provide a seal that is the least likely to leak. That kind of tight seal is not necessary for control tubing.

Glued fittings are used only with PVC pipe as indicated above. They are easy to use, but the installer must work fast as the glue takes affect within seconds when the two sections are joined and there is not much room for error or adjustment. Before using the glue (which in effect welds the pieces together as is dissolves the surface of the pipe and the fitting), the pipe and the fitting should be cleaned of burrs caused by cutting and wiped with a special cleaner that assures the glue will make a permanent joint.

There are too many holders and fasteners available for the mounting and fastening of pneumatic tubing systems in pneumatic control installations to mention here. Also, the proper way to install pneumatic tubing and test the installation could easily take up an entire chapter. Suffice it to say, there is an art to proper installation techniques, and only many years of training will allow one to become an expert.

Wiring Systems for Electric and Electronic Controls

Many old systems as well as new DDC and electronic systems use wire for the transmission of power and signals to the devices being controlled. Also, before there were electronic controls, there were electric controls that performed well in areas that could not use pneumatic controls and systems. Furthermore, there is the all important tie in between the pneumatic control systems and electric devices such as fans, pumps, and chillers that use electric motors and other electric devices. Without a knowledge of those items, the temperature control system, be it pneumatic, electric, or electronic, cannot be made to function properly even though the control engineer has specified all the components properly. This section will therefore discuss, among other things, low-voltage control wiring, line-voltage control wiring, and high-voltage and electronic low-voltage wiring and systems.

Before beginning the discussion, some definitions are in order concerning the hardware and the systems. The definitions generally are from official publications of organizations such as the Underwriters Laboratory and the National Fire Protection Association, and most of them are in a book called the *National Electric Code* (NEC), approved by the NEC committee of the United States of America Standards Institute (USASI).

For the purpose of discussion of control wiring, this chapter defines two types of wiring systems:

1. Low voltage: anything below 50-V ac or dc
2. Line voltage: anything above 50-V ac or dc

As a practical matter, no dc systems above 50 V are used in building controls. The term *line voltage,* although used with anything above 50 V, is usually meant to imply 120-V ac.

The types of wire that will be involved are generally classified as follows:

1. By gauge (Size)
2. By material in the wire (copper, aluminum, etc.)
3. By wire structure (solid or stranded)
4. By wire insulation (thermoplastic or fiber)
5. By shielding used with the wire

For a complete description and suggested usage of all types of wire, refer to the *NEC.* Also note that solid aluminum wire is seldom if ever used in the control industry. It is less expensive but too difficult to work with in the cramped areas involved in panels and so on. Thus, no stranded aluminum wire is available on the market. The only area where wire with a nonthermoplastic covering might be used is in low-voltage systems in residences; this wire is similar to that used with a 10-V bell transformer. Article 725 in the *NEC* covers the low-voltage wiring used in control systems.

As far as power wiring to commercial buildings is concerned, the electrical power supplied to most commercial buildings today is described as a four-wire three-phase service. This means there are three power wires and one neutral wire in the electrical distribution system. There are two basic types of three-phase systems in buildings—Delta and Wye. The Delta system provides 240 V across each phase and 120

V between two of the phases and neutral. A voltmeter across the third phase and neutral will read about 200 V. The Wye system usually provides 208 V across the phases and 120 V between any phase and neutral. The Wye system is also used to provide 480 V between phases and 277 V between any one phase and neutral. This system will not provide 120-V line voltage without an additional transformer. The electrical power to a building is usually pulled to the main power panel and from there to the subpanels as power wiring and lighting wiring. Power feeders usually terminate at power panels in the equipment room, and lighting power is distributed to lighting panels throughout the building and not necessarily to the equipment rooms.

Control system power is usually supplied from one of three sources:

1. From the load side of a motor starter. (See *NEC* articles 430-71, 72, 73, and 74)
2. From a circuit provided in a power panel for the control system
3. From a 120-V lighting panel

The latter is the least desirable source for the control voltage. In any event, a separate disconnect switch with circuit protection is a must, and the switch must be protected in such a way as to prevent unauthorized tampering that can shut off the controls and affect the entire HVAC system. According to the code, systems that use the normal 24-V ac power for the controls do not require circuit protection in the 24-V side of the transformer.

The most common low-voltage systems from the past as well as today use 24-V ac as power. Some people have referred to these systems as three-wire series 90 systems after the classification given to those systems by a major control company.

Basically, the controller and the controlled device were connected by three low-voltage wires, and the type that was used was standard 18–16-gauge solid or stranded copper wire with a fiber or thermoplastic cover. The gauge was increased from 18 to 16 in cases of exceptionally long runs. In many cases, raceways such as thin-wall conduit were not used unless specified by the consulting engineer. If there is a good chance damage can result from other workmen such as electricians and carpenters or from vibration or other movement, it is a good idea to provide some protection in the form of a raceway, even though the codes do not call for such protection. Low-voltage 24-V wiring is common in the residential controls area; as a matter of fact, all control systems that involve residential heating and air-conditioning use 24-V low-voltage systems. Seldom if ever will that style of wiring be done in raceways of any kind.

The wiring used for line voltage (120-V and above) must follow local and state codes and usually will mimic the *NEC*. The wire will probably be the same as that used for all other power wiring in either the solid or stranded form, with gauges that suit the length of the runs and the amperage draw in the wire. Line-voltage 120-V wiring is common in the wiring for controls connected to the fan motors, pumps, and so on, in a fan equipment room. Typical wiring is from a motor starter to an EP relay that allows air from the control system to get to the controllers when the fan starts and exhausts the air from the control system when the fan stops. This type of wiring is typically line voltage of 120 V or higher depending upon the system. The control system through clocks, relays and so on, can also be involved in the starting and stopping of HVAC devices such as fans and pumps, and that too usually comes under the classification of line-voltage wiring.

There are many many cases where the control system is interfaced with the starters and other devices that require the use of line-voltage wiring techniques. In all cases, the wiring must follow the *NEC* or the local codes that apply.

As far as electronic wiring for analog electronic and digital controls is concerned, an entire new science has come upon the scene in the last few years. Electronic controls use low-voltage (sometimes millivolt) signals that are sensitive to stray currents and voltages that can affect the quality of the control signals. As such, these wiring systems fall into a different category than low-voltage or line-voltage systems. Some signals are ac, some are dc, and they almost always require special treatment when it comes to wiring.

In the area of electronic wiring the terms *quiet* and *noisy* wiring are used, and the preferred wiring for electronic wiring systems is never near or in the same raceway with noisy wiring. Quiet wiring is free of transients. It includes binary inputs to DDC systems, analog inputs, and outputs, 24-V power, communications bus, and sensor wiring. Noisy wiring is subjected to transients because of its proximity to sources of electrical noise or because of the loads to which the wiring is connected. Examples are switched low-voltage loads, switched line-voltage loads, and line-voltage distribution. As such, noisy wiring should never be in the same raceway (conduit) as quiet wiring. In panels the noisy wiring, if possible, should be isolated from the quiet wiring with panel dividers and at least be at the other end of the panel from the quiet wiring. Shielding of sensor and communications wiring is a common practice in electronic control systems.

Typical wire includes Belden #18 AWG Beldfoil #8760™. In some instances the Beldon™ wire needs to be a minimum of #16 AWG depending upon the usage and the length of the runs from a panel to a sensor and so on. Other types of wire meet the specification, including a metallic shield around the conductors and insulation. The shield is grounded to either an earth ground or to another similar and suitable ground; an earth ground is preferred. This ground needs to be extended throughout the system so stray currents do not affect the control system. In most cases the easiest way to be sure of a proper ground is to ground the cabinets to an earth ground then connect the shields to the cabinets. The normal low-voltage wire used, for example, for the 24-V power to a panel, needs to be #16 AWG stranded. The information in this section is general and varies with manufacturer, so it is suggested that the installers and specifiers check with the manufacturer to get its wiring specifications.

To find out about specifications involving pneumatic piping and wiring of controls, check the Construction Specifications Institute (CSI) and *Specifications: For Architecture, Engineering, and Construction* by Chelsey Ayres, published by McGraw-Hill (1975).

Troubleshooting Systems

This section will discuss troubleshooting pneumatic control piping and systems. It will begin with a the testing procedure used after a pneumatic control system is installed.

At the end of each workday experienced installers check the work that has been done for that day by capping the newly installed lines and pumping them up to 30 psi so see if there is any appreciable pressure loss (no more than 1 psi/h is acceptable). If copper lines are installed directly into poured concrete, it has been suggested that the installer pump the lines up to 30 psi while the concrete is being poured to be sure the lines are not damaged in the process. Corrections can be made at that time before the concrete hardens. After the entire system is tested, then and only then should the instruments and controlled devices be hooked up.

When there is a suspicion of trouble with a pneumatic control piping system, the first thing that needs to be checked is if there is *water or oil* in the lines. This can be done by opening up a line at the furthest point from the compressor and letting the air blow into a white handkerchief. If oil or water is in the lines, it will show up. If oil or water is in the system at the furthermost point, it is in the entire system. Cleaning a system that is full of oil and/or water requires a qualified service representative. In some cases it may involve the complete replacement of all the instruments after the lines are purged of oil and water.

The search for leaks in a pneumatic system can be frustrating, especially if the leaks are in piping concealed in the walls or the poured concrete floors. Chemicals, such as oil of wintergreen, can be used; they give off distinctive odors that can be traced after they are pumped into an air line by a hand pump. Ether should not be used in the lines to detect a leak, since it is highly explosive and can cause injuries. Leaks can be found if the systems are isolated in a step-by-step procedure. Sometimes the problems can be a pressure loss in the system, which indicates that proper procedures were not used in sizing the air lines.

At one time it was thought that rodents would attack a certain color of polyethylene (plastic) tubing and chew on it to sharpen their teeth. Research has since proven that they will only chew on it when it is their way, and the color has nothing to do with it. Most control companies will encase their plastic lines in metal raceways wherever there is a problem of that type. Remember, the proper maintenance of the control system, in particular the air compressor, will go a long way toward solving problems that involve contaminants in the controls lines. Troubleshooting then becomes a matter of such things as calibration and rearranging the cycles or changing the components.

Troubleshooting electric control wiring and systems requires instruments that can read voltage, amperage, and continuity (resistance). Troubleshooting of electric control wiring requires a basic knowledge of electricity and a properly installed system. A wiring diagram from which to trace the wires is important, as is a system installed with color-coded wires so tracing the system is possible.

Think of electricity just as you think of water in a pipe. The voltage is the pressure and the amperage is the gallons per minute (gpm) of flow.

Troubleshooting wiring systems is just a matter of seeing to it that the electricity gets from one point to another as it is shown on the wiring diagram and that there are no opens in the wire caused by mistakes or devices that have the circuit open. The continuity tester (ohmmeter) is used to make sure a wire does not have any opens in it and that it goes from one point to another as indicated on the diagrams. There is no magic in tracing a wiring system if you remember that the electricity must get from the hot wire to the neutral wire through the device connected to the two wires, at the same time using that current to do something at the device. An example is a pair of wires to a motor with electricity in the wires that makes the motor run.

Two kinds of wiring diagrams are used in the control industry; point-to-point and schematic, or ladder, diagrams. In a point-to-point diagram, each wire is shown just as it is on the job, from terminal to terminal. In a schematic diagram, the only thing shown is how the hot power gets to the devices being controlled. Schematic diagrams are used mostly to show how motor controls work in conjunction with all of the devices needed to make the control system work automatically.

In the final analysis, successful control systems have complete wiring and piping diagrams for the installer and future servicepeople and troubleshooters. If a set of diagrams is not available, every action from then on is just guesswork.

In summary, this chapter discussed the heart of a pneumatic control system—the air compressor and all its accessories and components, with emphasis on keeping the control air clean and free of oil and water and keeping the pressure at the right level, while distributing the air where it is needed by the controls. It discussed the types of tubing used with pneumatic control systems and the wiring of electric and electronic control systems. The chapter also touched on troubleshooting for both pneumatic and electric control systems. The most important suggestion is to follow the diagrams as closely as possible, use logic in the thinking process, and be sure that all recommended safety procedures are followed.

Chapter

8

Electric and Electronic Control Products

This chapter will discuss electric and electronic control products and will touch on some systems controlled by those products. The chapter will also discuss residential controls, which are, with a few exceptions, electric controls and systems. It will explain the differences between electric and electronic controls and systems. Some discussion of the various voltages used in residential and commercial markets will be included.

The chapter will provide you with information to distinguish between analog electronic controls and digital electronic controls and will give you an understanding of the difference between the two types of electronic control products and systems. The digital products and systems discussed will be basically stand-alone types that are not connected with an energy management system. Energy management systems will be covered in other chapters of this book.

Electric Control Components and Systems

This chapter begins with a discussion of residential systems installed in our own homes, the control systems with which most of us have grown up.

Residential controls date back to the days when the furnace in a residence was fired with coal, and the only thing the owner of the home was able to control was the opening and closing of the air shutter below the fire pot and the damper in the flue from the furnace to the chimney stack. Whenever someone wanted more heat, he or she opened the air shutter at the bottom of the fire pot and fed more air to the coal, at the same time opening the damper in the flue from the furnace to allow the gases to escape up the chimney. In the cold climates where the furnace was in the basement, at first people had to go to the basement and operate the shutter and the flue by hand. Later, they were able to install a set of pulleys and chains so the control could be accomplished from upstairs. From that improvement it followed logically that a thermostat in the living space could do the job and operate a motorized set of levers to open and close the air shutter and the flue damper. The first motor was the forerunner of those used today on dampers and valves. When energized, it rotated through a set of gears, a crank arm that was attached with chains to the dampers mentioned above. By the time the first damper motor was invented, the concept of using lower than line voltage for a home door bell was already in existence. Therefore, it seemed logical to extend that concept to the control industry but at 24 V rather than the 10-V door bell systems.

Oil furnace

It soon became apparent that consumers wanted more than the dirty fuel provided by coal, and the next innovation was the oil furnace. By this time, the thermostat that was used with the chain damper motor had gone through some improvements and was ready for the oil burner market.

To burn oil in a furnace or a boiler, the oil must be a certain grade and must be mixed with air so that it is atomized into a fine spray that will burn efficiently. This required a fan and pump to force the oil from the oil tank into an atomizing nozzle. The fan and pump were line voltage, so the use of relays became standard, along with safety controls in the stack of the furnace or boiler to be sure the oil ignited and did not go up the stack in the fine explosive state that could cause problems. The ignition was accomplished with a high-voltage spark, similar to the spark in an automobile. All of the controls were (and are) low voltage except the oil pump, the igniter, and the fan that blows the oil mist into the fire pot.

The basic cycle of the thermostat, calling for heat which turns on the oil pump, fan, and igniter with the stack switch allowing continuation of the cycle as long as the oil ignites, has not changed much over the years, other than adding sophisticated electronic controls to make the system more reliable.

Natural gas furnace

After customers discovered that oil burners left a sooty residue, engineers began to look for a cleaner and simpler fuel to use. The next step was natural gas for heat and, in some cases, cooling as will be described later. Natural gas was considered by some a time bomb since a build up of gas in a furnace or a boiler that was not ignited by a flame could, if turned loose in a space, cause (and has caused) disastrous consequences. In some cases, complete homes have been destroyed from the resulting explosion when a small spark ignited a house full of gas. The concept of a standing pilot light and a system designed to ensure that the gas was shut off if the pilot light was not present to light the main burner was necessary. Thus was born the original Baso Valve™. The valve worked by having the thermocouple sense the pilot light flame and send a millivolt signal to a Baso Valve™ to hold it open as long as there is an adequate flame from the pilot light. The concept of two dissimilar metals sending out a millivolt signal whenever the two metals are heated has been used for many years. These thermocouple-operated gas safety valves became the backbone of the gas furnace and gas boiler business and allowed technological advances in that area. There were even controls and systems on space heaters that used the power of the thermocouple to power the safety valve, as well as the main valves operated by the thermostat on a call for heat. The gas safety control valve was also combined with a shut-off valve so one neat package could be used to control the flow of gas to the burner. Except for the completely self-contained systems, a 24-V gas solenoid valve is controlled by the thermostat and the burner lights after the safety valve is satisfied and open to the flow of gas.

Room thermostats

To prevent the blowing of cold air when the thermostat called for heat, the fan did not turn on until the heat exchanger was warm enough to send air to the space at a comfortable temperature. At the same time, since some of the fans were belt driven and an overheated exchanger would result if the belt broke and an open gas valve kept heating the exchanger without any heat going into the space, it was necessary to have a thermostat that shut off the gas valve in the event the fan did not blow the heated air into the space. Thus was born the high-limit and fan switch combination that controlled the gas valve and the furnace fan as outlined above. Experience has shown that the fan switch that turns on the blower when the chamber heats up needs to be set as low as possible, allowing the fan to run as long as possible to give an even distribution of heat to the space. If the fan switch is set too high, the fan will cycle on and off with short bursts of hot air causing much more discomfort than if the fan runs longer with reduced settings of temperature.

The room thermostats used in the gas furnace and small boiler residential systems are low voltage (24 V). Since they are two position (we are unaware of any modulating gas valves on residential furnaces), they generally have the same problem as commercial systems in that they tend to overshoot because the systems are oversized when they are operating in mild weather conditions, which is the majority of the time. To offset this problem, most of the thermostats in the residential market use a *heat anticipator*. This is a small heater built into the thermostat that turns on when the thermostat calls for heat. It is close to the bimetallic temperature sensitive element of the thermostat. This causes the bimetallic element to heat up and break contact before the heat from the system actually reaches the thermostat, thus the name *anticipator*. There is a variable resister on the heater so it can be adjusted to fit different requirements of anticipation (Fig. 8.1).

New innovations appeared on the market soon after the first thermostat was perfected. Among other items was the concept of *night setback*. This is the idea that the daytime temperatures set on the thermostat were not necessary when the occupants were asleep. Instead of setting the thermostat down when the occupants retired and having them wake up to a cold house, engineers decided to operate the thermostat with a clock. Thus was born the first clock-operated room thermostat. Since that time, exhaustive studies by Honeywell and others have shown that this system saves considerable energy. Today, that concept is promoted in all kinds of electronic thermostats that not only accomplish the simple cycle but have the ability to set different multicycles every day of the week.

When commercial air-conditioning systems were promoted for the ordinary residence and it became apparent that simple systems were needed, the hermetic compressor came into vogue, with the direct expansion (DX) coil in the furnace and the compressor and condenser coil and condenser fan exterior to the residence. Sometimes called a split system, it made the controls simple since the same room thermostat with the addition of a switch to reverse the action of the thermostat to turn on the furnace fan and hermetic compressor through a re-

Figure 8.1 Electric residential room thermostat. (*Courtesy of Johnson Controls Inc.*)

lay was all that was needed. In some cases, a cold anticipator is used on the cooling cycle, and most residences are operated as described. The problem with the danger of explosions as with gas do not exist, and home air-conditioning controls are pretty simple.

About the only place we might find line-voltage controls involving homes in the HVAC industry would be the thermostat in a window air conditioner designed to start and stop the compressor in the unit. The accuracy of that type of control leaves a little bit to be desired, but window units were never designed to be a precise method of air conditioning a space. Whenever a line-voltage thermostat is designed to handle a large amperage load, it must be built with rugged contacts and action, making it less sensitive to temperature changes. The most sensitive thermostats and controllers operate with low voltage through relays to control systems.

With most of the controls in a residence being low voltage (24 V), the wiring codes do not generally apply. You will see wire strung all over the place in a residence without being protected in any type of raceway. This can cause problems if a system is to be updated and air conditioning is added to a home that only has a heating furnace. This is because the wire cannot be pulled out of the walls and new wire added with the increased number of wires sometimes needed when adding air conditioning to a system designed only for heating. Most professional electricians pull wire with multiple conductors (five or more), color coded to allow for future changes such as adding air conditioning. Today's systems have many new electronic innovations, such as the spark ignition in place of standing gas pilot lights in order to save energy. All in all, however, the principles are about the same as years ago—low-voltage, simple two-position controls that do a fair job of controlling the temperature in a residence.

In climates that have extreme winter conditions, a residence might have a boiler for hot water or steam heat. The cycles are similar, with the exception that there may be a circulating pump in a hot water system that in effect takes the place of the fan on a furnace. A thermostat turns on the circulating pump whenever the boiler water is hot enough or operates the boiler at a constant temperature, turning on the pump through a relay. No circulating pumps are necessary with steam since steam circulates on the basis of a pressure difference and gives up its heat to the space, while at the same time condenses to hot water for the return trip to the boiler. Again, the controls are usually low voltage with the exception of the pump and the items on the boiler that are line voltage when the boiler is oil fed. In the case of a steam boiler, usually the pressure switch that shuts off the boiler is line voltage if the system is oil fired and low voltage if the system is gas fed.

Commercial Electric Controls

People in the controls industry divide systems into three basic classes: residential, commercial, and industrial control systems. Commercial electric and electronic controls are discussed in this chapter. They are those systems used for HVAC comfort control of people in buildings other than a residence. Industrial controls are those generally used for a process, not for comfort of the individual in a plant or other similar facility.

The first commercial electric controls used for HVAC systems were almost copies of the systems used for residences, along with some line-voltage devices such as line-voltage room thermostats that controlled steam boilers, starting and stopping them on a call for heat. As more complicated HVAC systems were born, the control industry responded with items to control the systems.

First was the oil-immersed motor used on dampers and valves with a reversing induction electric motor controlled from a low- or line-voltage thermostat. Most of the initial motors were strictly two-position or floating types. The line-voltage motors soon gave way to the low-voltage concepts. The controllers were single-contact, two-position devices designed, for example, to close on a temperature drop and open on a temperature rise. Some of the motors were coupled with a spring, so when the power to the motor was interrupted, the motor returned to a normal position. Some were operated by floating-type thermostats that had a center off position in the thermostat switch, a closed contact on a rise in temperature, a different closed position on a fall in temperature, and so on, in the space or medium being controlled. This meant that the thermostat made a set of contacts for heating, a different set of contacts for less heating (or cooling), as well as a center off position where no movement of the motor takes place. This allowed the motor to rotate in one direction, come to a complete stop as the thermostat contacts were opened, then to rotate in the opposite direction as the thermostat contacts were made again in the cooling or less heat direction.

It soon became apparent to the engineers and designers that two-position control systems would not suffice for the more complicated systems that were being designed. Thus, the concept of modulating controls with electric devices was born.

The three-wire concept of modulating electric controls was the first breakthrough in the attempt of the electric control companies to compete with the pneumatic control companies that were installing modulating systems of pneumatic controls. These first systems and devices were nicknamed Series 90™ (Fig. 8.2) after the number given to the devices by Honeywell when they were first introduced on the market. The principle of operation can be seen in the figure.

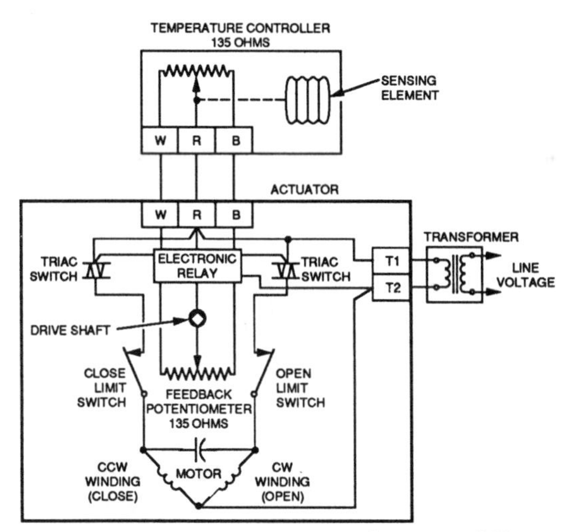

Figure 8.2 Series "90" systems. (*Courtesy of Honeywell Inc.*)

A room thermostat or other controller consists of a temperature-sensitive element (or pressure, humidity, etc., element) that drives a small 135-Ω potentiometer as the medium changes condition. The set of three wires from the ends of the potentiometer as well as the moving wiper in the middle are connected to a similar potentiometer attached to the shaft that rotates on the device (damper or valve motor) being controlled. At the same time, in the valve or damper motor there is a balance relay that senses the flow of current as the thermostat wiper moves to unbalance the whole circuit. The relay makes contact in one direction or another energizing the induction motor to move and rotate in the appropriate direction. As the rotation takes place, the wiper (potentiometer) in the motor is dragged along and rebalances the wiper (potentiometer) in the controller. If the upset at the controller is great and the potentiometer in the controller moves a large amount, the motor in the controlled device will also move a large amount to rebalance the three-wire system. When the system is again in balance the relay in the motor will seek the center position and stop the motor, ready to sense a change in current flow and unbalance again as the need arises. The sequence is repeated as, for example, the space or duct temperature changes.

One of the first problems recognized with the three-wire system was that during a power failure, the damper and valve motors stayed where they were when the power went off. With pneumatic control devices, however, the items returned to their normal position on a failure of air to the system. To overcome this deficiency of electric con-

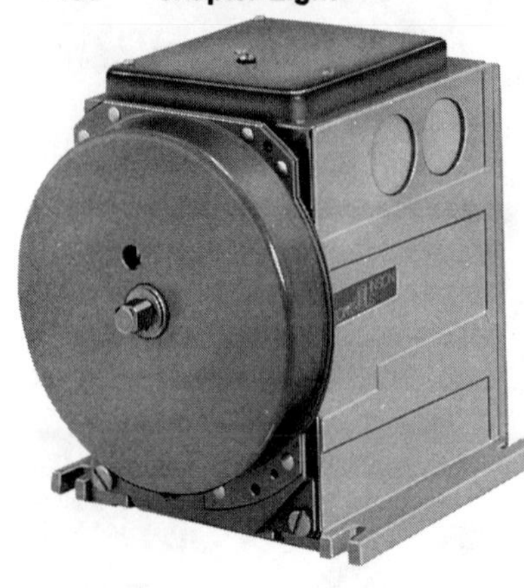

Figure 8.3 Return spring on electric damper motor. (*Courtesy of Johnson Controls Inc.*)

trols, the designing engineers built a coil spring (Fig. 8.3) into the electric actuators that wound up as the motor traversed, so that when the power was interrupted, the actuators returned to their normal position. A minor problem with that system was that the torque developed by those kinds of motors was reduced, as a majority of the torque is used to wind up the spring and cannot be used for useful work. Modern motors are sometimes provided with built-in batteries that return the motor to the normal position on a power failure without the use of springs. These batteries are the rechargeable type but must be replaced when they wear out and no longer accept a charge.

All of the Series 90™ and similar types of modulating control systems are low voltage (24 V) and as such can be used in various combinations with other devices in the system to affect just about all of the sequences specified in commercial HVAC control specifications. They are used with low limits, high limits, and heating and cooling switched systems. Controllers can be used to control more than one device, with one device acting as a master and the other a slave. When a system is installed that needs to reverse the action of the controller in winter versus summer, a double-pole double-throw switch is just added in the lines from the controller to the controlled device that accomplishes that task. The switch can be another thermostat with a built in set of contacts (DPDT), or the thermostat can operate a relay with the DPDT contacts. The same concept of a DPDT switch in the lines (three wires) from the controller to the controlled device can be used to allow the controller to control first one motor, for example, a heating valve, and control another device, for example, a cooling valve, when the switch is repositioned.

The only cycle and specification that gave the electric control manufacturers concern were the ones that called for sequencing of valves and damper motors. The specification for the pneumatic control people was not a problem, but it was almost impossible for the electric control companies. In some cases those providing electric controls and trying to compete with the pneumatic specifications had to furnish two different devices set at different levels to accomplish the same things pneumatic controls did with one device controlling two different actuators with springs set to sequence. Eventually, all manufacturers saw that there was a place in the industry for both types of controls. Commercial electric controls are still used today, although electronic commercial controls are gradually replacing them.

Electronic Control Components and Systems

Electronic controls are used today in many applications. The term *electronic* is used so often mistakenly that it has almost become a "buzzword" in any industry. Like so many other words that are misused, it has gotten to the point where no one pays any attention anymore to the phrase, and people expect all things to be *electronic,* so much so that engineers do not attach any importance to *electronic controls.*

Electronic controls came into vogue in the 1940s and 1950s when the systems used expensive clumsy vacuum tubes in the controllers. These controls were used because of their speed of response and accuracy. The size of the controllers limited their use to systems that could be mounted in equipment rooms. Not too much was known about the affects of long wire runs from the sensors to the controllers, and the controllers were not designed to operate controlled devices directly. Therefore, in the beginning electro-pneumatic transducers were used extensively. This required a pneumatic air compressor for the devices being controlled. The first designs used (and still do in some cases) the Wheatstone Bridge principle, which has been a known electrical concept for many years. According to that principle, the first applications of commercial electronic controls involved the used of indoor–outdoor controllers operating zoned systems. Examples are Honeywell's Weatherstat™ and Johnson's Duostat™. Those systems and all like them were deemed too costly and too large, in part because of the vacuum tubes that did not last long requiring maintenance that never took place. As a result they went out of fashion quickly, and were replaced by the transistor and the truly electronic age.

With the advent of solid-state science involving microprocessors, transistors, and printed circuit boards, we find electronic controls in every aspect of the controls industry, from the residential through the commercial to the industrial field. There are still, however, some misconceptions that need to be cleared up concerning electronic controls and systems.

First, electronic controls and devices use printed circuit boards with transistors, diodes, capacitors, and microprocessors and generally do not have moving parts. Electromechanical controls and devices can have some of the above but also have moving parts and items that are strictly electrical and not electronic.

Second, a distinction needs to be made within the electronic control systems concerning the difference between digital electronic and analog electronic controls. Digital control devices process information digitally just as computers do by recognizing ones and zeros that represent values such as temperature and humidity and massaging that information quickly. The programs that can be loaded into a small microprocessor in a controller are vast and allow for tremendous flexibility. The speed of the microprocessor also affects the ability for more accurate control. Experience has shown that electronic controls do not drift like pneumatic controls, which makes their repeatability a desirable attribute. There are some disadvantages to digital electronic controls, such as cost and the fact that the sensing of the elements to be controlled (temperature, humidity, etc.) still has to be presented to the controllers in an analog fashion and therefore must be converted to a digital signal at the controller through an Analog-to-digital (AD) converter. By the same token, since there are few known controlled devices that can accept a purely digital signal, the output of the microprocessor in the controller must be reconverted into an analog signal through a device known as a digital-to-analog (DA) converter (Fig. 8.4).

The use of electronic digital controls and systems has spawned another controls industry innovation—the use of computers and digital controls to manufacture and install *energy management systems*. These systems have existed since the 1960s. The reason DDC is available now is the need to decentralize control from the central computer and get it as close to the point of use as possible. This allows the central processing unit (CPU) to do its best job—gathering, storing, and reporting data. This chapter will not consider those types of systems but will be confined to stand-alone electronic digital controls.

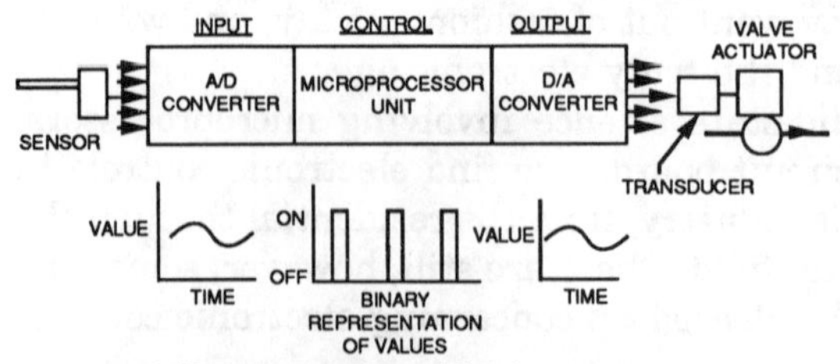

Figure 8.4 Representation of basic computer controls. (*Courtesy of Honeywell Inc.*)

Residential electronic controls

Examples of residential electronic controls are the new breed of night-setback residential thermostats (Fig. 8.5). They are vastly superior to the old electromechanical clock thermostats, which were only able to operate with one or two programs a day. The modern electronic residential thermostats can have up to six programs a day with a different schedule for each day of the week. Electronic controls are also used to eliminate the old standing pilot light of a gas furnace and to replace the slow-acting unreliable stack controls of the modern oil burner on a furnace or boiler. Residential split systems also use electronic controls to vary the speed of the condenser fan as the load changes, thereby saving energy. The energy efficiency ratio (EER) of the modern residential air-conditioning systems has gone from 4 or 5 to as high as 11 or 12 in the past few years due to the innovations of electronic controls and systems.

Commercial electronic controls

The area of commercial controls is where digital electronic controls have made the greatest gains. Their advantages of reliability, speed of response, accuracy have made users look twice at the idea of continuing to use pneumatic and straight electric controls. Two items, however, that have slowed the use of digital controls in the commercial field are the cost and the hesitancy by some to try something new. In addition, the input and output signals must be converted to analog signals. This requires devices that must convert analog signals to pneumatic signals for operation of the valves and dampers. This need

Figure 8.5 Residential night-setback stat. (*Courtesy of Johnson Controls Inc.*)

is particularly true when the valves involved are larger than, for example, 4 in. Most owners and engineers who are considering a retrofit of a commercial building control system will look long and hard at a system described above before switching to an all pneumatic or all electric system. Nevertheless, it seems electronic controls are being used more frequently in the equipment rooms of commercial buildings, where accuracy and speed of response are important. The terminal controls in the spaces may remain the conventional pneumatic and electric control items and devices.

There is a difference in the DDC area between the fully programmable DDC controllers and the ones that have some programs burned into the memory of the controller. The fully programmable types are typically used in the fan rooms to control an air-handling unit with a number of devices. These types allow flexibility in the programming and sequence changes. Fixed function units and controllers usually have a dedicated purpose, such as controlling a VAV box.

Although the emphasis in today's market is on digital electronic controls and systems, there is one other system of electronic controls that needs to be mentioned. It involves analog electronic controls. Analog electronic controls sense the items to be controlled the same way digital systems do. Analog controllers do not need AD or DA converters, as they accept the analog signals directly and have a variable gain analog output that is used directly by the devices being controlled. There are no microprocessors in the controllers. Most of them have Wheatstone Bridges in the sensing to generate a low-level analog voltage signal that is amplified through a variable gain op-amp to a signal that can be used by the controlled devices. Usually, the output signal is a voltage in the range of 0–20-V DC. DC voltage systems are used since DC current is easier to use in a system where it is desirable to have a plus and minus current for control. As with the digital controllers, the sensing of temperature is done with a wire bobbin, which consists of a wire that changes its resistance with an ambient temperature change (Fig. 8.6). The wire has a fixed resistance that can be measured as the temperature changes. The range, repeatability, and ohms per degree of the wire is a known value that is used in the analog and digital controllers. Examples of the wire are platinum, nickel, and alloys of those metals.

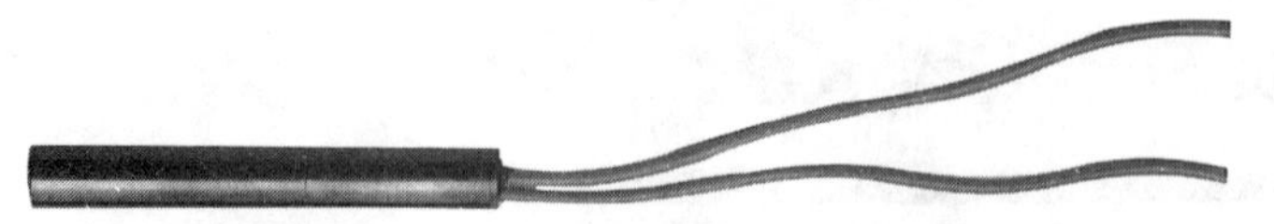

Figure 8.6 Bobbin RTD sensor. (*Courtesy of Johnson Controls Inc.*)

Since analog controllers do not have microprocessors built into them at the time of manufacturer, they are considered fixed program controllers and cannot generally be modified to function differently in the field as digital controllers can. They are, however, flexible when compared to pneumatic controls on a function to function basis. The accuracy, reliability, and flexibility of electronic analog controls has long been a factor in the industry where they have been used. In some cases they are also less costly than present-day digital controls. The weak link in the chain was and still is the devices being controlled. There are no pure electronic actuators, and thus as is the case with digital controllers, the actuators have been either pneumatic devices used with transducers or straight electric-operated valves and damper motors.

Before the advent of digital electronic controllers, the most impressive type of job was one that used analog electronic controllers with transducers and pneumatic actuators. Analog electronic control systems were first designed to compete with pneumatic systems, so they incorporated all of the features of pneumatic controls, including the relays and devices that act to modify the signals of a controller. Analog electronic controllers have voltage or amperage outputs that can be added, subtracted, multiplied, and divided just as is done with pneumatic output signals.

Analog electronic controls are used in the commercial as well as the industrial fields, but they are not generally used in the residential field since there are very few cases where modulating controls are used in residences.

As indicated above, the sensing portion of both digital and analog electronic controls are the same, but the outputs of the two systems are different. The sensors briefly mentioned above need to be expanded at this point to understand some of the controversy that has surrounded that area of control systems.

Temperature sensors are the easiest to understand, since they are wires that change resistance as the temperature surrounding them changes. The item that it not always understood is the difference between a resistance temperature device (RTD) and a thermistor, or a thermocouple. Thermocouples have been with us long before the advent of the RTDs and have been used to sense the pilot light in a furnace to prove the gas was on and would ignite when the main valve was opened. Thermocouples consist of two dissimilar metals that are welded together to generate a millivolt current when heated. Thermocouples, however, are not too accurate and dependable when it comes to controlling temperature in an HVAC system. Their signal range is small and spreads across a wide range. They are not generally used in modern HVAC electronic control systems. Their use is limited to high-temperature furnaces and laboratory instruments where they can be constantly calibrated and readjusted. They are, however, the least expensive sensing device.

Thermistors are solid-state or semiconductor sensors that are sometimes used with electronic control devices. They are not, in some cases, as accurate as RTD's, but there are applications where their use is suggested.

The best sensors for temperature are the RTDs, which are used in most systems today. Their accuracy is predictable and they do not generally drift over time. They are manufactured to very close tolerances and do not age as other sensors do. They come in a variety of types and styles and depending upon the quality required can cost much more than thermocouples or thermistors. Examples of metals used are platinum, nickel, balco (a nickel-iron alloy), tungsten, and copper.

The sensors used for sensing humidity in electronic controls can be of three basic types: the Dunmore element, the Pope Cell and the Polymer sensor. They are chemically based or electromechanical sensors that measure an electrical characteristic that changes with the relative humidity.

The Dunmore element is built using a dual element, an inert wire grid on a substrate that has been coated with a lithium chloride solution of a particular concentration. The conductivity of the element is measured with an ac Wheatstone Bridge and varies with the ambient moisture available. The Dunmore element is a narrow range element and must be clustered in groups to span the range needed in some HVAC applications.

The Pope Cell is a similar device, but the substrate is fabricated from polystyrene treated with sulfuric acid. This treatment results in molecules having highly mobile sulfates that attract free hydrogen ions from the moisture in the air. The Pope Cell has a wider range than the Dunmore element.

The Polymer sensor (Fig. 8.7) is fabricated using ammonium salts polymerized into thermosetting resin on an electrode base plate. The

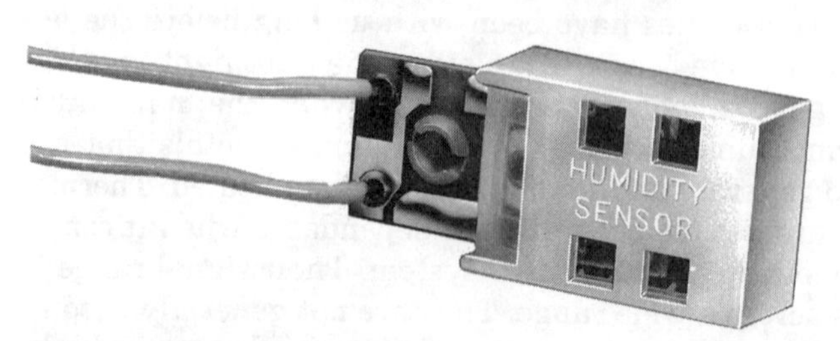

Figure 8.7 Polymer humidity sensor. (*Courtesy of Johnson Controls Inc.*)

conductivity of the sensor varies with the ambient moisture level that changes the concentration of mobile ions in the sensor. The electrical resistance of the sensor is changed in direct proportion to the relative humidity. One special type actually senses the buildup of moisture on a mirror that is chilled artificially with photo cells that sense the change in refractive light from the mirror. That type is more expensive than the standard polymer electronic types, but it is said to be very accurate.

Pressure sensors used with electronic controls are similar to the ones used with normal pneumatic controls, except that the bellows are connected to a variable resistance that changes as the pressure changes. That resistance is read as part of the control circuit just as with an RTD. There are very sensitive piezoelectric pressure sensors on the market that generate a small millivolt current in response to a change in pressure that can be read and used in an electronic circuit. They too are more expensive than the common bellows type; but if accuracy and repeatability are important, they are the types that should be used.

Specifying an appropriate system

Much has been said about where electric, analog electronic, and digital electronic control systems should be used. The important thing to remember is that each system or job needs to be analyzed to determine the correct systems and applications. Sometimes the answers will be obvious and other times not. The following discussion presents guidelines to help you select or specify the most appropriate system or devices for a general category of HVAC system.

Since residential systems are simple and always involve two-position control devices, they almost always use low-voltage electric controls. The only variation might be the fact that the new, day–night thermostats are electronic, but the devices they control are basic and electric in style. In all cases, you will find that residential controls are low-voltage except for the relays that run the fans and the compressors on air-conditioning systems and the pumps in hot water heating systems. One mistake frequently made when it comes to residential systems is not installing enough conductors in the wiring from the room thermostat to the furnace and/or boiler. When modifications are in order, such as adding air conditioning, the problem becomes acute if there are not enough low-voltage conductors from the stat to the basement.

When we start thinking of commercial systems, such as office buildings, schools, colleges, and hospitals, the picture changes, and the variety of controls available makes the decisions as to the proper type of system difficult. The following list presents items to consider. It is a place to start; you can add additional items to be considered when making that all important decision:

1. Size of building
2. Location of building (climate, etc.)
3. Use of the building
4. Occupants who will use building
5. Height of the building (number of floors)
6. Type of maintenance available
7. Type of owners of building
8. Cost of systems
9. Type of HVAC system planned
10. Energy management considerations

First, the size of the project is important. A good rule of thumb that has been used in the past is that any building over 50,000 ft^2 is a candidate for a commercial pneumatic control system. That rule, like all rules, needs to be tempered with the other items in the list. Location is a factor, since climate affects the type of control systems. In severe climates, engineers tend to design heating systems that attempt to take care of the skin of a building, using additional heating devices such as convectors and baseboard radiation, which in turn lends to modulating pneumatic controls and systems. Safety considerations in very cold climates usually require that on a failure of power the heating devices go to full heating. That is more easily accomplished with pneumatic controls on hot water and steam valves. In southern climates, in particular in desert areas, safety is not a consideration and electric and/or electronic controls are a good possibility. The building use and the occupants must also be considered, since a facility such as a hospital that operates on a 24-h basis would be different than a church that might be used only once a week. The height of the building sometimes determines the type of systems that will be specified, which can affect the type of controls used.

The most important item that must be considered, however, is the level of the service and maintenance that will be available when the building is ready for occupancy. If the owners and occupants of the building do not understand the importance of proper maintenance and service, the selection controls must be the type they are capable of working with and maintaining. There is an old saying in the control industry, Maintenance individuals will reduce the control system to their level of understanding no matter what the designer or installer tries to do. If, for example, the people who will be doing the service and maintenance are electronically oriented, it might be smart to use electronic controls for the HVAC system. If the owners and operators are interested in energy management, they would likely entertain the idea of digital control systems along with the energy management system. One of the most popular types of systems is the combination system of digital controls in the equipment rooms with pneumatic controls on terminal devices such as VAV boxes, fan-coil units, and radiators. At this point in the development of controls, it seems that inexpensive small electronic actuators for terminal equipment have not been fully developed. That situation may change in the next few years, but for the moment the facts indicate that where there is a large number of terminal devices to control, pneumatic controls are less expensive. Furthermore, an excellent job of controlling is done with the primary equipment in the fan room; the terminal controls will have less to do in the way of close or sophisticated controlling.

The cost of the control system always plays an important part in the selection. Since pneumatic control systems always require an air compressor, they are ruled out in many smaller systems. If the equipment is, as is the case many times, furnished with the controls as part of the package, it would be foolish to abandon the packaged controls and furnish something that has to be field mounted.

In summary, the selection of the types of control systems will depend upon a lot of factors that need to be analyzed, in some cases with the help of experts in the field. The designer needs to keep in mind that there is no one system that is proper and adequate for all buildings and the decision is always a compromise.

Chapter

9

Direct Digital Control

Definition and Historical Evolution

Direct digital control is the use of computers or microprocessors in conjunction with sensors and actuators to provide closed loop control. Direct digital control applied to HVAC is control of the heating and cooling processes used to maintain indoor conditions of temperature and relative humidity. Many types of control loops are involved in HVAC control. A *control loop* consists of a minimum of a sensor device, a controller function with a set point that creates an error signal, and an actuated device. In some cases there are multiple actuators to control different processes. In conventional control, the controller function is done by a hardware device. In DDC, the controller function with set point and error signal are calculations done by a computer or microprocessor. Analog-to-digital interface from the sensor device provides values to the controller calculation. Digital-to-analog interfaces take values from the controller calculation and provide variable voltage signals to the actuated devices. In most DDC hardware designs, a single analog-to-digital converter scans many control loops to update input readings, and a single microprocessor scans each loop updating the controller calculation. This time-sharing of hardware greatly reduces the hardware device costs.

DDC became practical in HVAC control when microprocessors were put to use in smart data-gathering panels in building automation systems. This put the computational power for controllers into the field panels physically connected to sensors and actuators. This distribution of intelligence to remote panels gave the speed of response and reliability required for successful DDC control of HVAC. Transmission to a central computer and the reliance of all controls on a central computer was avoided by this use of distributed intelligence.

Functionality with Examples of Applications

The functions common to HVAC control have evolved over the years and are generally proportional controls with limit actions and interlocks to the starting and stopping of equipment. The basic building block of HVAC control is represented as a control loop. The most prevalent example of HVAC control is the control of a fan system that provides heating, ventilation, and air conditioning in sequence to control temperature in a space. This is considered a single control loop because the single variable of space temperature is the primary controller of the three processes in a coordinated manner. The control loop used in this case involves numerous other sensors, limits, and interlocks. Figure 9.1 shows the functional flow diagram that represents all the elements of a complex control loop.

A sequence of operation describes in further detail how this fan system is controlled. Some people might refer to this control scheme as multiple control loops with interaction between them. The interaction needs to be further defined, making this the most complex approach. Other people might specify individual control loops for each of the processes being controlled and not address the issue of interaction between loops. This would give far different control results with serious energy implications. This approach should be avoided.

Sequence of control for Fig. 9.1

Space temperature subject to discharge low-temperature limit modulates in sequence heating closed, ventilation open above a minimum and cooling open on a rise in space temperature above set point. The sensing of outside air enthalpy higher than return air enthalpy will close the outside air damper to a minimum position. The mixed air low limit will modulate outside air damper closed as necessary to maintain the coldest mixed air sensed above 45°F. Fan shut down will close the outside air and exhaust air dampers and open the recirculated air damper. It will also close the cooling valve and open the heating valve. This sequence of control and the accompanying functional flow diagram specify how the fan system is to be functionally controlled.

This sequence of control could be fulfilled by DDC, pneumatic, or electric and electronic control devices. The interactions of heating, cooling, and ventilation are ensured by this sequence of control. If each process of heating, cooling, and ventilation were specified as a separate loop, the sequential action and prevention of heating and cooling at the same time would need to be ensured by some other means and would become complex.

The functional diagram in Fig. 9.1 shows functional relationships as well as a method of implementing DDC. The implementation of the

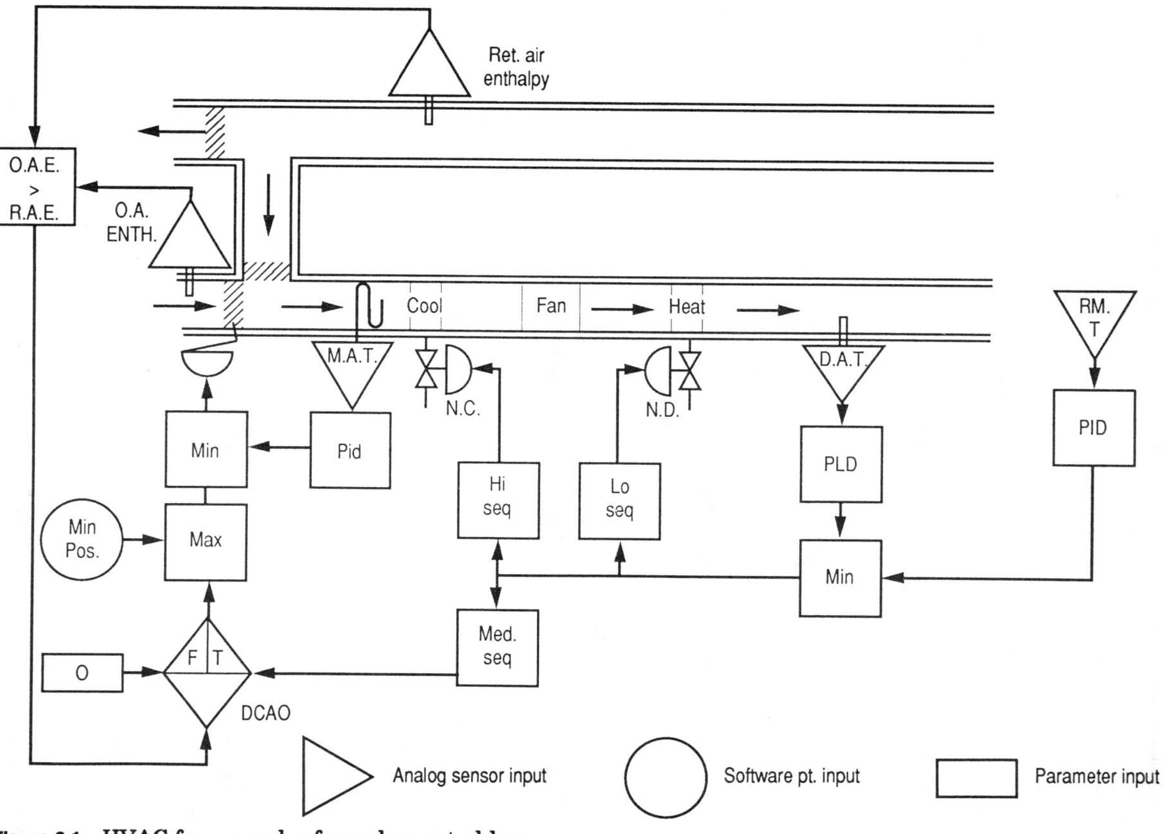

Figure 9.1 HVAC fan example of complex control loop.

functional diagram by DDC is accomplished by the use of software modules called *operators*. These operators provide the calculation of each type of function block shown in the figure. The results of each calculation are passed on to the next operator (function block) in the flow diagram, as indicated by the directional arrows.

Commonly used operator (function block) names and definitions are listed below:

PID. Controller that compares a sensor reading to a set point value and calculates an analog output control signal. The calculation of control signal is proportional and may have integral and derivative effects added. Adjustments are provided for each mode of proportional, integral, or derivative action.

SEQ. Sequence that provides for a full output signal change over a defined range of input change. This allows the operation of different actuators in sequence from a common control signal.

REV. Reversing provides the ability to reverse control action to accommodate a required normal (no power) position of an actuator or to provide a required action related to a limiting control signal.

MAX. Maximum provides a selection of the higher of several input values as the output value.

MIN. Minimum provides a selection of the lower of several input values as the output value.

RATIO. Ratio provides a predefined output scale proportional to an input scale of predefined range and direction.

ACDO. Analog controlled digital output provides a change in digital output state when an analog input reaches predefined values for on and off.

DCAO. Digital controlled analog output provides a selection of one of two analog inputs dependent upon the state of a digital input.

TD. Time delay provides a delay in change of output state after an input state has changed.

Besides these operators, which are specific to proportional control, there are arithmetic and logic operators common to the calculations and logic used in general-purpose programming. These are also used in many control applications and should be available in a DDC language.They are as follows: add, subtract, divide, multiply, equals, greater than, less than, square root, and if–then–else.

Because DDC calculations are continuously reexecuted to simulate continuous analog control and an output calculation is expected in every program cycle, GO-TO or DO-UNTIL programming actions are not necessary. Because programming errors could stop the DDC controller, these programming actions are not desirable in a DDC language.

HVAC control is standardized where possible, but there are so many variations in plant configuration and design that there is usually a significant degree of customizing of control sequence and/or points and parameters. Therefore, configuration of DDC control programs is usually to specific project specifications of job requirements.

The functionality of any control system includes not only what actuators and equipment are automatically controlled but also what information is available to people concerned with the HVAC system and its controls. In conventional hardware control systems, visual indicators are a part of many of the devices, such as controllers and thermostats, that are not present in DDC systems. Set point indicators and control signal gauges or meters are examples of these sources of information that are taken for granted on hardware jobs but not necessarily present on DDC jobs. Figure 9.2 shows a control functional diagram of converter control that includes software points of information of interest to a systems operator. The decision of which type of points should be available to the operator should be considered; software points accessible to the operator should be specified to help run the system.

In addition to points available to the operator, there are tuning adjustments and initial setup values that normally should not be available to the operator but must be available to the technician doing the initial commissioning and follow-up tuning on the project. These values can be described as parameters, and a restricted method of viewing and modifying them should be provided. Examples of parameters are controller adjustments of throttling range, integral timing, the start and stop points of sequence relays, and scaling ratio relays. There is variation from project to project in the type of information appropriate for the operating staff. Therefore, thought must be given to what if any tuning parameters or set points should be accessible and the responsibility of the system operators. This will be covered further in the discussion of specifying automatic control by DDC.

Variations in Programming Methods

There are probably as many programming methods as there are manufacturers of DDC equipment because the method depends on the equipment design and the tools. The programming of DDC has two major tasks. The first is establishing data files; the second is providing the coded instructions that constitute a DDC program.

The first major task of establishing data files is to represent the identity, location, and nature of all points used in the DDC programs.

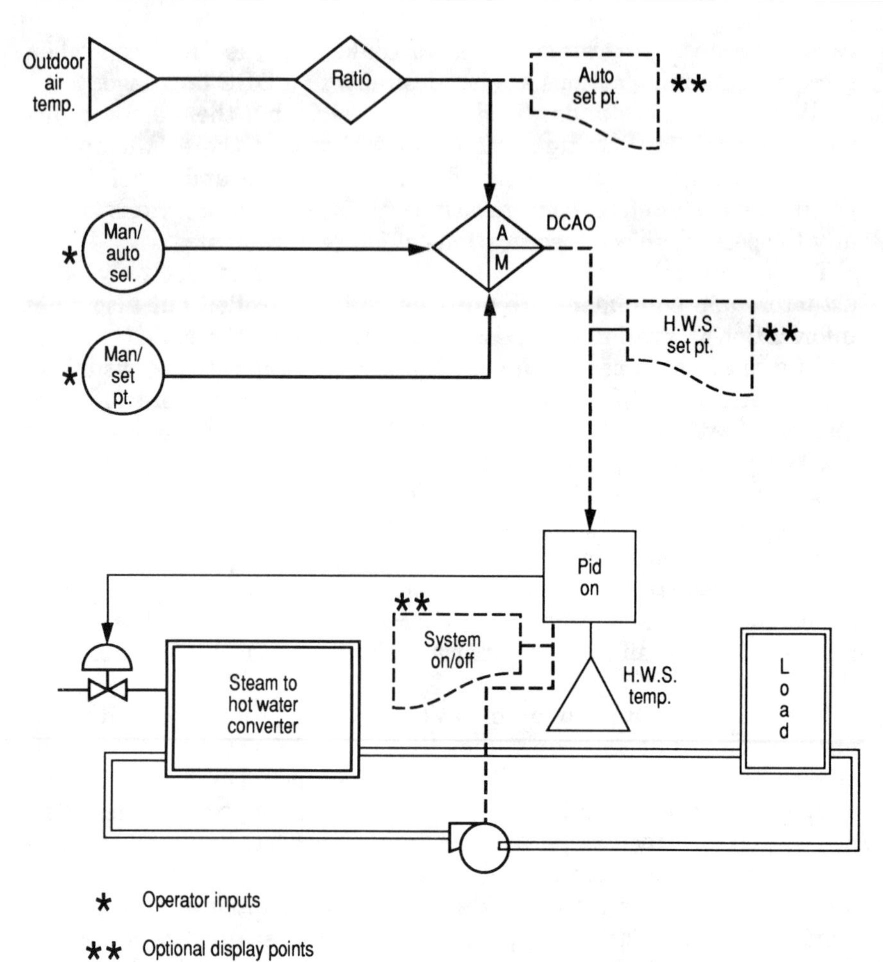

Figure 9.2 Converter control system with operator access software points.

These points include real hardware points that are the sensors and actuators that sense and cause changes in the real world HVAC system. The other type of points are software points that are a means of communicating between programs and the operator of a system, such as set points for the operator to command.

The identity of a point includes its physical identity (if it is a physical point rather than a software point) of the sensor or actuator plus its application identity of the HVAC system it is in and its function, such as discharge sensor or mixed air sensor. The location of a point includes both a physical terminal location and a logical address location that gives access to its file information. The physical terminal address is used to tell the controller how to process the signal measured on those terminals. The software file address is used to give access to the value for display or DDC calculation.

The nature of a point is defined by all of its attributes, both physical and application. These include analog or digital, input or output, scale and engineering units, plus any special functions such as alarm limits or initiation of programmed events.

The second major task is programming the DDC code that defines and implements the control logic and calculations. Several major types of programming are used. The first is with a high-level language that defines the logic and calculations. A compiler translates these instructions into machine instructions specific to the microprocessor, and the compiled program is loaded into the controller. The compiled program is then executed, based on a time schedule, and the resulting commands are executed.

A second major type of programming is with an interpretive language, which is interpreted then executed line by line. A third major type of programming is with registers to stack individual instructions in the order in which they are to be executed, with results being passed from one stack to another until the final results are achieved and executed.

Each of these types of programming has strengths and weaknesses. The power and understandability of a high-level language is its strength, whereas the need to recompile the program when any changes are made is its weakness. (This could also be considered a strength because of the error checking done in the compiling process.) The ease of changing a single line of code in interpreter code is convenient when making changes, but the limited power of using only single line instructions according to a fixed format recognized by the interpreter could be a limitation. The compact memory usage and simplicity of operation of a register language is an advantage in lowering hardware costs, but the difficulty in putting together and understanding the flow of logic is an added application cost.

The generality of the programming type should not be the only factor used in decision making, but it can help in comparing the capabilities of different programming methods. In judging capabilities, the questions to ask are, Who will do the programming? How will it be documented? Who will make changes? How will changes be tested? How will changes be documented? Who needs to understand the documentation? What tools are provided to do programming or to make changes to programs? Comparing alternative products and programming methods should then be done with these type questions in minds.

The specification of a project that will be installed and commissioned by the control manufacturer makes the programming method evaluation a minor subset of the evaluation of the total capabilities of the control manufacturer–contractor. If the project is planned to have programming done by the owner's engineers, the evaluation of the programming method becomes a major consideration and should involve the HVAC engineers who will do the programming. They should be familiar with the required sequences of control as well as the type of programming to be used. The major questions then become, does the language have the functions needed to implement the sequences of control required? Are the language and documentation understandable to the people doing the programming? What is the relative usability and costs of alternative products and programming methods?

Energy Management Functions and Interfaces to DDC

Although energy management system functions were initially implemented in the central processors of building automation systems, many of them are now implemented in the smart remote panels that are also the location of DDC controllers. The energy management programs that associate with a single fan system are logically located in the panel that also has the DDC control of that fan system. The distinction between the two types of programs is not only that they serve different purposes but also that they typically have different characteristics. The energy management type of program saves energy and is a complex but standard type calculation best implemented as a standard program. The DDC type of program is for basic control and is tailored to a particular project and mechanical subsystem.

The types of energy management programs distributed to local panels are as follows: optimum start–stop, duty cycle, enthalpy control, night cycle, night purge, and load reset. The interfaces between these two types of programs are that some common sensor values are used and some common actuators are controlled. The starting and stopping of an HVAC fan system is normally done by an EMS program, and DDC controls the individual processes. In some cases, the EMS programs reset the set points of DDC control loops. Here, the set point in the DDC program must have a software point identity that is then referenced in the EMS program data file to be commanded by the EMS program.

Other EMS programs are global in nature in that multiple fan systems are controlled in coordination with each other. An example is a demand control program where a total building demand is controlled by cycling fan systems in rotation. Another example is where a central chilled water supply temperature is controlled to just satisfy the greatest demand fan system, as indicated by the most open cooling valve on all fan systems being 95 percent open. A global EMS strategy may be controlled from the remote panel where DDC controls the process, but communication with all the other panels being coordinated is needed.

Stand-Alone DDC Controllers versus DDC as Part of a BAS

Whether or not a DDC controller communicates with a central processor or with other controllers is what determines if it is a stand-alone controller. If a DDC controller does not communicate, it is a stand-alone controller. If it serves as a data-gathering panel as well as a DDC controller, the added functionality of this communication capability can make it part of a building automation system. To be a full-scale building automation system, there should be a central operator interface. When considering stand-alone DDC versus DDC as part of a BAS, we need to recognize that there is a middle ground, where there can be a system that is neither a stand-alone DDC nor a BAS. To discuss this subject we need to examine system architecture from the viewpoint of both function and communications.

Functional considerations

There are different sizes and types of DDC controllers designed for different uses. Sizes range from small, single-loop controllers controlling 1 or 2 actuators to large multipurpose panels controlling 16–32 actuators or relays. The smaller DDC controller is typically used as a zone controller for single-zone HVAC units such as variable air volume boxes. These zone controllers can be application specific with limited programming or can have custom programming capability to meet a variety of needs. They can also be stand alone or have communications capability and be part of a BAS. The larger, general-purpose DDC controllers can be described as system level controllers. They have full flexibility of programming and normally control one or more HVAC subsystems such as fan systems.

Communication considerations

Larger DDC controllers can also be stand-alone controllers or have communications capability. They can, in some cases, communicate both up to a central and down to zone level controllers. This capability can be an efficient way to collect data for a BAS central. It is also a natural hierarchy of function for system optimization when all the zones supplied by a fan system communicate to the fan system controller. Then, all the fan system controllers communicate to the chiller system controller. Figure 9.3 shows this type of architecture.

The consideration of present and future needs will, in some cases, lead to selecting equipment for present use as stand-alone control that at a future time can communicate to a central unit and become part of a BAS. It is also possible that a network of zone controllers could communicate with a fan system controller to accomplish system optimization but not have a central operator interface to make it a complete BAS.

The selection of stand-alone controllers or a BAS with DDC will be guided mostly by the benefits to be obtained by having BAS versus the added costs. In the situation where stand-alone DDC controllers are chosen, there still must be a provision for some level of operator interface for such things as control set points, on–off commands, controlled point status, and analog values of controlled conditions and actuator positions. This type of operator information can be obtained on some systems by an operator's terminal plugged into the DDC controller. In some cases, this may be the use of a single portable operators terminal (POT) usable on all DDC controllers and providing information in a fixed format. In other cases, a display can be incorporated with the

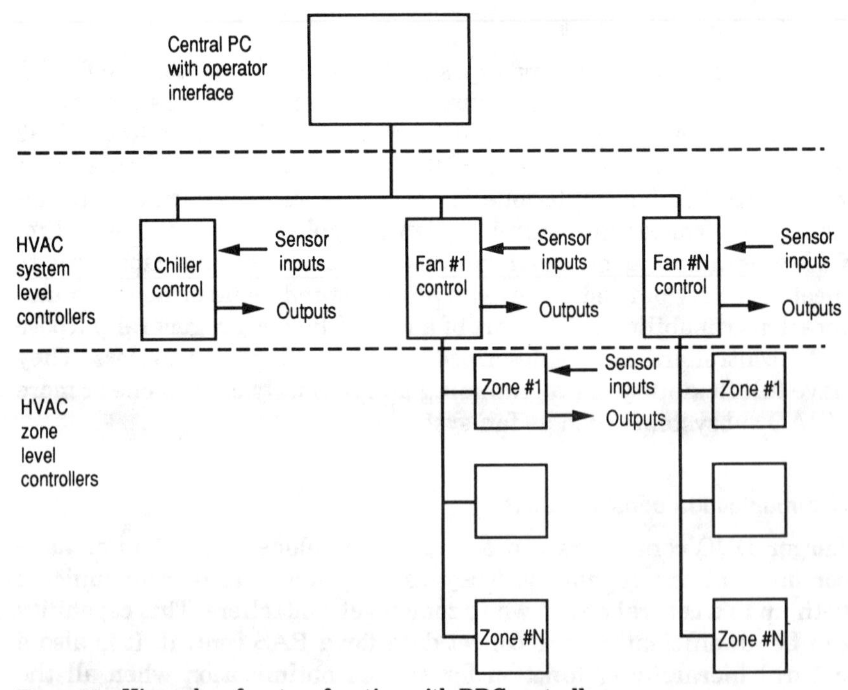

Figure 9.3 Hierarchy of system function with DDC controllers.

controller. In either case, documentation specific to the project must identify the points by function such as space or discharge temperature to give meaning to the information. In a BAS, this information can be incorporated into a custom configured display representing a particular HVAC subsystem.

In the case of stand-alone DDC or DDC as part of a BAS, provisions for changes in DDC programs should also be considered. A portable programmer's terminal (PPT) is the normal means for a stand-alone DDC controller. A BAS with DDC would normally have central programming capability and the means to downline load programs.

Specifying Automatic Control by DDC

The main points to be covered in a specification are as follows:

Sequence of operation for each type of HVAC subsystem

Definition of the points to be in the operator interface as information or command capability specific to each sequence of operation and definition of the function of those points within that sequence of operation for a specific HVAC subsystem

Definition of whether the system is to be stand-alone DDC controllers or a BAS with DDC

Definition of how operator interface information is to be displayed and commands implemented

Documentation requirements for submission and approval before installation

Requirements for commissioning and acceptance procedures to be followed, as built documentation of completed job and training

The first step in defining what DDC control will be on a project is the communication of the plans and specifications. The second step is information from the supplier that shows in more detail how the specification was interpreted and how the DDC is to be accomplished. This information could be a functional flowchart like Figs. 9.1 and 9.2 or high-level language code. In either case, a more detailed sequence of control should supplement the information to ensure understanding by the approving specifier. The information should also identify inputs and outputs and adjustments. Adjustments should include displayed software points for use by normal operators and nondisplayed parameters that are adjustments to use in initial setup and follow-up tuning by control technicians. Figure 9.4 shows operator-used software points and technician-used tuning parameters for the system in Fig. 9.2. In this example, there is provision for the operator selection

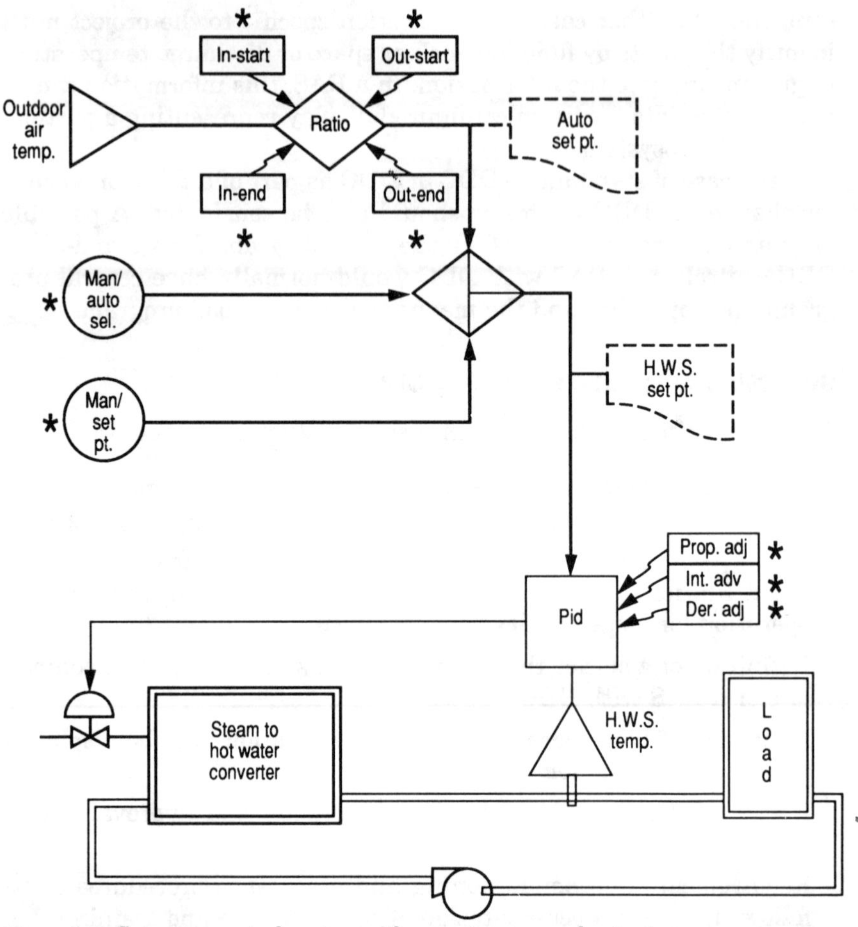

Figure 9.4 Converter control system with operator access plus tuning parameters.

of discharge water temperature or automatic reset of discharge water temperature by outside air temperature. This would be appropriate for a situation in which the converter is supplying an outside zone of radiation and an operator can adjust supply water temperature and monitor the results for proper temperature in typical spaces. The example of tuning parameters for the control technician are shown as the set point and range of the outside air rest plus the throttling range and integral time settings for the discharge controller. The specifier should define what is needed for the operator, and the control supplier should define the adjustments needed for set up and tuning by the control technician. In this example, it is possible that the adjustment of the outside air reset be made available to a well-qualified operator. Whether a parameter is to be used by an operator depends upon whether the operator has the technical understanding of that aspect of the system. The review of the information with the adjustment and tuning parameters listed gives the specifier the opportunity to see if any of the parameters should be made available to the operator. Giving more information than the operator understands should be avoided, since the operator may then bypass the system.

Checkout, Commissioning, and Acceptance

Checkout of hardware and software are two independent functions to be done before their combination as an operating control system is checked. Checkout of hardware is a matter of checking that each sensor and actuator is properly wired to the correct terminals of the controller and that manual commands from the controller are properly executed. Also all sensors are checked for proper calibration to ensure that actual, measured conditions are accurately represented by the controller to the operator interface. Checkout of software is typically done by fixing (manually setting) the values of inputs and looking for the proper output. This process is done for each logic path through a program, and several values of input are tested to ensure that the direction and amount of change in output are correct. If the DDC is part of a BAS the printout of each test run's input and output values can be used as documentation of the checkout.

When these hardware and software checkout steps are completed, the dynamic control of the HVAC subsystem in operation can be tested and tuned as necessary for stable control of each loop. If the DDC is part of a BAS, trend logs of the controlled variable and the controlling actuator position can be an aid in testing, as well as documentation of the dynamic checkout of the system. If practical, this dynamic checkout should be done under both light and heavy load conditions for a long enough period of time to show stability of control after an upset in control.

Commissioning of a DDC system should include demonstrating the system to and training the owner's representatives using the as-built documentation and operating instructions that will become a part of the project deliverable. The as-built documentation should include the normal terminal wiring diagrams, the schematic diagrams that show control device locations in the mechanical subsystems, as well as the functional flowcharts and/or high-level language code that represent the software performing DDC. It should also include the sequence of operation written to include all input and output points identified as they are found in the operator interface. These should include software points that are operator commandable set points or DDC calculation results of use to the operator.

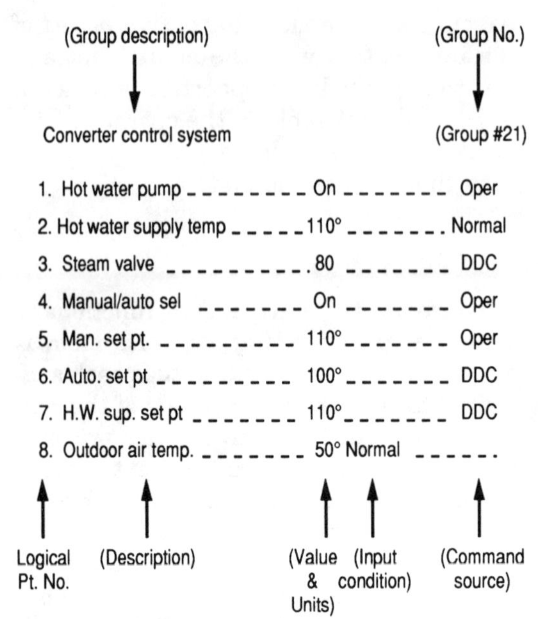

Figure 9.5 System display of points for converter control of Fig. 9.2.

If the DDC is part of a BAS, there should be a subsystem display that identifies primary control loops with their set points and measured results as well as actuator position and the on–off status of the process being controlled. If the DDC controller is a stand-alone type, it should have documentation that represents all points in a system both in the format they are presented in the operator interface device and with the functional description used in the functional flow diagram and sequence of operation. Figure 9.5 shows the system display of points in a BAS that represents the converter control subsystem defined in Fig. 9.2. In a stand-alone controller these same points should be covered in whatever fixed format is available, and the documentation should provide the information not given by the operator interface (such as description and engineering units).

Chapter

10

Air-Handling Units

The term *air-handling unit,* AHU, generally means the items (fans, duct work, coils, filters, etc.) that comprise the system that delivers conditioned air to a space for the purposes of maintaining either comfort or other environmental conditions. These units can be factory manufactured or field fabricated. They can be used to maintain conditions for human comfort or for preserving conditions in a laboratory or a freezer plant atmosphere. They can be very small in terms of cubic feet of air delivered or very, very large. The cabinetry that surrounds them (sheet metal) can be finished, as in the case of a unit in an occupied space, or it can be rough plain galvanized sheet metal ductwork. Thus, anything that delivers air that is changed by passing through the unit is considered an air-handling unit.

Units generally, but not always, involve dampers, fans, filters, coils, and some form of ductwork or sheet metal casings. This chapter will discuss the history of air-handling units as well the various types of units that have evolved over the years. It will also discuss the control cycles and different types of pressure used with air-handling units. Maintenance and troubleshooting will also be an important part of the chapter. The chapter will *not* concern itself with the terminal devices that sometimes are considered part of certain air-handling units, as these are discussed in other chapters, such as the one on "Terminal Units."

Packaged versus built-up units will be explored and the advantages and disadvantages of each type will be covered.

Evolution of Air-Handling Units

The initial heating systems used in residences and buildings years ago depended upon gravity (i.e., the fact that hot air rises) to transport the warm air to the spaces that needed it. If we go back far enough, we can see that convection as well as conduction and radiation were the only way heat transferred from the source of the heat (e.g., a flame) to the human body to create a feeling of warmth. The early systems tried to improve upon fireplaces by moving warm air from the fireplace to the other parts of the house or building by various means. As an example, homes had a fireplace in the middle of the house so the transportation of the warm air was simpler. Where the furnace or fireplace was on the first floor, holes were cut in the ceilings so warm air would traverse to the upper floor by the natural ability of hot air to rise.

Some innovative early systems that used the natural tendency of hot air to rise used steam boilers to furnish steam to radiators as well as to cast iron coils placed in a horizontal position at the bottom of a two-or-three-story duct shaft that fed the upstairs rooms. Thus air heated by the cast iron coils rose up the shaft to the upstairs rooms and returned to the basement, which was colder than the air being supplied to those rooms, by gravity.

The coils were not coils as we know them today. They looked almost like cast iron radiators placed on their sides. They did the job, however, even though they were inefficient. Steam boilers by the way were prevalent in those days, and the idea of heating with gas had not been approached at all. Coal and oil were the fuels, and about the only improvement was the addition of a stoker to the boiler so the operators did not have to constantly feed coal to the boiler. A stoker is a device that fed pulverized coal at a slow rate to a boiler through a hopper and a screw drive. The rate was adjustable, and the steam rate of the boiler could be controlled that way.

The first fans were nothing more than glorified paddle wheels that could move air. Some were steam driven, since the technology of electric-motor-driven fans had not advanced yet. The first squirrel cage fan, as it was called, was fabricated by bolting pieces of metal together to form a fan and hoping it would not throw itself apart as it rotated because it was out of balance.

Eventually the technology advanced and all types of fans appeared on the scene. Today the science of moving air with fans has advanced to the point where there are fan curves, we know exactly how they operate, and we can classify fans for various jobs for which they are needed.

There are class I, class II, and class III fans; the classifications are based upon the way the fans are constructed and the pressures they are designed to operate against. They are built with forward curved, backwardly inclined, and airfoil blades (Fig. 10.1). Unlike the first fans, they are slow speed as well as high speed with ratings involving low pressure, medium pressure, and high pressure. There are propeller fans, vane axial fans, tube centrifugal fans, and standard centrifugal fans. There are

Figure 10.1 Typical centrifugal fan. (*Courtesy of Twin City Fan Co.*)

fans for almost any purpose we can think of in the industry. There are fans for moving not only air but all the other gases.

The first breakthrough involving modern air-handling units came when it was discovered that air could be heated by blowing it through a finned tube coil that had steam inside the coil in a set of tubes. Thus was born the first real air-handling unit.

Just as the first air-conditioning systems consisted of separate pieces (the compressor, the cooling coil, and the evaporative condenser all piped together in the field), the first air-handling units were field fabricated and constructed of fans, coils, piping, filters, and duct work (Figs. 10.2 and 10.3). Often, the specification called for a fan room, which was built of concrete block or poured concrete with openings left for the duct work to get the air to and from the fan room. The fans, coils, filters, and so on, were shipped to the job and assembled in the field to complete the system. That concept required quite a bit or coordination, and many times the pieces did not fit as planned. An interesting point is that some of the supply duct work consisted of tunnels and other poured concrete raceways, and the concept of varying the *amount* of air to a space was tried even back then, with the use of

Figure 10.2 Air-handling cooling coil. (*Courtesy of Buffalo Forge Co.*)

Figure 10.3 Air-handling heating coil. (*Courtesy of Buffalo Forge Co.*)

dampers controlled from a room thermostat in the space. 7
was that the fans that were available were not capable
down to the degree we have today, and the speed controls tha.
used could not operate the fans the way they can today. For that reason, as well as the fact that the diffusers dumped the air into the spaces when the amount was throttled back as the requirements changed, the jobs that were tried turned into failures.

Those first attempts at air-handling units resulted in new and innovative methods to deliver the air to the space that was heated and in some cases cooled by the refrigeration systems that were around.

The first systems provided only ventilation air to the spaces; heating was done by perimeter radiation. Those systems only had to temper the air, which was a mixture of fresh air and return air that was filtered and heated to a temperature close to the space temperature. As the idea of cooling the air took hold, the engineers looked for new and better ways to deliver the air to the space and also looked forward to the day of the all air system.

At one time, the thought was to send both the cold air (tempered or cooled as the case may be) as well as the heated air directly to space and mix it just before entering the space with dampers controlled by the zone or room thermostats. Thus was born the first *double-duct system*. Today we call that and other systems like it *dual-path systems,* as opposed to the single-path or series system. Single-path systems, as the name implies, pass air through the heating and cooling coils, filters, and so on, in a single path so the same cubic block of air that is heated can be cooled or acted upon by a humidifier. Dual-path systems, on the other hand, are so arranged that the air passing through the units have dual paths, and at the same time as some of the air is being cooled, other portions of the total air stream are being heated.

Modern air conditioning did not just happen. It required many years of research and development of products to get to where the industry is today. Just one example will illustrate the point.

The first steam coils were brass and copper tubes with aluminum fins pressed on the tubes to make a tight fit. The headers were plain cast iron, with no effort to distribute the steam evenly through the tubes. After many freeze ups in cold climates, it became apparent that a steam distributing headers, as well as a method to make sure the steam did not condense prematurely in the coil and freeze were necessary. After much trial and error, it was discovered that the control valve not only throttled the *amount* of steam that went into the coil, it also created a pressure drop to operate, which lowered the pressure in the coil. At times a vacuum was created in the coil, subsequently holding up the condensate in the coil so there was freezing. Another problem that had to be avoided was cold air stratification that can allow bypassing air at high velocity through part of the coil (usually the wrong part). This can cause freezing of stagnant coils downstream or at least nuisance alarms from freeze stats. To help correct some of these problems, the condensate trap was increased and work was done on the coil with internal distributing tubes so that today we have steam coils that can be used in cold climates along with the proper controls that will operate satisfactorily. Even with the best control systems, however, the years have taught most engineers that steam coils are not the best item to use in extremely cold climates without other methods of protection.

All of the modern systems, such as VAV and single-path and dual-path high-pressure systems, are the result of experiments and trial and error of years ago. Most of what we see today in the up-to-date air-handling packaged units were tried as built-up units at one time or another.

Packaged Units

The factory assembled units (Fig. 10.4) used in buildings and on rooftops today are a result of the work done with field-fabricated units years ago. Manufacturers recognized that given the competitive pressures of the market place, the company with the best mousetrap would get the most sales. The effort was concentrated in the area of standardization, flexibility, compactness, and cost reduction, including minimum field labor, with shipping weight a large part of the equation. This learning by experience sometimes resulted in units shipped to the job that were totally inadequate to do the job, and more importantly they were units that were almost not controllable. An example is dampers that not only slipped on the actuator shafts (which was not apparent from the outside of the unit) but also leaked.

Packaged units are available as both single-path and dual-path units. They are placed inside buildings and hooked up to the duct work or are used as weatherproof units on roofs of buildings to feed the duct work within the building.

The advent of the packaged unit went a long way toward the development of the roof top units we see today. Generally, however, rooftop units are not used on buildings that are more than three stories high, although there are exceptions. The rooftop unit was developed as an answer to the architect's lament about the rentable space that was being used by engineers designing a normal job. Packaged air-handling units can be as small as 300 cfm, for example, a fan-coil unit in a room providing conditioned air just for that room. Usually, however, we think of air-handling units in the range of 1000 cfm to hundreds of thousands of cfm.

Figure 10.4 Packaged air-handling unit. (*Courtesy of York International.*)

Single-path units

Single-path units, as we have seen, allow the cubic foot of air that is being conditioned to be worked upon in a series of actions taken by the dampers, coils, filters, and so on. Single-path units can be field fabricated as well as packaged units. A typical single-path unit—a *single-zone unit*—is shown in Fig. 10.5. In this case, it is only a heating unit with the unit supplying air to one space and the room thermostat operating the heating coil that is heating a fixed amount of outside and return air mixture. It is called a single-zone unit since it feeds only one area. This unit is basic and easy to build.

Figure 10.6 shows the same unit with automatic outside air and return air dampers controlled by an economizer system. This portion of the unit can be considered a module and might be called the OA–RA control module.

Economizer control of the outside air and return air dampers is one of the most common basic controls specified for that portion of the system. The principle behind the system is the use of outside air only when it is economical and the air does not need to be heated or cooled to satisfy the space requirements. This means the controls operate the

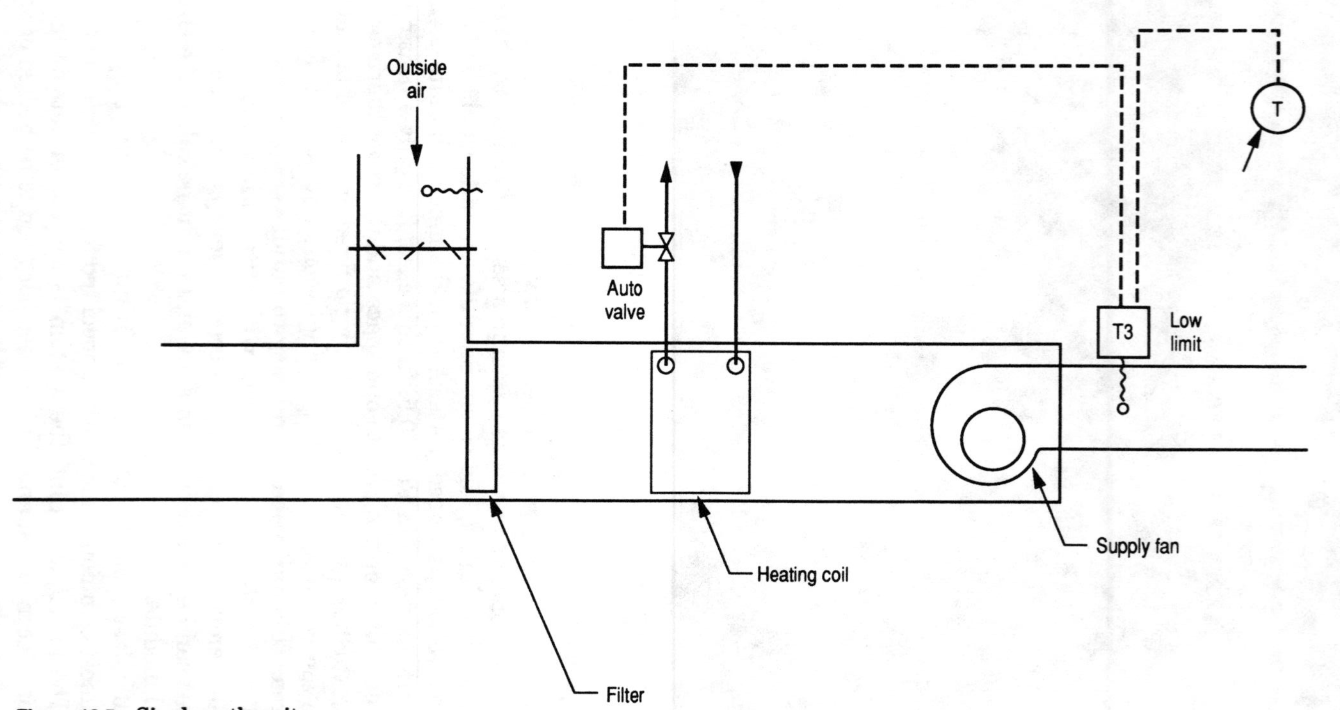

Figure 10.5 Single-path unit.

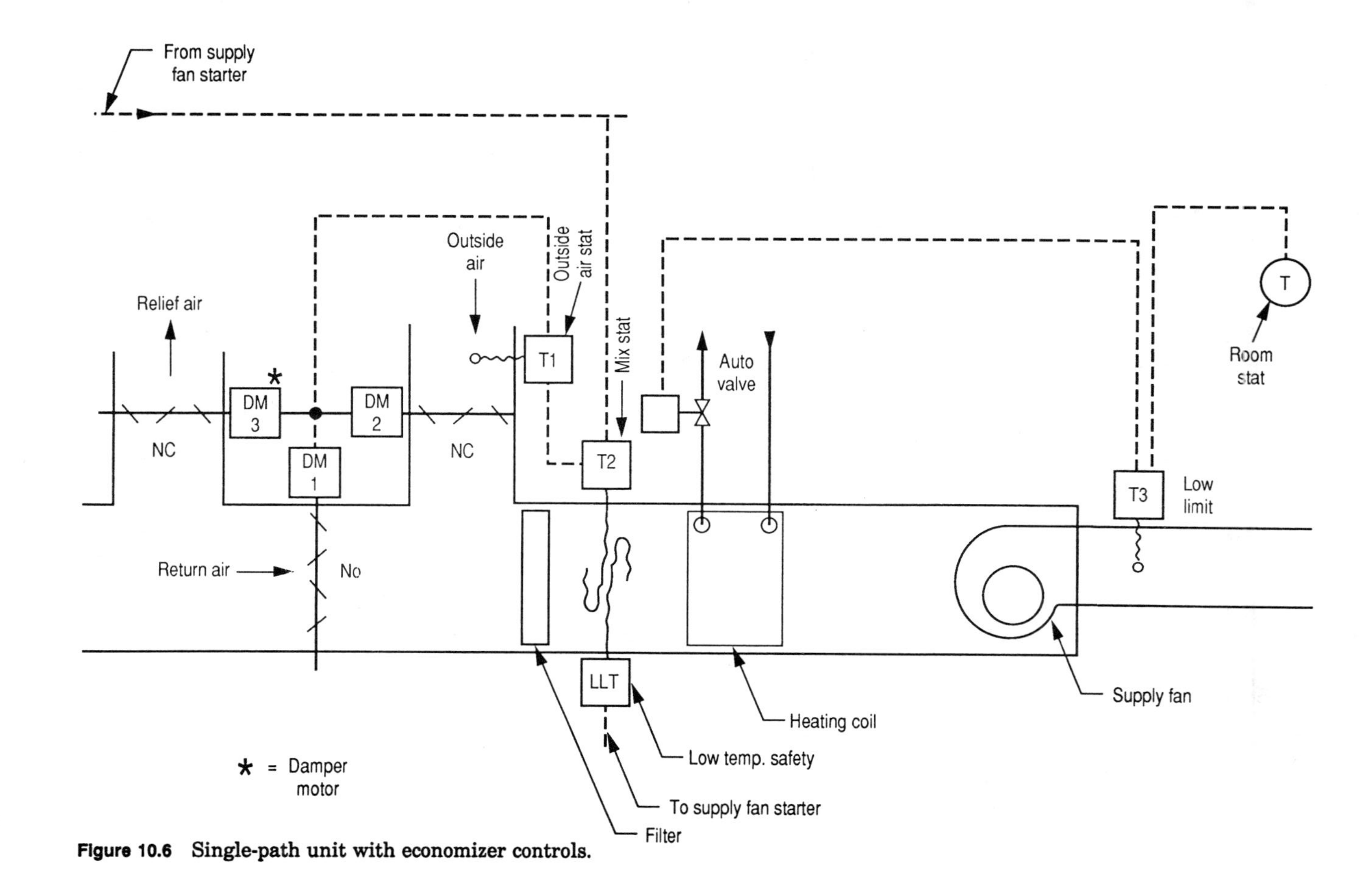

Figure 10.6 Single-path unit with economizer controls.

outside air and return air dampers to allow for 100% outside air and no return air when the temperature is ideal and reduce amounts of outside air as the temperature outside becomes more severe (either too hot or too cold). One of the ways to do this, as the controls show in Figs. 10.6 and 10.7, is to sense the mixed air temperature and operate the dampers to maintain a temperature of between 50 and 65°F as the ideal temperature of air that neither needs to be heated or cooled. Another way is to have a thermostat in the mixed air controlling the dampers with a limit thermostat sensing the outside air temperature that will return the dampers to a normal position when the outside air temperature is no longer a source of cooling.

Figure 10.7 shows the addition of a cooling coil to the system and the control cycle controlling the heating and cooling coils in sequence to maintain space conditions. We have now added a cooling control module. Filter modules, relief modules, and return air fan modules can also be added. To complete a complicated single-path system, including such things as minimum position switches for the dampers and auto override from the room thermostat on the economizer control system as shown in Fig. 10.8. The important thing to recognize is that fan control systems are a *combination* of different modules or dedicated control schemes to make up the entire air-handling unit. Single-zone units can be controlled in a variety of ways depending upon the desires and requirements of the designer. For example, a unit may use face and bypass dampers to control the cooling as opposed to a valve on the cooling coil. Generally, if the cooling coil is a DX coil, we would find face and bypass dampers the rule for control since it is almost impossible to control a DX coil with a valve.

Note the position of the cooling coil in Fig. 10.8 as opposed to the heating coil and recognize that the position in the air stream is an arbitrary one, usually decided upon by the type of service and the climate where the unit will be used. Sometimes the heating coil is after the cooling coil, in which case it can be called a *reheat coil*. This is to say, its specific purpose is to reheat the air that has been cooled in order to dehumidify it. In some cases, since the system may feed many zones with one cooling coil, reheat coils are used as zone reheat coils to correct for the unbalance that may occur because only one cooling coil can be used and it must be controlled to address the needs of the area with the greatest demand for cooling. In severe winter climates, a heating coil can precede the cooling coil and follow the cooling coil. The coil that precedes the cooling coil is called a *preheat coil* and, as the name implies, preheats the air and protects the cooling coil from freezing. Various methods of control of the preheat coils have been used, including steam coils that use a control valve as well as face and

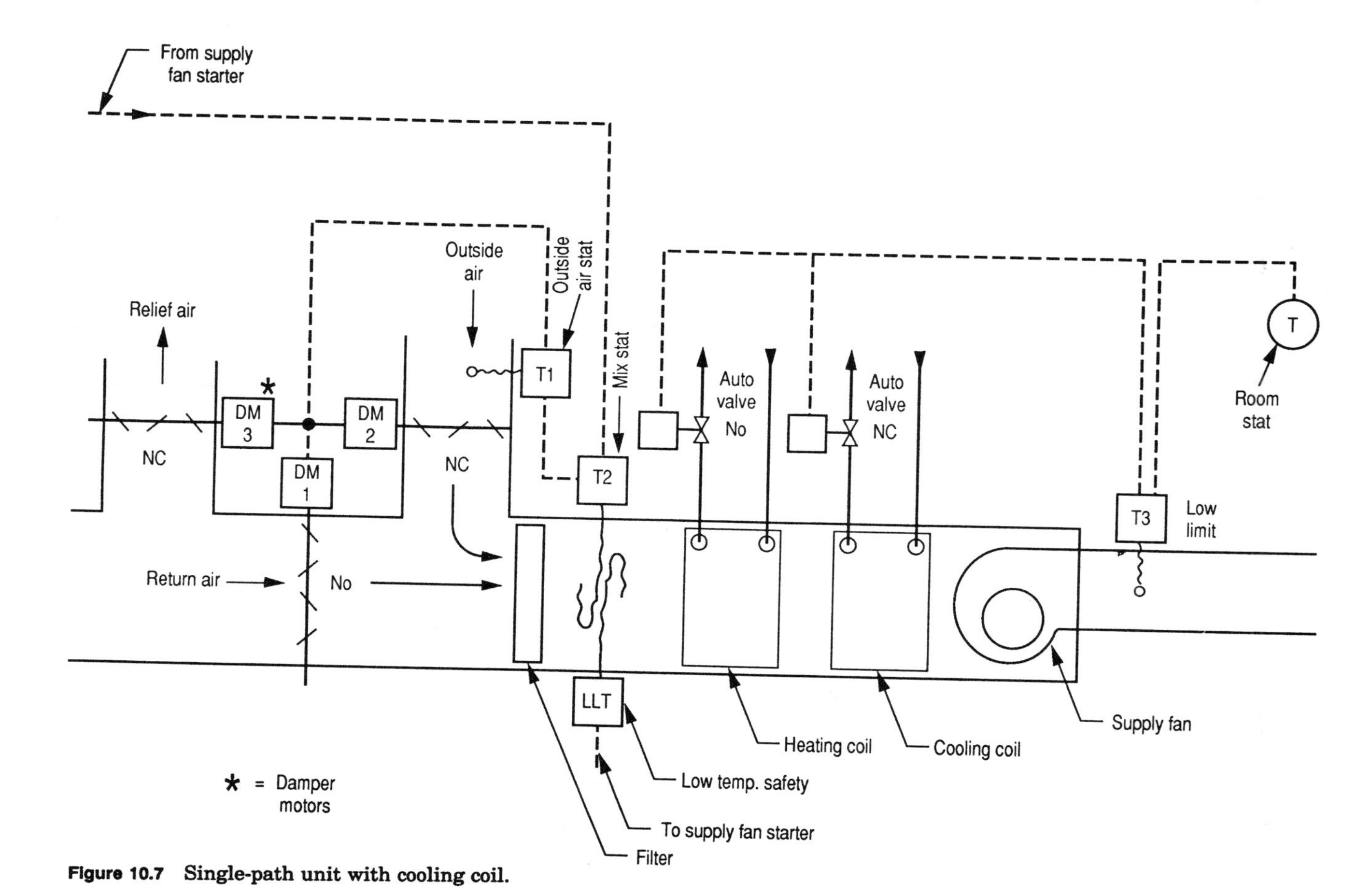

Figure 10.7 Single-path unit with cooling coil.

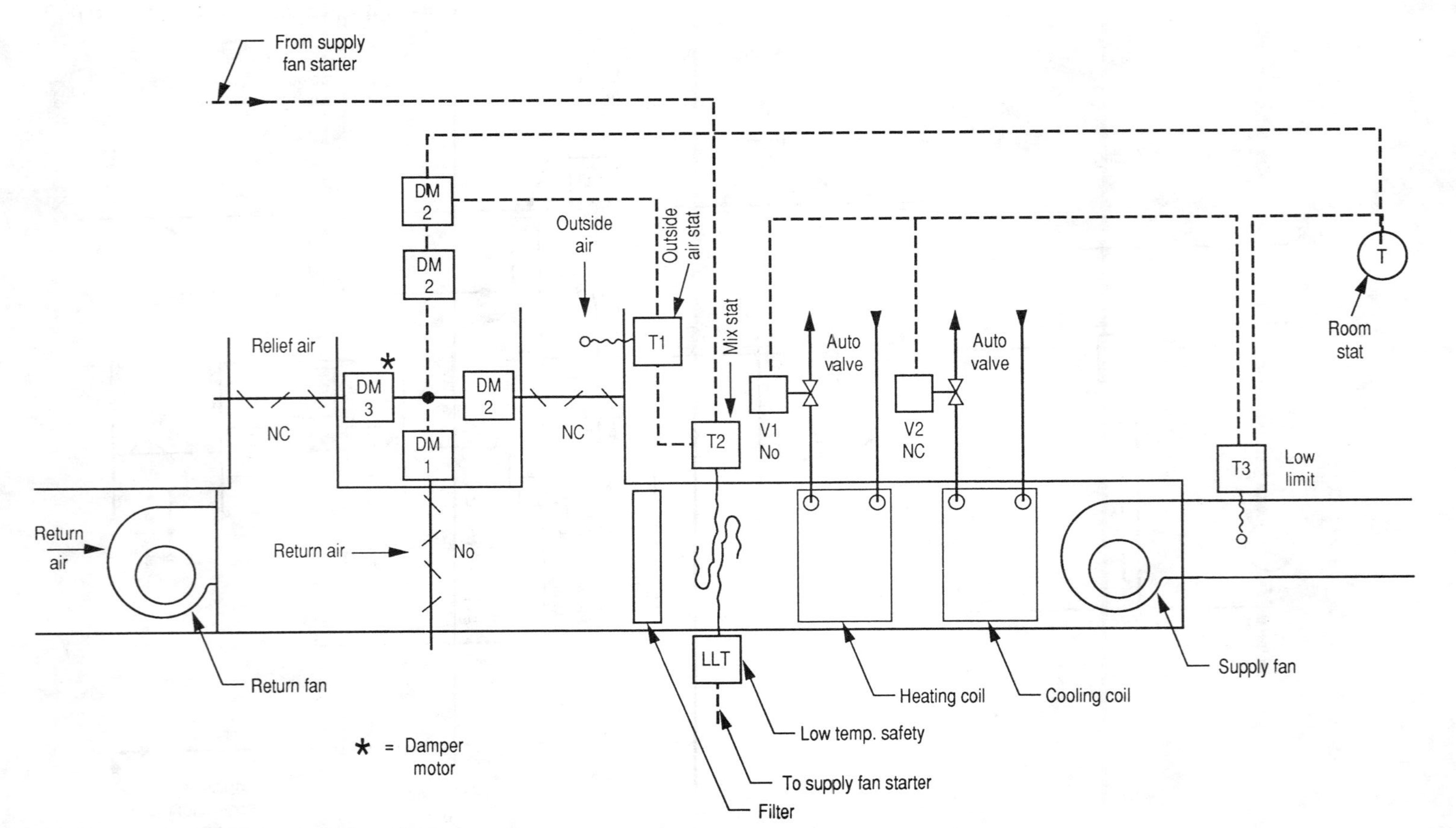

Figure 10.8 Single-path unit with outdoor air, and so on.

bypass dampers to protect the coil from freezing and still temper the air to the system. The cycle is simple in that a thermostat sensing the outside air temperature opens the steam valve wide open when the temperature reaches 35°F, and the discharge temperature of the air-stream is maintained by controlling the face and bypass dampers. The theory is that if the coil is full of steam under pressure (no valve to take a pressure drop), its chances of freezing are less.

There are other methods of controlling preheat coils, in particular the coils that use hot water instead of steam. Some years back it was discovered that *moving water under pressure* will not *freeze*. Therefore, if we control the temperature of the preheat coil by the *amount* of water going to the coil, it can freeze. If we control the air stream temperature after the coil by *temperature* of the water that is constantly circulated in the coil, it will generally not freeze. Tests have shown that a velocity of about 3 fpm in the coil will protect it from freezing way below its normal freeze point. Figures 10.9 and 10.10 show how the cycle and the controls work.

It is also possible to change the solution to the coil and add glycol to the water as an antifreeze. Note too that a three-way or a two-way valve can be used. In all cases with the temperature control method, however, a separate unit pump must be added in addition to the system pumps. In some instances such as operating rooms of hospitals, the units must be by code, 100% outside air units with no return air allowed to the unit. In this case in severe climates the preheat or tempering coil, as it is sometimes called, is a must.

Humidification is a requirement in some areas due to the very low humidity that exists in cold climates and the damage that can occur to furniture, fixtures, and paper for copy machines, not to mention the affect on human health. The human skin tends to dry out at low humidities, and even the nasal passages are affected in the winter. A quick look at a psychometric (psyc) chart will show, for example, that when outside air at 35°F is brought into a building with a humidity of 50% and heated to 70°F, the humidity drops to about 12%. The way to correct the problem is to add moisture to the air stream before it gets to the space. One way to do that is to add steam from the system boiler. The device is called a steam jet humidifier and it can be controlled with a humidistat and steam valve as seen in Fig. 10.11. It can also be done with a pan-type humidifier that allows moisture to be generated in a pan of domestic water that is heated either by steam through a coil or an electric coil controlled from the humidistat. This method overcomes the complaint of the smell of live steam. Plain water sprays can also be used, but they require eliminators to prevent water carryover into the spaces unless the atomizing system injecting the water into the air stream is efficient enough to prevent water

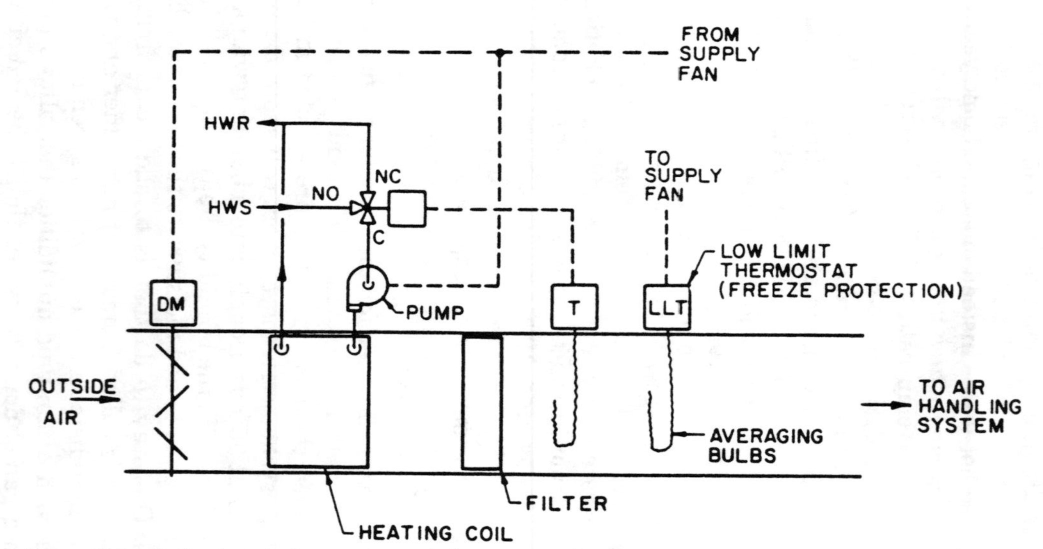

Figure 10.9 Preheat secondary pump and three-way valve. (*Courtesy of ASHRAE.*)

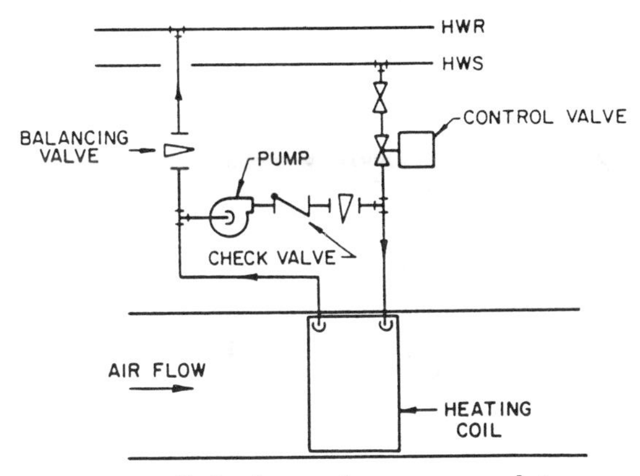

Figure 10.10 Preheat secondary pump and two-way valve. (*Courtesy of ASHRAE.*)

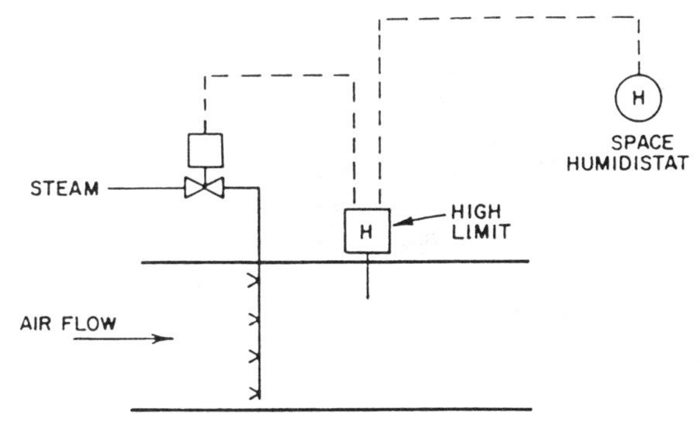

Figure 10.11 Steam jet humidifier. (*Courtesy of ASHRAE.*)

carryover. Thus, humidity controls can be considered another module of the air-handling unit. Some units have it, some do not.

Sometimes humidification control is accomplished by the use of a sprayed coil system that allows for saturated air for the system at all times. If you examine the psyc chart in Fig. 10.12, you see how the system works to control space humidity. Follow the psyc chart to see that the system supplies a constant dew point temperature produced through the use of the wetted coil. At that point, the control is through the dry bulb control device to maintain space conditions.

Some single-zone units (although not very common) can be operated with return air fans as well as supply fans; they will be discussed in paragraphs involving VAV units.

At this point, we discuss the need for return air fans, where to use them, and where not to use them. Space is at a premium, and the mechanical engineer has been charged with using as little space as possible in the design of his mechanical system. That means that if a sys-

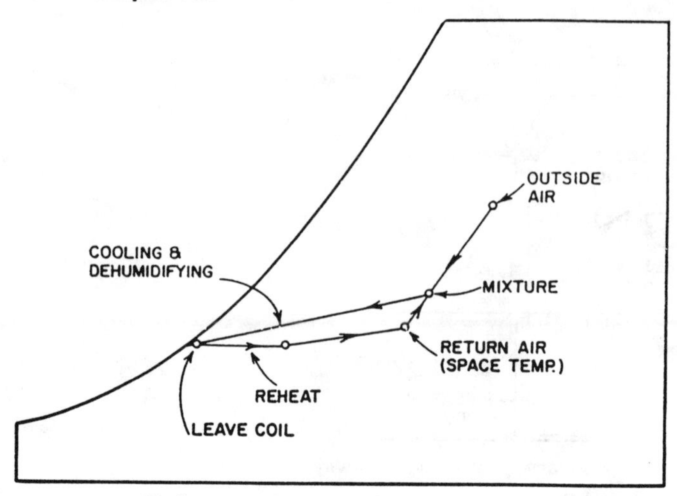

Figure 10.12 Skeleton psychometric chart. (*Courtesy of ASHRAE.*)

tem requires a total of 100,000 cfm to heat and cool a space, it might be wise to divide the system into two 100,000 cfm fans, one supply and one return fan. The single fan concept, with the attendant large pressure drops that can occur, can cause a negative static pressure in the duct work where the air returns to the fan or is exhausted to the outdoors when the system is in the economizer mode. This situation can occur because of high duct pressure losses in some very large systems with minimal sized duct work and high velocities. The challenge with the two-fan concept, which will be discussed further in subsequent chapters, is the controllability of two fans in one system and the ability to have the fans track each other.

As other types of single path and dual-path units are discussed, reference will be made to the modules of the basic single-zone unit. The reason is since all systems are built from the parts of the single-zone unit, we can look at complicated units as combinations of the modules.

The units shown in the above figures are constant volume units; that is, the amount of air being pushed and pulled through the duct work did not change, only the temperature of the air changed with its passage through the air-handling unit. Those temperature changes took place in the unit or the duct work after the unit, as is the case with zone reheat systems as shown in Fig. 10.13. Unless zone reheat systems use recovered energy, they are energy wasteful since the air is cooled to take care of the worst zone and the rest of the zones are reheated up to their zone requirements. As a matter of fact, reheat systems are not according to code in some areas unless they are modified to meet the energy efficiency codes.

Since the 1970s, when every HVAC engineer was concentrating on saving energy, the concept of varying the *amount* of air circulated in

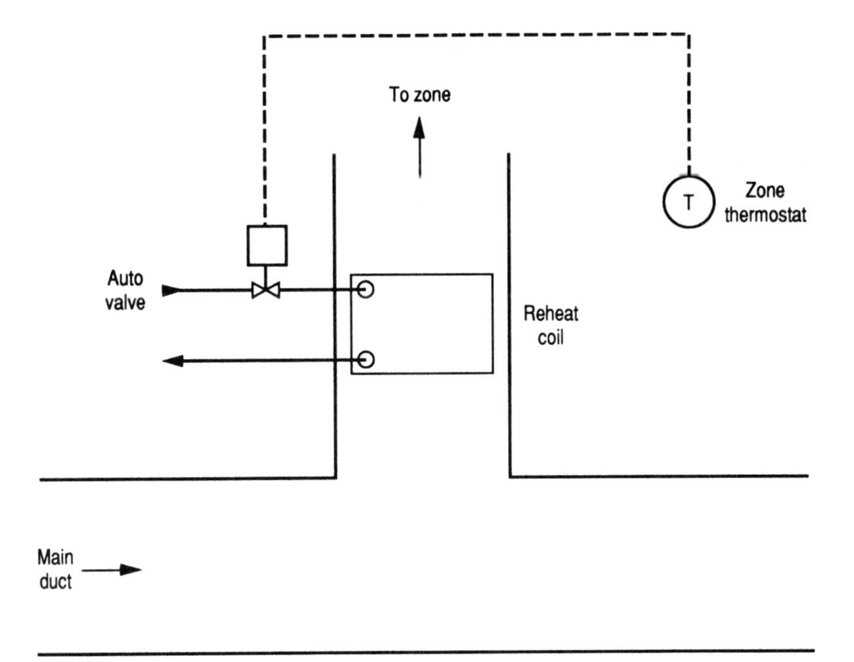

Figure 10.13 Zone reheat coil control.

the air-handling unit has been added to systems we see today. The fans, diffusers, and control systems that were available years ago did not make the variable air volume systems practical, and people were not concerned about conserving energy. Today, the story is different, and with air diffusers, controls, and fans that are available today, the concept makes a lot of sense.

Basically, as the requirements for heating and/or cooling change in the spaces, the fan delivery is reduced to meet the reduced requirements of the system. This can be done by inlet dampers, axial vane pitch, or fan speed controls. A close look at Fig. 10.14 shows what happens to the horsepower requirements of the fan when the fan is slowed down to adjust the cfm delivered through the air-handling unit. An examination of the fan laws shows the power savings that can be accomplished as the fan is slowed down from maximum cfm to a 50% reduction. See Figs. 10.15 and 10.16.

The terminal units used with a VAV system are discussed in another chapter, and the VAV air-handling unit itself can be made up of any or all of the modules mentioned previously. That is, the units can consist of the OA–RA, tempering coil, cooling coil, heating coil, reheat coil, humidifier, and other modules used in the single-path constant volume units.

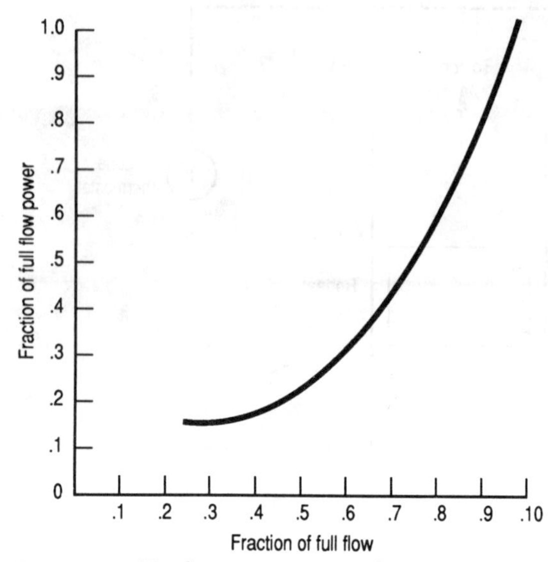

Figure 10.14 **Fan horsepower versus flow.**

For All Fan Laws: $\eta_{t1} = \eta_{t2}$ and $(\text{Pt. of Rtg.})_1 = (\text{Pt. of Rtg.})_2$

No.	Dependent Variables	Independent Variables
1a	$Q_1 = Q_2$	$\times \left(\frac{D_1}{D_2}\right)^3 \left(\frac{N_1}{N_2}\right)$
1b	$P_1 = P_2$	$\times \left(\frac{D_1}{D_2}\right)^2 \left(\frac{N_1}{N_2}\right)^2 \frac{\rho_1}{\rho_2}$
1c	$W_1 = W_2$	$\times \left(\frac{D_1}{D_2}\right)^5 \left(\frac{N_1}{N_2}\right)^3 \frac{\rho_1}{\rho_2}$
2a	$Q_1 = Q_2$	$\times \left(\frac{D_1}{D_2}\right)^2 \left(\frac{P_1}{P_2}\right)^{1/2} \left(\frac{\rho_2}{\rho_1}\right)^{1/2}$
2b	$N_1 = N_2$	$\times \left(\frac{D_2}{D_1}\right) \left(\frac{P_1}{P_2}\right)^{1/2} \left(\frac{\rho_2}{\rho_1}\right)^{1/2}$
2c	$W_1 = W_2$	$\times \left(\frac{D_1}{D_2}\right)^2 \left(\frac{P_1}{P_2}\right)^{3/2} \left(\frac{\rho_2}{\rho_1}\right)^{1/2}$
3a	$N_1 = N_2$	$\times \left(\frac{D_2}{D_1}\right)^3 \left(\frac{Q_1}{Q_2}\right)$
3b	$P_1 = P_2$	$\times \left(\frac{D_2}{D_1}\right)^4 \left(\frac{Q_1}{Q_2}\right)^2 \frac{\rho_1}{\rho_2}$
3c	$W_1 = W_2$	$\times \left(\frac{D_2}{D_1}\right)^4 \left(\frac{Q_1}{Q_2}\right)^3 \frac{\rho_1}{\rho_2}$

[a] The subscript 1 denotes the variable for the fan under consideration.
[b] The subscript 2 denotes the variable for the tested fan.
[c] P = either P_{tf} or P_{sf}

Figure 10.15 **Fan laws.** ***(Courtesy of ASHRAE.)***

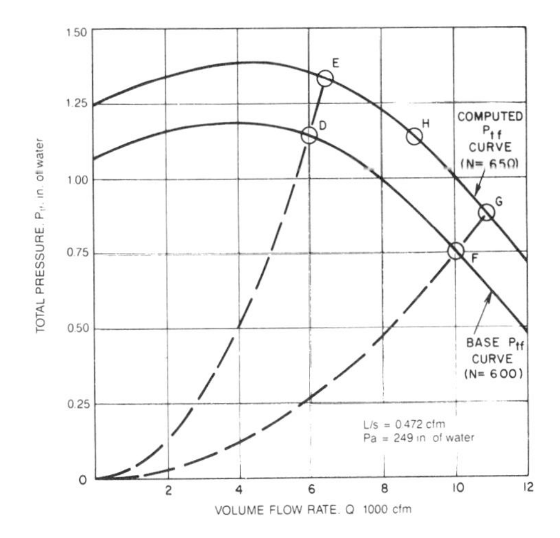

Figure 10.16 Example of an application of the fan laws. (*Courtesy of ASHRAE.*)

The control of the fan speed is now an accomplished art in that it can be done in a variety of ways. There are frequency modulation devices on the market that can easily change the speed of the motor without excessive heat losses. There are variable speed drives used on the motor and fan sheeves. There are inlet vane dampers on some fans that can change the cfm output of the fan by imparting a swirling motion to the air as it enters the fan chamber. Some systems use a bypass damper on the fan as well as a fan discharge damper and systems where the pitch of the fan blades (vane axial fans only) is changed to accomplish the cfm change requirements.

All of these methods of controlling the cfm output of the fan in the AHU require a control system that can sense the need to change the fan characteristics as the terminal units throttle down from maximum requirements. This is done is almost all cases by sensing the static pressure in the duct work at about 66% of the way down the duct work from the outlet of the AHU. Some of the control schemes can be seen in Figs. 10.17 to 10.21. The static pressure controllers should always incorporate PID algorithms, since any tendency to drift that is inherent with proportional only controllers can cause upset in a VAV system.

Note that in all figures showing VAV units, the return air fans are shown, and, as discussed earlier, the control schemes are designed to allow the return air, RA, fan to track the supply fan as closely as possible. The adjustment of the controls for these systems can be complicated, and a competent controls engineer and system commissioning expert should be used. It is also a good idea to use the best PID controllers (preferably DDC types) for the fan-matching control systems.

VAV systems operate in the low-to-medium pressure ranges but can be used with high-pressure systems, although the chances for noise

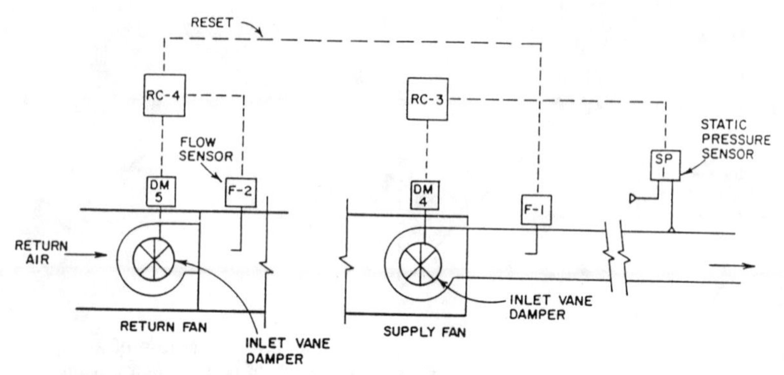

Figure 10.17 Fan volume control flow measurement. (*Courtesy of ASHRAE.*)

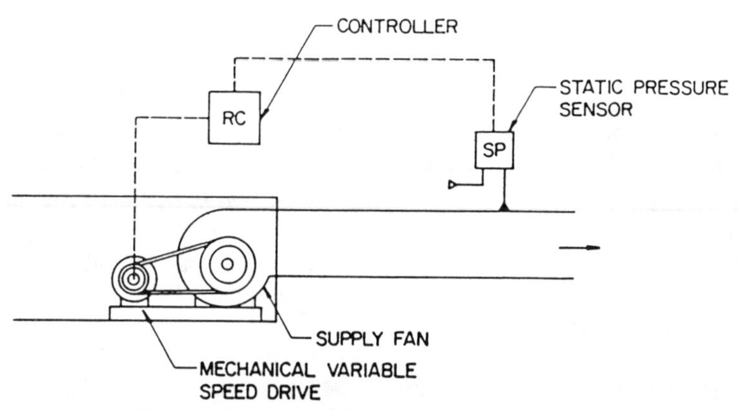

Figure 10.18 Fan speed control—variable speed drive. (*Courtesy of ASHRAE.*)

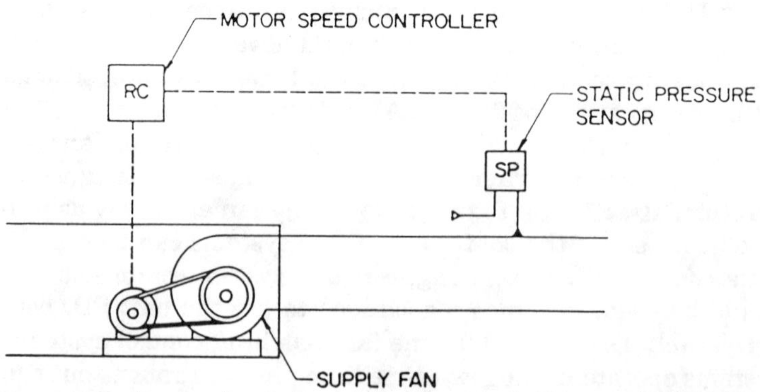

Figure 10.19 Fan speed control—variable motor speed. (*Courtesy of ASHRAE.*)

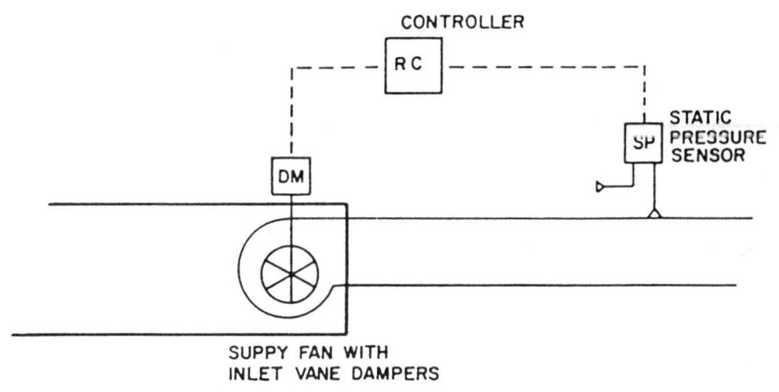

Figure 10.20 Fan volume control—inlet vane dampers. (*Courtesy of ASHRAE.*)

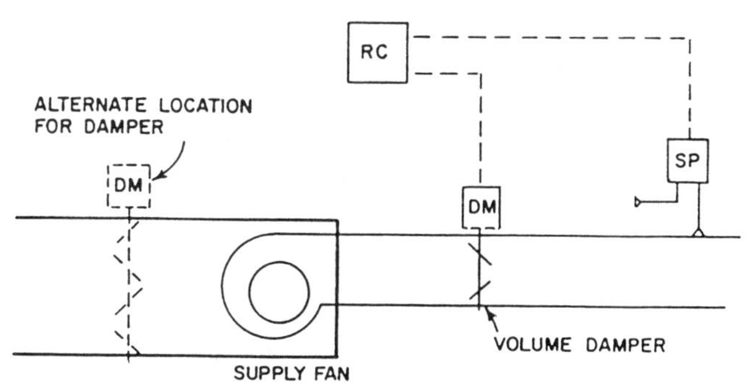

Figure 10.21 Fan volume control—volume damper. (*Courtesy of ASHRAE.*)

problems are increased if VAV systems are designed around the high-pressure concept.

As indicated in the chapter on terminal units, single duct VAV systems can be used with all types of VAV boxes on the market today. These include the standard boxes, ones that use reheat coils, induction boxes, as well as fan-powered VAV boxes used for special or building perimeter systems.

One system involves the use of induction terminal units. Here, the primary air dehumidifier (PAD) unit is used to cool and dehumidify the air sent to the induction units and to allow the inductions units in the spaces to do the sensible cooling required in the rooms and zones. The ratio of PAD air to air being induced in the space is about 1 to 6. In most cases the PAD units are 100% outdoor air units, so they must be carefully designed and controlled in severe climates. Some have been designed with return air capabilities, but they are in the minority. Quite a few have used sprayed coils as humidity control methods. Induction systems using PAD units were more popular in the past and have been replaced with the VAV units, are more energy efficient.

There is nothing new in the control cycles used with the PAD units, as can be seen in the figures. Just note the modules that are part of the PAD unit as they relate to the modules in other units.

Another single-path unit is the makeup air unit, which is generally used in factories and industrial complexes where there are large amounts of air being exhausted due to the processes involved. Usually, they are only heating units when used in those types of complexes. Since they are almost always 100% outdoor air units, the control cycles must take the severe weather conditions into account. One innovation that has come along in the past few years to protect the units in freezing weather is the use of direct fired units. These units use natural gas for heating that is *not* fired in a heat exchanger. The gas flame is used right in the airstream with the air sent to the building being heated as a result of the direct contact with the gas flame. These units are efficient from a BTU standpoint, and when the flame is properly adjusted, there is no danger of noxious gases being introduced into the airstream. Complete combustion is accomplished in a properly adjusted unit. These types of units are popular in facilities in very cold climates that require vast amounts of makeup air as a result of the exhaust systems used. They are *never* used in normal comfort control situations such as office buildings and schools.

Makeup air units are also used with heating coils and evaporative cooling coils in areas that require some minimal cooling and heating, such as restaurant kitchens, where it might be impractical to cool with refrigerated air because of the heat generated by the stoves, ovens, and so on. Here the theory is to move a lot of air (more than normal systems) and create at least a sense of comfort for the occupants as a result of the air movement.

There are other single-path industrial units designed to do a specific job. For example, there are units that move some materials through the manufacturing process. These are materials that are light enough to be carried by fan systems and are difficult to move through the plant any other way. An example is sawdust that is moved through a large saw mill using fans. Paper mills use fans to move products around the plant. Air-handling units are used in some places to move gases. One special air-handling unit has been used on a large flat bed truck to cool down the electronics of the military jets after they return from a mission. These types of air-handling units can also be seen in trucks as portable units that cool the commercial jets as they sit at the airline terminal gates.

Dual-path units

Dual-path units, as the name suggests, are units where the air being heated or cooled is acted upon in parallel paths so not all of the air supplied by the fan is heated or cooled. For example, half of it is heated and half of it is cooled. The percentages stated are not always correct, but the concept of parallel paths is the one to remember.

Some modules that were discussed previously, such as the OA–RA module and the tempering coil module are the same for dual-path units.

One typical dual-path unit used in the past and used on a limited basis today is the double-duct unit. In this case the two paths of air (warm or hot air and cooled air), pass through the air-handling system and continue down the duct work in two separate duct systems to be mixed just before entering the space through a device known as a double-duct mixing box. The operation of these boxes is described in the chapter on terminal units. These air-handling units are usually of the constant volume type, (although not always the case), so that the control cycles are generally the same used with all constant volume systems as shown in the previous figures. These types of air-handling units are usually either high pressure or medium pressure, although some have been used with normal low pressure concepts. The theory here is that the air is being delivered right to the point of usage before being mixed. Allowing better control of the two streams at that point. Static pressure control is required since there are times when one or the other of the two sections of duct work is trying to supply all of the air being fed to the boxes. Also, there is the problem of balance since the cfm of cold air for a space is considerably different than the volume of hot air needed for the same space. An inspection of the control scheme shown in Figs. 10.22 and 10.23 shows one of the methods suggested to control that apparent difference in air volumes. Velocity pressure along with total pressure and static pressure can measure cfm of airflow and adjust the heating and cooling coils to compensate for the differences in flow in either duct. It will be clear when the subject of master–submaster control is discussed why that cycle is shown in almost all of the cycles discussed thus far. It is a common system used in almost all single- and dual-path systems.

Some systems that use the dual-duct method to transport the two types of air to the space do not use a mixing box and just use dampers in the sections of hot and cold ducts to mix the air before it enters the room.

Dual-path systems are also used in medium- and high-pressure VAV systems. The VAV boxes are described in the chapter on terminal units, and the AHU itself is controlled as shown in Figs. 10.22 and 10.23. The control systems are no more complicated than the systems used for the single-path VAV systems. There is one innovative type of dual-path VAV system using two fans that has been seen on a number of jobs. It allows the hot duct to use only the *return air* and the cold duct to use the mixed air (return and outside air) (Fig. 10.24). This system is reported to be more energy efficient than the standard system since the return air requires less energy to heat it than the mixed air.

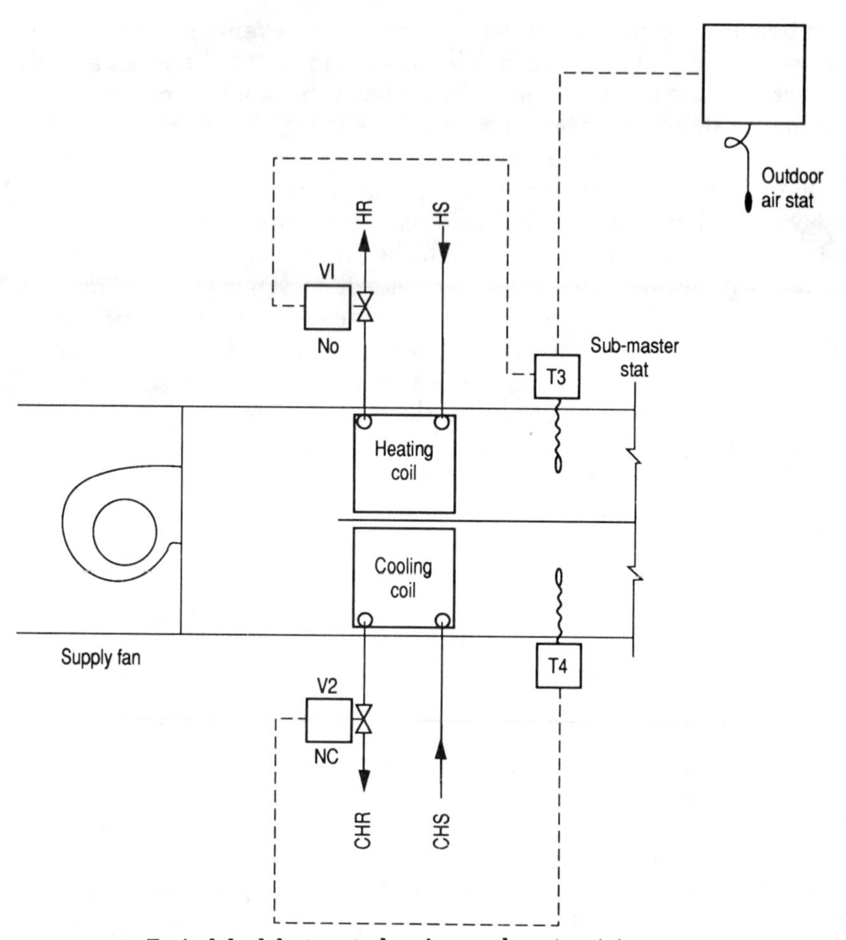

Figure 10.22 Typical dual duct controls using a submaster stat.

One of the most common and oldest dual-path systems in use today is the *multizone air-handling unit* (Fig. 10.25). With this air-handling unit, the air travels two parallel paths at the unit as it is heated and cooled. The dampers in the multizone unit mix the two air streams and send that air to the zone or room being conditioned. The control system for the coils and the dampers in the unit are shown in Fig. 10.26. There are limitations on these types of units as to the number of zones that can be incorporated in them. The modules of the OA–RA dampers, tempering coils, and so on, are the same in this case as in the case of all the units discussed so far.

These types of units were and still are popular for some applications and are almost always packaged units. The result is that some manufacturers have been known to skimp on the designs of the unit. There

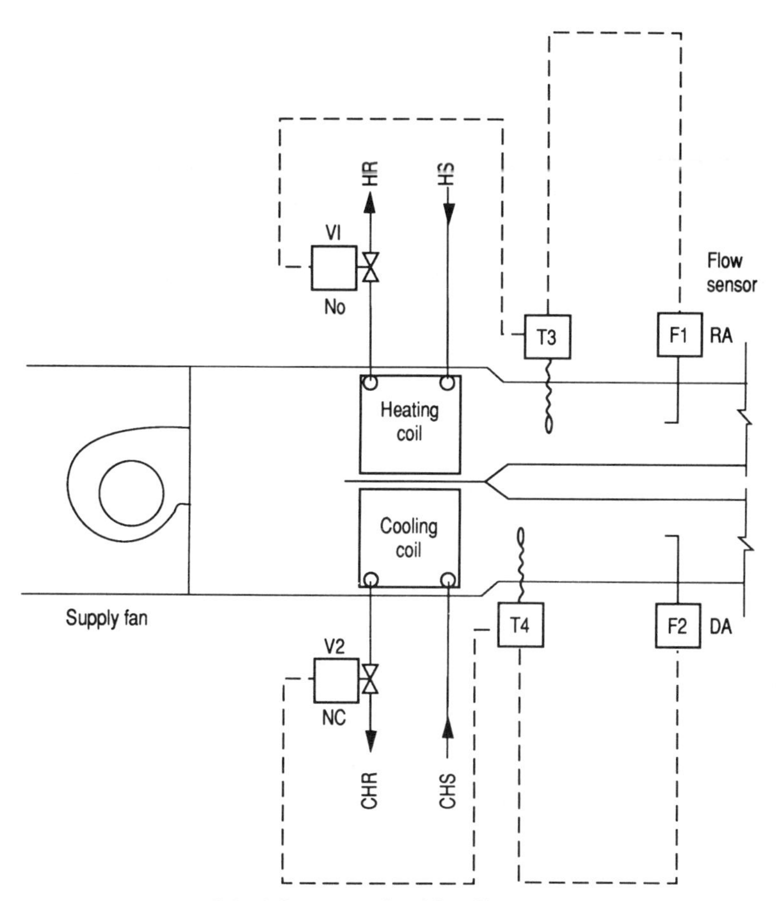

Figure 10.23 Typical dual duct controls with a flow sensor.

is a special three-duct multizone unit available that is more energy efficient than the standard units, but it is impossible to retrofit existing units to it (Fig. 10.27). Multizone units use steam heating coils, hot water heating coils, chilled water cooling coils, and direct expansion cooling coils, the latter being the most difficult to control since DX coils are either on and very cold or off and not cold at all. There is little or no in-between for DX coils.

Multizone units can sometimes be retrofitted into VAV units if the duct work, fan, and other items can be reasonably changed without too much cost. This involves removing the zone dampers and changing the fan to a variable speed or other controllable fan system. The quality of the dampers on multizone units has always been a bone of contention as far as the control manufacturers are concerned, and at times, consulting engineers have specified that the dampers be pro-

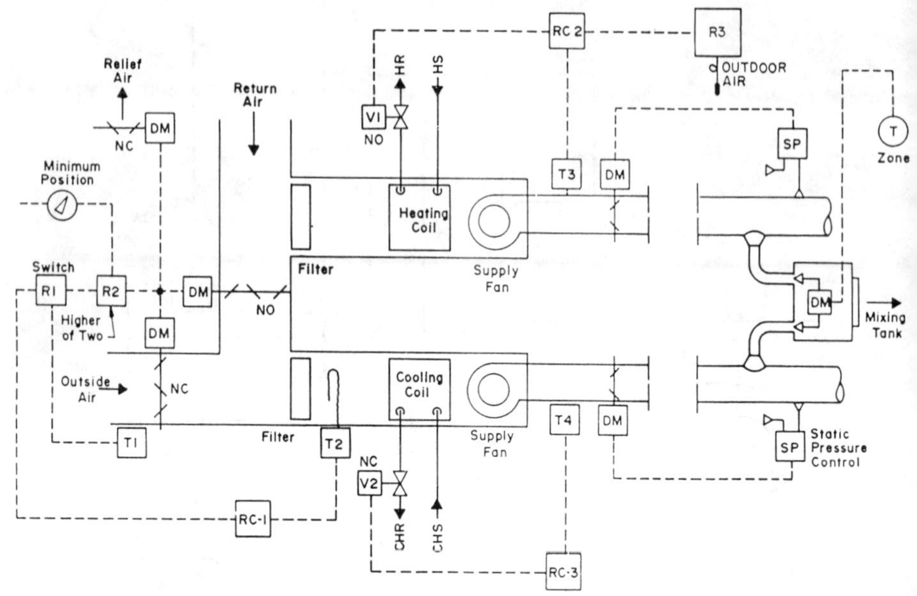

Figure 10.24 Dual duct air-handling unit with two supply fans. (*Courtesy of ASHRAE.*)

vided by the control subcontractor to ensure the quality of the dampers and prevent leaks from upsetting the controllability of the unit.

General Comments

Next, we provide general comments that apply to just about all the units, whether they are field erected or packaged units. The comments apply to low-pressure, medium-pressure and high-pressure units.

First, remember that the simpler the control system, the better chance it has to work. Adding controls on top of controls to correct a poorly designed system will accomplish nothing and only frustrate the controls engineer trying to decipher the system. If one controller will do the job, use one not two.

Second, most, if not all, air-handling units supply conditioned air to some sort of terminal device that reconditions the air just before it enters the space. The best systems are the ones that try to modify the mixture of the OA and RA so that it is close to what is required by the space before the terminal device does its work. That is, the unit should try to bring the air as close to the needs of the building in terms of heat loss or heat gain as possible *before* the final modulation is done by the terminal device.

Figure 10.25 Multizone unit. (*Courtesy of Trane Co.*)

Third, one way this is done is by resetting the hot deck temperature and, in some cases, the cold deck temperature, in accordance with the outside air temperature. The concept is sometimes called master–submaster control and is used in almost all single- and dual-path units. The concept of resetting one controller from another is used in other applications also. For example, it is used when the controller that is controlling a steam valve of an air-handling unit coil is quite a distance from the valve.

There is a problem of what is sometimes called old man delta T (time lag), which will cause constant cycling no matter what the proportional band is set to and even with PI and PID incorporated into the algorithm of the controller. The problem is that there is just too much time that elapses from the change in the airstream temperature to the time the sensor in the space feels that change. If the sensing bulb of a duct controller is placed right after the coil, the time lag is minimal and the duct controller can do a good job. The concept of master–submaster comes in to determine what the set point of this controller is to be. The room controller at a great distance resets the control point of the controller in the duct work that is doing the actual controlling; we then get stable control.

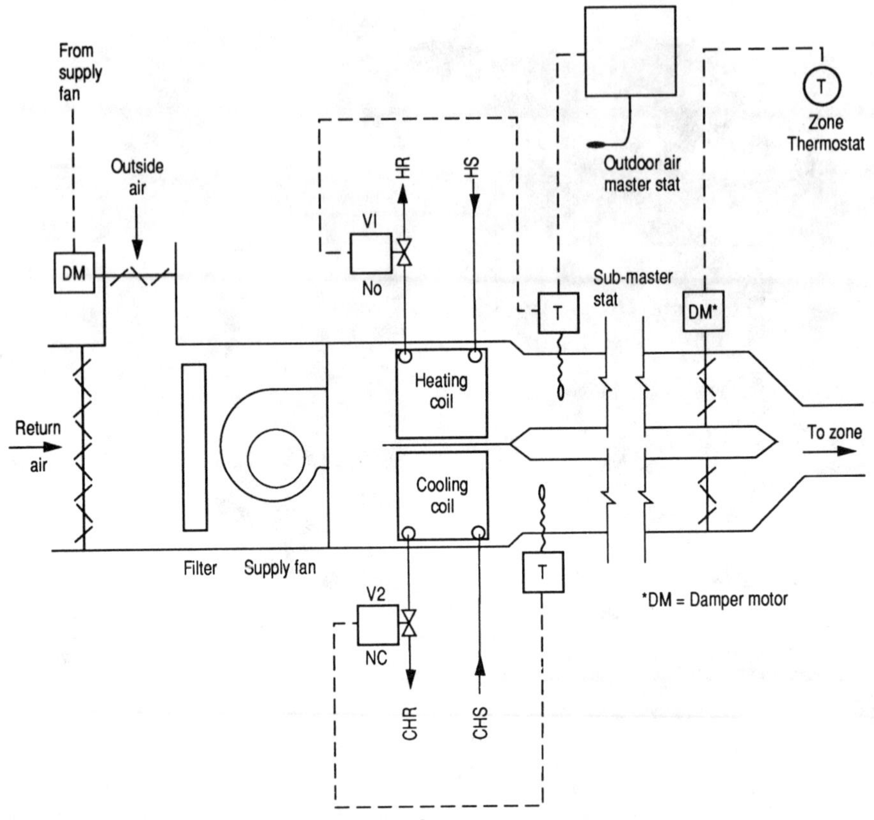

Figure 10.26 Typical multizone unit control system.

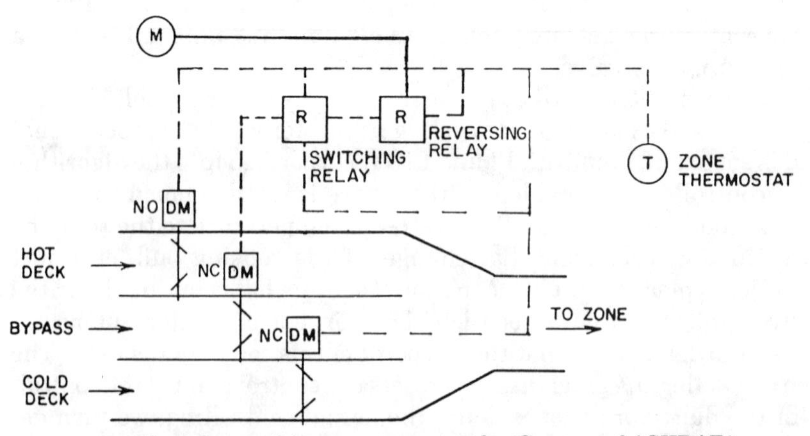

Figure 10.27 Three-duct multizone system controls. (*Courtesy of ASHRAE.*)

Other applications of the concept of lag are in a complete closed loop control system that prevents actions from being instantaneous and in a situation in which air passing over a liquid-filled capillary of a remote bulb controller does not affect the liquid inside the bulb immediately. There is a lag due to a wall of metal between the liquid in the capillary and the temperature of the air passing over the bulb. Other lags in closed loop control systems affect the ability of the system to control properly. Generally, the most unstable condition exists when the additive time lags in the process after the control corrections are about equal to the time lags in sensing the change and calling for another change in control signal. Conversely, the most stable condition exists when the control correction and resulting process is slower than the sensing lag and change in control signal. The more stable the timing, the larger the gain (or smaller the throttling range) needs to be to achieve stable control.

The control systems shown in the figures in this chapter have generally been drawn as if the systems were all pneumatic systems. This is not to infer that analog electronic, electric, or digital control systems will not work on the systems described. The reason for using the pneumatic concepts is that the majority of commercial systems have been pneumatic, and most control engineers can identify with those types of controls. In the past few years, however, the trend has been to convert and even start out with either analog electronic or electronic digital controls. The use of straight electric controls in the larger projects has been diminishing of late. There are basically no commercial control systems that cannot be changed from pneumatic to electronic (either analog or digital). Remember, however, that even with the electronic digital systems, in many cases it is more practical for the operators or actuators to be pneumatic as that still is to this date the best all around technology when price, simplicity, ease of operation, serviceability, and so on, are taken into consideration. Most digital jobs use electronic sensing, electronic controlling with pneumatic operation. That scenario is quickly changing, but at this writing the facts are that is the way the systems are being installed and sold.

Troubleshooting

The problems that can develop with air-handling units generally involve the moving parts of the unit and the items such as the coils and filters. The parts of the unit that are nothing more than sheet metal seldom cause problems that need attention. The moving parts of a typical air-handling unit are such things as belt drives, belts, electric motors, fan wheels, fan bearings, dampers, as well as coils, filters, other specialized items.

The first item to be aware of is the noise level of the unit when it is running. If the initial noise level changes drastically, there is a good chance that something is wrong. The first thing to check is the condition of the fan and the electric motor bearings. If everything in that area seems alright, it is possible that one of the automatic dampers is stuck in one position or another. Also check the pressure drop across the filters, since they may have reached their peak and need to be changed. In the case of automatic filters, it is possible they became jammed in one position. Look also for a splitter damper that may have come loose and is crossways in the duct work. That would also be possible with some turning vanes in the duct work, so they need to be checked. On systems with multibelt sheeves, check for broken belts that have changed the noise characteristics of the system. Something could also be jammed into the blades of the fan. This can happen if objects get by the filters or are introduced after the filters when an access door is left open. Noise changes can also be generated if access doors are left open even a crack or noise can result from a broken vibration isolator on the fan itself or on the motor driving the fan system. Sometimes the belts dry out and need belt dressing to stop the noise.

If the system does not seem to be delivering the capacity it did when it started, the solution can be quite varied. For example, if the electric motor was recently repaired, look to see if the motor is running backward. Three-phase motors can be wired to run backward by reversing just one wire, and a forward curved fan will put out about one-third of its rated capacity even running backward. The filters can be clogged. The fan blades can be choked with dust and debris, which will reduce the capacity of the fan to move air. This is particularly true of forward curved fans. The coils, like the filters, can be clogged with dirt and need to be cleaned. The capacity of the coils can be reduced by the dirt in between the fins, which is a big problem with a cooling coil that is six or more rows deep. The fins of a coil can be damaged and bent, reducing the capacity of the coil. This can be caused by inattentive workers who may be inside the unit doing something else. For example, in an attempt to save money, a building owner placed blocks of ice in the plenum area just before the cooling coil on a large air-handling unit supplying a St. Louis department store with cooling. After a couple of weeks, the fins on the cooling coil were so flattened the coil had to be entirely replaced before the system would work.

In the case of heating coils, look for steam traps that need servicing, since they can be preventing the steam heating coil from functioning properly. In the case of water heating coils, look to see if someone reset the balancing valves or turned the isolation valves either open or closed as the case may be. In the case of a DX cooling coil, the capacity reduction system on the compressor may have malfunctioned and the coil will not be getting the benefit of the total system capacity.

The controls can be malfunctioning, not allowing the system to deliver the maximum capacity. This situation can be caused by many items discussed in the chapters on specific controls. In general, check the calibration and make certain there has not been a failure at the air compressor (oil or water contamination). Also remember that controls that have not been checked for a while may need adjustments. Checking controls can be done by changing the set point above and below the measured condition and observing the resultant control action.

Automatic valves have a way of causing problems if the packing wears out and if the stem is in a vertical position with the operator portion upside down with a leaking packing nut.

Sometimes the use of recorders to check the actions of the air-handling unit and the controls is in order. Recorders are available with all kinds of ranges and drives, as well as with remote bulbs that can be placed in the duct work at strategic positions to monitor the temperatures and pressures of the system. If there is an energy management system that has logging capabilities, that capability can be brought into play. The secret to the use of that kind of a system is the access to individuals who can *interpret* the data that comes from that logging.

Maintenance of Air-Handling Units

Just as the moving parts and other items are the ones that can be involved with troubleshooting, they are also the ones that usually require maintenance. Everything that moves usually requires service and maintenance from time to time. The important thing is not so much knowing *what needs to be maintained* but is *setting up the schedule of maintenance and following it to the letter.*

To start, all bearings that need to be oiled or greased per the manufacturer's instructions need to be part of the maintenance schedule. Some items, such as damper blades, are *not* supposed to be oiled, as they tend to attract dust causing the blades to bind and not function properly. These filters need to be on the schedule and maintained according to the manufacturer's instructions. Throw-away filters need to be replaced periodically. Coils need to be looked at to see if dirt and dust are starting to accumulate on the fins. When that happens, a company that specializes in coil cleaning needs to be contacted. Some of the pan humidifiers need to be checked since scale can form in the pans and on the coils due to the water hardness and needs to be cleaned off. Whether the steam and the condensate of the system will keep the scale and dirt off the pipes and in the pans depends on the water treatment program of the building and its contents. The jet type humidifier can also clog up after a while due to inadequate water and steam treatment. The fan blades need to be cleaned periodically, as they can be clogged with dust and dirt. The belts need to be checked and replaced as recommended by the manufacturer. Unfortunately, the use of belt guards, although necessary, tend to make the belts victim of the out of sight, out of mind syndrome. The controls are also a part of the system that needs to be checked periodically, and at the first sign of air line contamination, the system should be shut down and cleaned.

When checking the controls, the operators are often frustrated by the lack of gauges and other devices to see how the control systems are operating. In that situation, operators should make every effort to have the necessary gauges and indicators installed to permit proper servicing and maintenance of the systems.

Design Considerations with Air-Handling Units

The use of air-handling units in a building brings with it the responsibility to use those devices correctly, and often times engineers and designers do not take a few important items into account when they are planning on the location and use of air-handling units. This section will discuss things to do and not do in regard to the out-door intake location, the unit location, the type of air-handling units, the size of units, and so on.

One of the most important considerations for an engineer is the selection of the location of the unit, in particular the location of the outside air intake. Too many systems have failed and have created disastrous results from improper location of outside air intake for the air-handling unit. American Society of Heating Refrigerating and Air Conditioning Engineers' research and new standards have tried to spell out what quantity and quality of fresh air need to be introduced into a commercial building. The problem is that with a poorly designed air intake duct system, all the standards in the world will not solve the problem of the quality of the air in the space. Recent studies have shown that a poor location affected by predominant winds, exposures, updrafts, snows, heat from exhaust fans, and other factors will ensure that not only is the air-handling unit not taking in the required amount of outdoor air, the air can be traveling backward. Added to that is the problem created by a system with the usual exhaust fans that are not functioning properly as a result of the poor control over the amount of air entering the building through the outdoor air intakes. Some have suggested that the way to be sure the amount of air that was designed into the system is actually getting into the building is to install flow-monitoring and control devices in the outdoor intake that will adjust the system accordingly. Building static pressure controls that are properly engineered can also be used to ensure that the correct amount of outdoor air is coming into the building. The problem with that type of system is that, as with all static pressure controls, there must be a stable reference point, and the only reference point that can be used is the outside air. The location of that reference point and the tip that senses it is a matter of conjecture. If a good location can be found for that tip and the static pressure controls are stable, the system will be able to control through static pressure the amount of outdoor ventilation air entering the building. All of the above not withstanding, the location of the outdoor intake is a matter that requires some study. Furthermore, the location should not only take into account the wind, and so on, it should consider the predominant winds, the snows that may occur, the sun's affect, and the controls that will be applied to the automatic dampers used in the cycle.

The classic example of *what not to do* is the situation in a retrofit job where the outdoor intake is installed in an old window frame just because the window is available. Many times the hole created by the removal of the window or windows is either too large or too small and typically it is extended to the air-handling unit's full size with the outdoor air automatic damper installed full size in that duct work. Any good control engineer knows this is the most difficult control situation one can imagine. The outdoor air damper *must be* sized in accordance with the standard recommendations of the control manufacturer. If that means the duct work in that area needs to be reduced with transition pieces, so be it. At times it would be more expedient to install a fresh air fan on the roof of a building that feeds the air-handling units on lower floors with a positive pressure to ensure that the unit or units receive the proper amount of fresh air. There is a better chance that such a fan on the roof, if its intake is not restricted, will get the job done and not be affected by the winds, and so on. The intake louvers with their bird screens also need to be considered in the design of the outdoor intake. Many of the systems that have been looked at a few years after they were installed were found to have intake louvers that were totally blocked by leaves and other debris because of their locations. That is, they were in a location that could not be checked easily by the operating personnel. The bottom line is that the location, size, and other factors in the design of the outside air intake as well as the duct work connected with it are very important and must be of prime consideration in selecting the units.

Where to put the units is a question often asked by the designing engineer when first faced with the initial design plans of the architect or the owner. Too often the compromises that result from the first meetings result in locations that cause problems throughout the life of the job. The architect and the owner want to save space. They need to remember, however, that if the system does not heat and cool properly, the space *can never be rented or sold.* There must be space for the operators to maintain the equipment, and if they cannot get there, the unit will not get looked at and maintained. If an electric motor, as an example, is on the side of the air-handling unit that is up against a wall so tightly that someone cannot squeeze into the space, you can be sure the belts will never be checked and the motor bearings will dry up of lubricant long before they burn out. If the installers and contractors involved in the construction of the project have to use a shoehorn to get the duct work and the units in place, you can be sure that the cost of the project is going to go up.

The selection of the types of units, be they VAV, multizone, single zone, high pressure, or whatever, will depend upon factors too numerous to mention. The selections, however, should not be geared to any one item. They need to take into account energy considerations as well as factors such as the intended use of the building and the owner's future plans for the building. Too often compromises are made that come back to haunt the designing engineer. The engineer who is too proud to ask for help when there are special problems with special building requirements is bound to make mistakes in the design.

In summary, this chapter presented the history of air-handling units in the HVAC field, with some information on the types that have been installed and that are being installed and used. The chapter looked into the control systems that can be used with the various types of air-handling units and warned of some of the problems that can result with improper selection and design of air-handling systems.

Chapter

11

Terminal Units and Systems

In this chapter the discussion will center around terminal units. A *terminal unit* is any heating or cooling unit, with or without any kind of fan, and in some cases, with auxiliary devices such as a heating or cooling coil used to condition the air entering an occupied space just before its entry to the occupied space being conditioned. The chapter will include units currently in use as well as units that have been popular in the past but have been phased out because of changes in energy codes and other reasons. The proper controls and cycles will be discussed and illustrated. The problems and solutions that can arise in the use of the various types of terminals and how adding or taking away control devices and systems will correct the problems will also be discussed. The chapter will clarify and expand on the control cycles that were used and then abandoned and the reasons they were abandoned.

A main part of the chapter will be a discussion of the importance of proper control with the primary systems and their affect on the terminal units, with emphasis on the selection of the proper cycles in the primary unit controls.

The items that will be covered include

Unit ventilators

Fan-coil units

Double-duct mixing boxes

Induction units

VAV boxes (variable air volume)

Other items such as convectors and radiators, etc.

Before discussing terminal units, a discussion of the difference between the primary systems and the terminal units is in order. A primary system consists of the boilers, chillers, fans, coils, duct work, piping, dampers, automatic valves, filters, louvers, electrical motors, controls (both electrical and temperature controls), and so on, involved in supplying the mediums such as hot water, steam, and chilled water to the terminal units in the building that conditions the space. Basically, it is the equipment in the boiler and fan rooms that supply the mediums to the terminal equipment such as the VAV units and other units in the space or adjacent to the spaces.

An important point that needs to be stressed about the primary and terminal systems and units is that these two systems must operate in conjunction with each other and to be sloppy in the design, installation, or operation of one or the other is to miss the design criteria completely. If the controls that control the primary systems are *not* operating to maximum efficiency, the terminal systems will have more work than they were designed to do. The whole concept of the two systems is based on the idea that the primary system does most of the work, with the terminal or secondary systems taking the rough edges off the mediums provided because the spaces do not all have the same heat loss or gain. If the system is properly designed and operated, the primary system will almost be able to do the entire job for the building by itself, except where there are differences due to exposure, people load, and so on, in the various spaces.

What this means to the control engineer and operator is that time and effort must be spent to control the primary part of the system as closely as possible so the terminal units can do the final work. Here is an analogy: Years ago buildings were heated by providing 212°F 5 psi steam to cast iron radiators in the space while asking the room thermostat to cycle the valves on the radiator and maintain the space conditions. It was apparent that the room thermostat could not do that job with the medium (steam) as hot as it was, and the temperatures in the space had wide swings. Today we might do the same control system with hot water that is adjusted as the outdoor temperature changes. The same concept will apply in *all* types of primary controls. If the control system involved in the primary portion of the entire system is properly designed, operated, and maintained, the terminal units can do the job. If not, the designers and operators *must* look for the reasons the primary system is *not* providing the properly controlled mediums to the units.

As is often true in other things, the design of the system itself is often at fault, and *no amount of controls will correct the problem.* The emphasis here is to be sure that the kiss (keep it simple stupid) principle is used. Too many times controls are added on top of controls that do nothing more than cause problems and compound existing problems. An analysis of the *system* will often show what the problems are.

Unit Ventilators

To understand unit ventilators and how and where they came into being, we will look at the history of the HVAC industry before the advent of affordable air conditioning. Back in the days of mostly steam radiator heating systems, there was not much concern about overheating rooms as the outside temperature varied during the day from morning to noon to afternoon, *except* in school classrooms. The usual heating systems in classrooms could not control the overheating when the rooms went from *no occupants to 30–35 occupants* in a matter of minutes. Coupled with that was the fact that lighting experts insisted that classrooms have large windows to let in natural light. Thus, the average classroom started out in the morning with the ambient temperature anywhere from 0°F to 40°F outside. After the students arrived, the room temperature rose and was way above the comfort level by noon; it stayed high the rest of the day up until 4:00 p.m. What did the teachers do to try and control the overheated classrooms? They *opened the windows.* That in turn compounded the problem since energy went out the windows and janitors had to pour more coal into the boilers. An added problem of overheated classrooms was that students fell asleep, while the teacher was trying to teach. Many times in the past, one could walk by an elementary school in the dead of winter and see half the windows open during the day and even into the night when the sun was set.

Then, an engineer decided that if we need to cool the classrooms after the students arrive, *why not use the best cool air available, outside air.* Thus was born the classroom unit ventilator (Fig. 11.1).

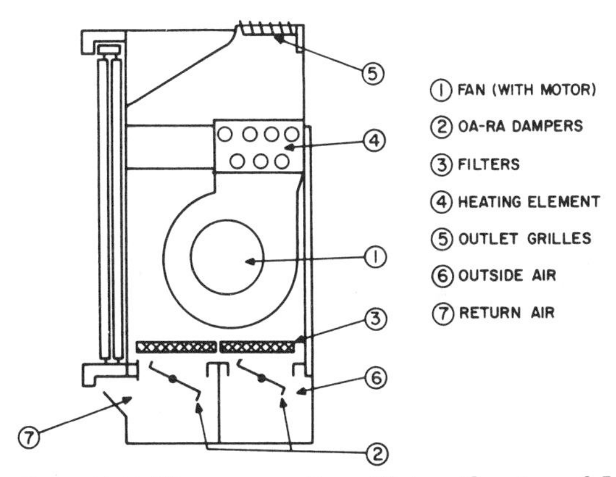

Figure 11.1 Classroom unit ventilator. (*Courtesy of Johnson Controls Inc.*)

Outside air will cool the classrooms in the winter and keep students awake, as well as eliminate the need to open the windows in the middle of the winter in northern climates. Further study indicated that the same apparatus could be used in other areas of high concentrations of human bodies, especially where the areas were empty one moment and full of people the next. Thus, unit ventilators were and are still used in gymnasiums, auditoriums, churches, and anywhere people congregate and give off more heat than is necessary to maintain a balance between the heat loss and the outside air temperature. Experience has shown that on an average winter day in a northern climate, the outside air temperature may start out cold enough to require heat in the space, but after the sun comes up and the temperature moderates, a classroom with 30 students can easily become overheated.

Basic cycles of operation

Soon after a few manufacturers started producing unit ventilators, it became apparent there was a need for standards so users did not purchase a variety of units with different features and cycles. About that time ASHRAE set about to create standards and eventually came up with three basic cycles of operation for unit ventilators. At first the ASHRAE cycles were known as Cycles A, B, and C, but today they are known as Cycles I, II, and III, with an extra cycle known as W. These will be explained in detail since they are the basis for all the controls.

To begin, all of the cycles use some common items. These include an EP switch, or solenoid air valve, wired to the fan motor so the dampers (outside and return air) return to their normal positions when the fan stops. There is always a low-limit thermostat to prevent excessively cold air from discharging into the room and causing discomfort. There is a room thermostat in the space to control the unit and maintain a comfortable space temperature. In all of the cycles, some of the units are heating only and some are heating and cooling; in those cases there can be further subdivisions in that the heating can be provided by steam coils, water coils, or electric coils. The cooling can be from DX coils or chilled water coils.

Cycle I (Fig. 11.2) can be described as follows: The unit operates with 100% outside air at all times that the room thermostat is satisfied. This means that during warm-up, the unit is controlled by the low-limit thermostat, which controls both the heating source (the steam or water valve) and the dampers. In all unit ventilators, the dampers are linked so that as the outside air opens, the return air closes and vice versa.

Cycle II (Fig. 11.3), which is the most common of the three cycles, operates with a fixed minimum of outside air (usually anywhere from

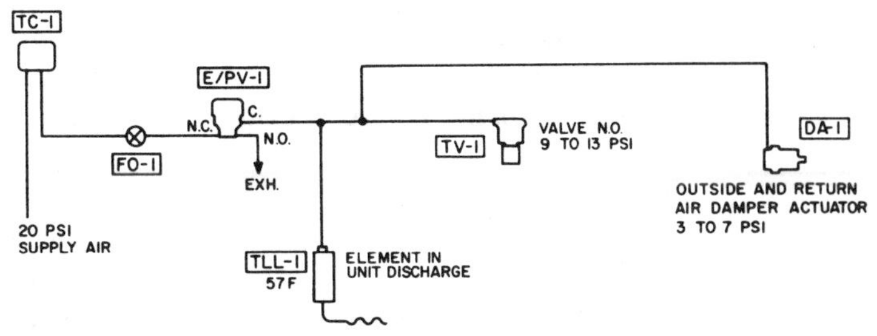

Figure 11.2 ASHRAE Cycle I. (*Courtesy of Johnson Controls Inc.*)

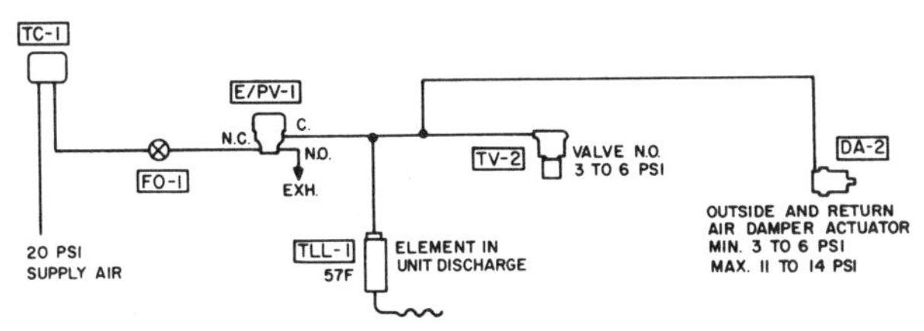

Figure 11.3 ASHRAE Cycle II. (*Courtesy of Johnson Controls Inc.*)

15% to 50%) admitted during the heating and ventilating period. That percentage is gradually increased to 100% if required to counteract the tendency of the space to overheat. There is a low-limit thermostat in the fan discharge that prevents the discharge temperature from going too low, even when the unit is trying to admit 100% outside air for cooling the space. That low-limit thermostat controls the dampers and the heating valve.

Cycle III (Fig. 11.4) is different in that there is a thermostat in the mixed air chamber that only controls the dampers to maintain a fixed temperature to the coil, with the room thermostat controlling only the heating medium (the coil valve).

Cycle W is the same as Cycle II, except that the low-limit thermostat controls only the dampers and not the heating valve. The room thermostat controls only the heating valve.

There are variations of the cycles in that in some units the control of the coil is through the use of face and bypass dampers not the usual

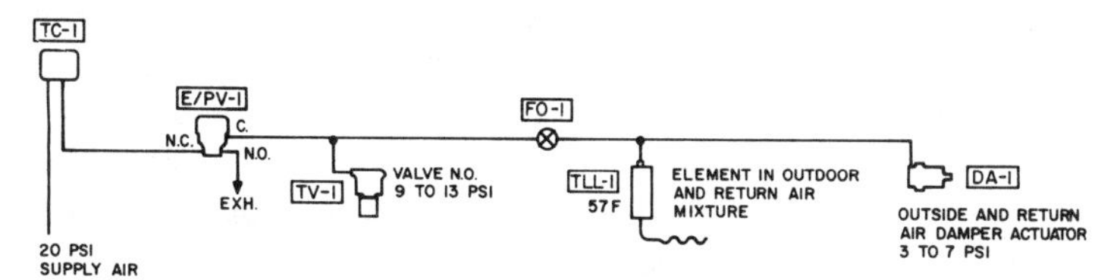

Figure 11.4 ASHRAE Cycle III. (*Courtesy of Johnson Controls Inc.*)

valves on the coil. Some units use the face and bypass dampers *along with* valves on the coil. During the warm-up period all of the units operate with 100% recirculated air to bring the temperature close to the set point.

Over the years there have been some innovations to the standard units to try and combat the down drafts that are often a problem with large expanses of windows as in a classroom. These innovations include using a long grill under the entire window length to draw in the return air to the unit, while taking very little air from the floor of the room, using a long grill under the entire window blowing the air up against the window to counteract the drafts, and using a long window length grill with baseboard radiation in the cabinet to warm the windows and prevent the cold down draft off the windows.

The introduction of these three types of window draft preventers required the control companies to design additional controls for units with the baseboard radiation under the windows. Most of the time the basic control of the units is from a room thermostat in the space controlling the dampers and the heating mediums in the unit. It is possible, however, to use a return air controller that has its bulb installed in a sampling chamber in the unit to sense the return air. An analysis of Cycle II will show that there is a period when the damper motor controlling the dampers must hesitate and stay in one position even as the output pressure from the thermostat rises. This is done on a pneumatic system with a set of two springs in the damper motor. Each spring is calibrated and has a different spring range.

Electric unit ventilators are the same as the ones described, except that they have electric coils instead of the steam or hot water coils used in the majority of the standard units.

Over the years the control manufacturers have made standard arrangements with the manufacturers of the units to have some of the control items installed at the unit manufacturer's factory. These control items include the damper actuators (outside and return air as well as face and bypass), step controllers used on electric coils in electric units, and return air–sampling chambers.

After the advent of practical and affordable air-conditioning systems, some engineers decided that schools that held classes when mechanical refrigerated cooling was needed should have unit ventilators equipped with an air-conditioning coil (either chilled water or DX) and should extend the usage of the unit to those times. In the case of units with DX coils, a complete system, including compressor, is installed in the unit itself. The cycles are the same for the heating, the only difference is when the unit switches to cooling where a fixed amount of outside air is used during that period.

Unit ventilators can be used with night setback thermostats in the same way as with other heating and cooling systems, and as a matter of fact they are often used in schools with just such cycles. In those cases, some changes in the cycle usually take affect. The operation to maintain the night setback temperature is done by intermittent cycling of the unit fan *without* any outside air. Sometimes a zone thermostat controls a group of units, cycling them on an intermittent basis to maintain the minimum temperature set on that thermostat.

In general, all of the cycles described up to this point were designed around pneumatic controls. It should not, however, be inferred that electric and electronic controls that perform the same functions with pneumatic control devices are not available. As a matter of fact, some control companies manufacture electric damper motors that have built-in terminals for the low limits and other devices necessary to comply with the standard cycles of ASHRAE. The companies that manufacturer electronic controls also have devices that can be used to control unit ventilators, with the results of the standard ASHRAE cycles being just about the same as with the pneumatic control systems.

In summary, unit ventilators can and are being used in many applications where the internal gains due to people, lights, and machinery are rapid and excessive in comparison to the heat losses of the building.

Fan-Coil Units

The term *fan-coil unit* (Fig. 11.5) is used to describe a multitude of heating and air-conditioning units, such as the ones that have a fan and a coil to deliver conditioned air to a space needing to be heated or cooled. In this section, however, we will limit the discussion to units in the same cfm range that are similar to the unit ventilator except that they generally do not take in outside air under a controlled condition and use almost all recirculated air.

The fan-coil units usually run in the neighborhood of 200–800 cfm and are often found in motels and other facilities where there is little or no need for ventilation and inexpensive systems is the watchword. They are mostly used with heating *and* air-conditioning systems installed on the perimeter of a building to counteract the heat loss at that location. By comparison with other types of systems, they are simple, inexpensive, easy to maintain, and easy to control. They come in basically two styles of cabinets with a fan and a fin-tube coil. The vertical units are used under the window in offices or motels, and the horizontal units are used in a furred space above a hallway or a bathroom in a motel. The horizontal units have less cabinetry than the vertical ones, but the principles are the same for both types.

The control schemes for all types will be discussed in this chapter, but before continuing the discussion of fan-coil units we look at some of the advantages and disadvantages of using these units.

Figure 11.5 Typical fan-coil unit. (*Courtesy of Carrier Corp.*)

The advantages of fan-coil units are as follows: They allow individual room temperature controls of various cycles and schemes. There is little if any recirculation of room air from room to room since the units are not connected to other units. There is minimum or no duct work, except where the horizontal styles are used above the furred ceiling in motels and other such places, and even there the amount of duct work is small. The under-the-window installations are ideal for small rooms, and sometimes the units can be adjusted to wipe the window surface and prevent cold down drafts.

The main disadvantage of fan-coil units that use a common coil with hot water in the heating season and chilled water in the cooling season is that when the mild time of the year arrives, in the morning the units require hot water and by noon they need chilled water. Since these units generally have a common piping system, it is almost impossible to switch from heating to cooling and vice versa easily. Added to that is the danger of switching an entire system from heating to cooling with the return mains full of hot water. Chillers have been known to blow up if they are faced with trying to cool return water that is being supplied at about 120°F. The same problem exists when the system is switched from cooling to heating in the fall or spring. Boilers designed to heat water from 100 to 200°F cannot accept water at 50°F, and have been known to lose tubes that have shrunk as a result of the change: Fan coil systems that have a common supply and return system for both the hot water and the chilled water do not have the ability to switch from one medium to another easily. It is possible to arrange the piping so the system is zoned (Fig. 11.6), and zones can be switched to account for the different building exposures in the fall and the spring. Even those systems can cause problems if individual rooms require cooling when there is only hot water in the piping. There are methods using the multiple-piping systems to alleviate the above problems, and they will be discussed under multiple-pipe systems.

As far as piping arrangements are concerned, fan-coil units are classified as two-pipe and multiple-pipe (three or four pipe) systems. As

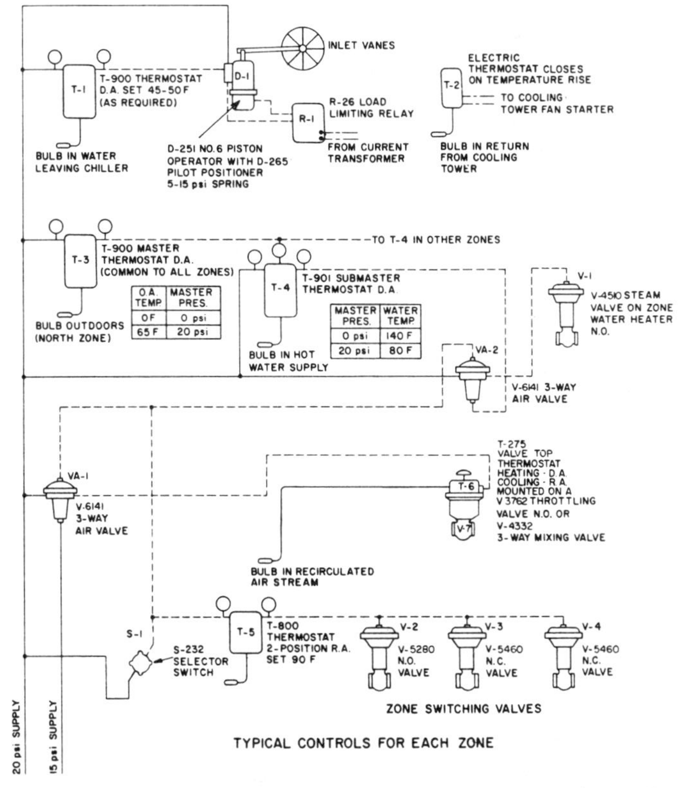

Figure 11.6 Controls for a fan-coil system. (*Courtesy of Johnson Controls Inc.*)

for the controls on the common two-pipe systems, there are several ways to control the fan-coil units. The simplest and most economical way is the line-voltage thermostat (Fig. 11.7) that cycles the fan motor to maintain the space conditions. Since the unit fan must turn on on a call for heat in the heating season and also turn on on a call for cooling in the cooling season, the thermostat must be switched from direct acting to reverse acting whenever the units are being supplied with different temperature water. This is done in a couple of ways: One method is using a switch at the thermostats that reverses the action of the thermostat labeled summer–winter. That type of thermostat usually has a fan switch on the thermostat that also controls the speed of the fan motor and allows the occupant to turn the fan off. Another method is a strap-on thermostat that senses the water temperature and does the switching. These types of thermostats are usually straight line-voltage electric thermostats, although they can be low voltage with relays in the units to do the controlling.

The next step is to use control valves on the coils of the units that are either modulating or two-position controlled from a room thermostat or a return air unit thermostat. Again, the action must be reversed as the mediums in the coil change from hot to cold. With pneumatic controls that can be done by changing the supply air pressure to the thermostats. Some companies have developed special heating–cooling thermostats (Fig. 11.8) that mount on top of the valves in a side compartment of the fan-coil unit with a small remote capillary sensor to sense the return air temperature in the space and control the unit during summer and winter. The units are designed so the dial of

Figure 11.7 Line-voltage room thermostat. (*Courtesy of Johnson Controls Inc.*)

Figure 11.8 Fan-coil valve top thermostat. (*Courtesy of Johnson Controls Inc.*)

the thermostat is visible under a small access door on the side compartment. The small valves on the units that are controlled either from a room thermostat or a unit thermostat with the thermostat a part of the valve or as a separate unit can be either two way or three way, depending upon the design of the primary water system supplying the fan-coil units.

The controls required for the primary system, which involves supplying the fan-coil units with hot or chilled water, will be discussed in the chapters on piping systems for various terminals as well as other systems.

The main disadvantage of two-pipe, fan-coil units is that there is a common supply of hot or chilled water in the same piping. This can be overcome with either a three-pipe or a four-pipe system of chilled water and hot water. The three-pipe system is more versatile than the two-pipe system, since the units are supplied with both chilled water and hot water year around. Each unit functions independently of the others in the building, and there is a special valve (Fig. 11.9) on the unit that looks like a three-way valve but is in fact a valve supplied with hot and chilled water with a dead-band position where neither hot nor chilled water is supplied to the coil. This is accomplished with a special spring arrangement inside the valve. In the case of the three-

Figure 11.9 Fan-coil three-pipe valve. (*Courtesy of Johnson Controls Inc.*)

pipe system, the returns from the units are common and the return water flows to the proper item (chiller or boiler) in the equipment room. In this case, it is important that the coils have high delta T on the water side, and it is advisable that both the hot water and chilled water supply temperatures be adjusted by the outside temperature so the return water is never too severe for the chillers or the boilers. Delta T (ΔT) also can have a large affect on system economy when hot and cold returns are mixed.

The four-pipe units are similar to the three-pipe systems except that the return systems are not common so the water systems are completely isolated. These systems use valves on the supply line to the coils and valves on the return lines where there is a single coil in the units. In these cases, the return line valve is a diverting valve and care must be taken to ensure that pressure differences between hot and cold systems do not exceed valve close-off ratings. Some units use a special split coil and even two coils (one for the chilled water and one for the hot water). In those cases, the four-pipe systems use two valves only on the supply line to the units, and there is no chance of mixing hot and cold water.

With the four-pipe systems, the room thermostats can be direct acting and do not need to be reversed in summer and winter. The same is true of the unit type thermostats mounted in the unit compartments.

The three- and four-pipe systems are more complicated and more costly to purchase and install. They also require more maintenance and qualified personnel. The bottom line is that two-pipe systems can be economical to purchase and install, but they do not always give the best comfort conditions because of climate and the problems with changing weather in the spring and fall.

Double-Duct Mixing Boxes

Double-duct mixing boxes (Fig. 11.10) are discussed in this chapter on terminal units since they have been used to deliver properly controlled air to a room or space.

Some years ago when building designers were trying to reduce the size of the duct work that went into commercial buildings, engineers designing the HVAC systems and major equipment manufacturers decided to try increasing the static pressures in the systems that delivered the air to the rooms, thereby delivering the same amount of air in smaller duct work. At the same time, the concept of making the hot and the cold air available just before entering the space to give good individual as well as zone control of the rooms was considered. Thus was born the *double-duct mixing box* that was used in those first systems. The box systems discussed in this chapter are the ones that mix hot and cold air (either air from an economizer system or mechanically cooled air) just before they deliver the air to the space. Although the emphasis in double-duct box systems is on the high-pressure types, there were lower pressure systems that also used the double-duct mixing box concept. Remember when reading this section that the *volume* of air from the fan system in the case of dual-duct mixing box system is *not* varied.

Figure 11.10 Double-duct mixing box. (*Courtesy of Anemostat Co.*)

In the 1950s and 1960s there were many manufacturers of these boxes, and, as usual, the control companies were required to come up with designs of controls and systems that allowed the boxes to function.

In any discussion of double-duct as well as VAV systems, there needs to be a discussion on the ways to control the static pressures in the duct work. That subject will be discussed with VAV systems. This section will discuss the types of boxes themselves, their controls, and their problems and the solutions.

When the designers of the higher pressure systems started to see those systems installed, they began to realize that the higher pressures brought additional noise problems to the system. The increased noise level was most apparent at the boxes themselves, especially when the majority of the mixing boxes were using mostly hot or mostly cool air and the other boxes in the system were mixing so the primary system was not limiting the static pressure. This problem was solved with static pressure controls in the boxes themselves.

There are and were various types of double-duct mixing boxes manufactured and used in double-duct systems. The most common is the one that uses one damper motor operating two hot and cold dampers or valves, as they are sometimes called. The damper motor is controlled by a room thermostat to maintain space conditions. In some cases the noise level at the boxes is controlled by a mechanical constant volume system supplied by the box manufacturer (Fig. 11.11). The systems usually consist of spring-loaded and calibrated valves, or shades, that closed as the pressure increased. Further refinement consists of a locally mounted static pressure controller sensing the pressure drop across a baffle and controlling, through a highest-pressure relay, one of the damper motors on a two-motor box. In this case, the room thermostat controls both the hot and the cold valves, with the static pressure controller overriding the room thermostat as the static pressure changes.

Some manufacturers furnished mixing boxes that used the pressure in the duct work itself to power the controls for operation of the boxes. One problem involved with those systems was the fact that the system air used as the power was never as clean as the air used in a pneumatic control system. Another problem was that the pressure in the duct work was really too low to provide any real power for system operation. As a result, system powered boxes never really took off.

In some cases manufacturers provided boxes with built-in reheat coils for special applications. Also, double-duct systems were controlled by both pneumatic and electric control systems. Since the terminal controls were not very complicated, pneumatic, electric, or electronic type of controls were used.

Today, since double-duct systems are not very energy efficient, they

Figure 11.11 Mechanical constant volume controller. (*Courtesy of Titus Co.*)

are used less and less in modern commercial buildings. In some cases a VAV version of a double-duct box is being produced or retrofitted. The double duct VAV box varies flow either hot or cold down to the minimum needed for ventilation or distribution. This type of double duct box requires VAV control of the system. The double duct VAV box type of terminal is explained in the sections on VAV.

Induction Units

In the 1950s and 1960s induction terminal units (Fig. 11.12) were a fad in the HVAC market. Here again, the attempt to save space was one reason designers started using induction unit terminal units.

The concept is simple and can be described as follows: A central fan system that usually used only 100% outside air supplied tempered and conditioned air to terminal induction units in the space. The air that was supplied to the induction units was heated in the winter and cooled and dehumidified in the summer. The supply system to the induction units was called a primary air dehumidifying (PAD) unit. The unit's purpose was to supply the induction units with warm ventilation air in the winter and provide dehumidified ventilation air in sum-

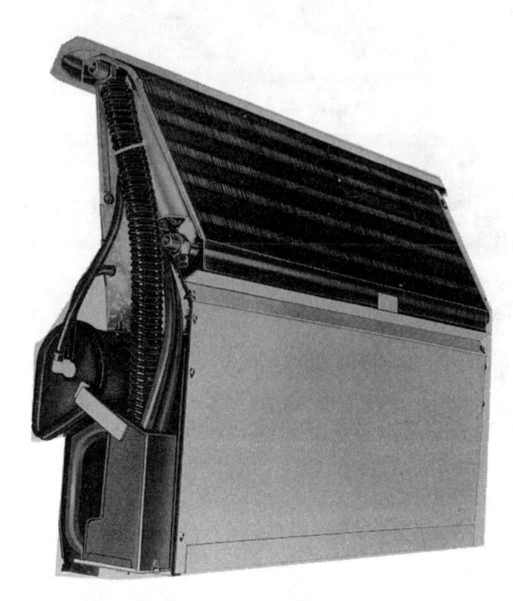

Figure 11.12 Typical induction unit. (*Courtesy of Carrier Co.*)

mer. That air had most of the latent heat removed from the ventilation air before supplying it to the room units. The theory is that the outside air contains most of the latent heat in the summer, and the cooling job in the rooms is mostly *sensible.*

The terminal induction units get their name from the fact that the PAD air to the unit in the space (usually on an outside wall under the windows) is supplied at a high pressure and fed to a set of nozzles to create a venturi effect and induce air from the room that mixes with the PAD air on about an 8 to 1 ratio. The mixture of the PAD air and room air is what cools the space. The air also has to be cooled in the unit in the summer and reheated in the winter, since the amount of heating and cooling the PAD unit does with the PAD air is not enough to take care of the entire load in the space. This means that the induction units had coils in them. These coils were sometimes like the fan-coil unit coils in that they were supplied with chilled water in the summer and hot water in the winter. There were also induction units like the fan-coil units that had split coils and separate coils for the heating and cooling.

The same problems that plagued the industry concerning fan-coil units were present with induction units: the units with the common coil for both heating and cooling. As far as controls are concerned, the control companies basically use the same controls for induction units that are used on the fan-coil units. At one time, one manufacturer promoted and sold a system that used the system air as the power to operate the controls and furnished all the control devices for the induction units. That in effect locked the control companies out of the picture. The idea did not last too long due to the same problems that confront most engineers trying to use the power of the system air to operate the controls.

The concepts that were discussed with fan-coil units apply here. The controls used on the PAD units are standard controls and cycles that could be used with any 100% outside air-handling unit, and they will be discussed in the chapter on those fan systems.

Most of the problems with induction units are the same as those found with fan-coil units, in that the switch from heating to cooling and vice versa was affected by the temperatures of the return water in the systems. The concept of three-pipe and four-pipe was also found to eliminate some of those problems, but others persisted. The PAD unit itself, which usually used 100% outside air, had problems caused by the severe temperatures in some climates, requiring the use of tempering coils as well as other items to protect the chilled water coils from freezing.

Most of the induction units were mounted under the windows, but some special ones were installed in furred ceiling spaces. One reason the induction units did not survive and become more popular is that in today's commercial building most of the cooling requirement of the space is not *sensible*. The facts do not support that concept.

Variable Air Volume Terminal Boxes

Variable air volume systems represent the majority of HVAC systems in use today, at least as far as the systems that were specified and installed in the last 15 years. To understand why these systems are so prevalent today, we need to look back at constant air volume (CAV) systems and their advantages and disadvantages.

To begin, CAV systems were energy hogs, and the simplest way to design a system was to supply the air to the spaces as coolly as possible, then reheat the air just before it got to the space to satisfy the space requirements. This meant there was a great deal of business for the people who supplied the booster coils that reheated the air before it got to the room. At that time, however, no one was concerned about energy as it was relatively cheap. Furthermore, the concept of varying the *amount of air* to the space was not considered because of the noise problems caused by increased static pressure and the fact that the controls available for the fans were crude at best when it came to slowing down the fans and adjusting the static pressure in the duct work.

Then, in the energy crunch of the 1970s engineers searched for ways to save energy. The fan manufacturers in cooperation with the electric motor and control manufacturers began to develop fans, motors, motor drives, and controls to allow the VAV systems to function without all the noise problems of the 1950s and 1960s. Also, air diffuser designs suitable for variable air volume were developed. These designs allowed large reductions in air volume before distribution broke down and dumped air into the space. With the development of new and better fans, controls, and drives and with the fan curves showing the amount of energy that could be saved with VAV systems, there was and still is a switch to VAV systems.

As is the case with induction and fan-coil units, the discussion of the fans and systems supplying the air to the VAV boxes will be covered in another chapter. This chapter will limit itself to the discussion of the boxes themselves (Fig. 11.13).

Remember, as is the case with all systems, the requirement for full cooling as well as full heating only takes place about 20% of the time. The rest of the time the systems are operating at less than full capacity. This means that the need for full airflows to the spaces to offset the maximum heat or cooling requirements is never needed all the time. For that reason as well as others, VAV systems are filling the fill for efficient modern HVAC systems.

As the definition indicates, the modern VAV box adjusts the *amount* of air that enters the space or room to maintain the temperature both in the heating and the cooling season. In its simplest form, a VAV box

Figure 11.13 Typical VAV box. *(Courtesy of Anemostat Co.)*

is a valve that closes off the air with a damper motor that is operated by a room thermostat trying to satisfy the space requirements.

The usual VAV box supplies only cool air to the space, and as a result, these boxes are usually confined to interior spaces where the requirements are for cooling year around. If they are used in exterior spaces, there is usually supplemental heating equipment that takes care of the skin loss. VAV boxes are usually classified as pressure dependent and pressure independent. In a pressure-dependent unit, the air-throttling device (air valve or damper) is controlled directly by the room thermostat. In a pressure-independent unit the valve or unit damper is controlled by a flow or velocity controller that is reset from the room thermostat. This means, in effect, that the pressure-dependent unit has no way of sensing changes in the system pressures that may change due to a number of boxes throttling down and reducing the airflow of the system. These types of boxes depend entirely on the ability of the central fan system to adjust the flow as the boxes close off or open up. The result is that boxes near the fans get the majority of air during heavy loads or pull down periods. The pressure-independent boxes, on the other hand, can maintain closer room control by varying the position of the damper in the unit to compensate for changes in supply duct static pressure. Also, since the flow control or velocity device has a low limit set point, the lack of acceptable air distribution is not a problem. This is especially true in light of the new codes and the problems with indoor air quality that have surfaced in the past few years. Those problems are not going to get better and may get worse, so air distribution is on the minds of all engineers involved with the design of commercial VAV HVAC systems.

The most common VAV box involves the use of the pressure-independent units with a reheat coil used to reheat the air when additional heating is needed after the cooling is reduced. Usually these types of boxes have a minimum amount of air that flows to space, and it is that minimum amount of air that is heated up by the reheat coil. The coils can be steam, hot water, or electric. The sequence is such that the reheat coil is operated in sequence after the room controller in conjunction with the velocity controller throttles the damper to the minimum setting. Electric reheat coils should never be used with pressure-dependent VAV boxes as they require a certain minimum airflow to prevent the coils from burning out.

Dual-duct VAV boxes that mix the hot and cold air being supplied in two ducts by the central fan system are also available. The central fan system supplies hot air in one set of duct work and cold air in the other. They qualify as VAV, since they vary the amount of hot and or cold air being supplied to the space. In theory, the *amount* of air (either hot or cold) being supplied to the space is the same at all times. These units are similar to the standard dual-duct boxes mentioned above but those systems do not attempt to vary the fan speeds, and so on, to save energy and VAV systems do.

Sometimes the dual-duct VAV boxes will be furnished by the manufacturer with flow controllers that are used in conjunction with the room thermostat to provide different cycles to satisfy the requirements of the designing engineers. Those cycles might have full shut off of the hot and cold dampers and no mixing of hot and cold air. In this case, the flow controllers control two damper motors, while the room thermostat resets the flow control point of the flow controllers. In addition to the above, the cycles might have minimum settings for the hot and cold dampers. Or they may have a total airflow controller that senses the hot or the cold duct flow and controls the cold duct motor, with the room thermostat controlling the hot duct motor through a selector relay. This is in effect a constant volume sequence with both dampers *normally open* and a mixing of the hot and the cold air. In another cycle, the room thermostat resets the control point of two flow controllers that control the hot and cold dampers that are both normally open. The hot damper motor is also controlled through a selector relay so there is mixing at the minimum settings of the dampers and there is flow control of both dampers at maximum flow.

Other manufacturers can provide other sequences on special order, and some promote special sequences. One manufacturer, for example, lists 60 different pneumatic sequences for controlling single-duct and dual-duct VAV boxes.

Some common VAV units that help with the ventilation problem (Fig. 11.14) use fans in the VAV boxes. For example, one takes return air from the return plenum when the single motor box is at its minimum setting

Figure 11.14 Typical VAV box with reheat coils. (*Courtesy of Anemostat Co.*)

so that there is still circulation. The cycle is as follows: As the unit goes from full cooling to full heating and the damper is modulated to its minimum position, the fan will start and return air will be mixed with the minimum supply air from the central fan. This type of unit generally has a reheat coil in conjunction with the unit, and if the space requires more heat than can be supplied by mixing the return air, the room thermostat modulates the reheat coil valve open. This way there is still some circulation of air in the space or room.

Another type of box with a fan has the fan mounted in series with the damper or air-throttling device. Like the previous unit, this terminal unit also operates with a reheat coil. The control cycle operates as follows: The cycle is the same as the unit with the fan that supplies return air except that the fan runs constantly. This unit provides the same airflow regardless of the actions of the central system. This type of unit uses more total fan energy, but it permits the best circulation and the diffusers work better with the constant flow of air.

In the final analysis, the selection of a VAV unit requires a thorough study of the particular building and an energy analysis to see which types are the least costly from an energy consumption standpoint. There are many instances in the past where engineers have retrofitted constant volume systems to VAV systems in an effort to save energy. These low-pressure constant volume systems can, in some cases, be converted easily; in other cases, the costs involved make the conversion prohibitive. Each case needs to be analyzed to see if the energy saved will have a reasonable payback.

Some problems with the use of VAV systems involve the ability to balance the system after it is installed and working. To balance the system, technicians may have to open up all the boxes to maximum flow as they attempt to adjust the flow at each box. This procedure can upset the users of the building if it is partially occupied. In some cases (usually the pneumatic type), the flow controllers can lose calibration and cycle the damper or air valves at a rapid rate causing pulsing in the system. Complaints from the occupants can involve the lack of circulation in some of the types of boxes. In those cases, it might be advisable to retrofit that box with a fan-powered box.

Noise problems are an issue that is often considered when VAV systems are specified, and the box selection is an important part of that issue. Some boxes are better constructed than others, and the use of inexpensive boxes is certainly not recommended.

Electric and electronic controls have come a long way in the area of VAV boxes with the advent of practical and inexpensive damper motors for the boxes. Some of the new digital controllers have impressive capabilities as well as flexibility at controlling VAV boxes. They have the added feature of being able to have their set points, and so on, checked with handheld devices that can read all the information necessary while the technician is standing on the floor under the ceiling VAV box. This can also speed up the time required to check the system. In fact, systems are now being specified and installed that have the ability to send all that information through an energy management system to a central computer for monitoring, reset, and control. There is a cost involved with these additional features, and the designers need to determine whether the costs are justified. The standard *electric control systems* are not being used much in modern VAV box control applications.

Other Terminal Units

Other devices that can classify as terminal units are those installed on the perimeter of a building that attempt to heat the skin of the building. Such units are classified as radiators, convectors, baseboard radiation, and fin tube radiation. These devices use either steam or hot water and are used extensively in northern climates to offset the cold wall effect. The control of these devices is as follows: A room thermostat controls a valve at the unit to maintain the space conditions. This is also an area where the self-contained valves are used successfully, even though they do not control as close as the ordinary room thermostat and valve. This is an area where the greatest experience has been gathered by the control companies, since it is the oldest form of comfort control.

The problems involved with this type of control are in the area of steam or water distribution and not with the control devices themselves. The proper use of traps in steam systems, as well as the design of the piping systems with steam, has caused more problems than can be mentioned here. Anyone designing a steam distribution system today who has not had experience with that type of design needs to study the available books from places such as ASHRAE before attempting the task. Water piping systems have also come a long way, and the guide and data books from ASHRAE have tried and tested methods that can be used in the design of those systems. Some of those types of designs will be covered in other chapters, and the illustrations will depict the various types of radiation that are available and being used in modern commercial buildings.

In summary, this chapter has brought you up to date on the various types of terminal equipment that were and are available. It discussed the control cycles as well as problems that can result from the use of the equipment. A discussion of the central fan systems that supply the terminal units will be in other chapters.

Chapter

12

Primary Supply Systems

Primary supply systems are central subsystems that convert prime energy, such as electricity or fuel, to thermal energy. These subsystems include chillers, steam and hot water boilers, and alternative sources of energy conversion such as solar collectors, cogeneration, and thermal storage. The general approach for all of these supplying subsystems is to control the rate of thermal energy supply at a value that satisfies the total thermal load. In cases in which there is no storage, the load to be supplied is the instantaneous load imposed by the distribution system that is supplying all user subsystems. In cases with storage, the load to be supplied is normally a 24-h total load. In all cases, the goal of the control strategy is to meet the load needs at the lowest cost. The control approach differs with the characteristics of each type of subsystem.

Heating Supply Systems

Boilers

Boilers are the most prevalent form of heating supply; they are either electric powered or combustion of some form of fuel. The capacity control of an electric-powered boiler is typically to vary the number of stages of heater elements and cycle the last one on to maintain a temperature level in the boiler. The capacity control of a combustion boiler is to control the firing rate. On smaller boilers, the combustion is cycled on and off as necessary to meet the load. On larger boilers, the firing rate is changed from a minimum to maximum as necessary to meet the load. Some boilers have a low and high fire rate, with one of them cycled as necessary to meet the load. Other large boilers have a modulated control of firing rate between the minimum and maximum rate. The implementation of this modulating control can vary from simple linkages between fuel and air valves to complex metering and flow control systems. The accuracy of the fuel to air mixture affects efficiency of combustion. This will be discussed further under optimizing control. The capacity controls of boilers are normally applied by the boiler manufacturer.

Combustion boilers require combustion safeguard controls that are applied by the boiler manufacturer. On smaller boilers (up to 400,000 BTU/h) the combustion safeguard controller is called a *primary control.* The primary control checks for the proper sequence of start-up. Essentially, these controls prove the existence of combustion air, purge the combustion chamber, prove a burner flame is established, then supervise the flame during combustion. Any failure causes a safety shutdown. On larger boilers, several more steps are included in a proper start sequence, and the combustion safeguard controller is called a flame–safeguard–programming controller. The added steps are a check of minimum purge flow rate, ignition and proof of a pilot flame before starting the main flame, and proof of a main flame. These controls are always applied by the boiler manufacturer because they are closely related to the boiler and burner design.

Optimizing control for a single combustion type boiler essentially optimizes the combustion process by controlling the amount of combustion air to match the fuel injection rate and provide efficient combustion. The percent efficiency is highest when there is 0% excess air and goes down sharply when there is not enough air because of incomplete combustion. Efficiency goes down more gradually when there is excess air because thermal losses are increased. These effects are shown in Fig. 12.1.

Controls for excess air are called *trim control* and are appropriate for large boilers that have modulated capacity control. These controls may be installed by the original equipment manufacturer or added

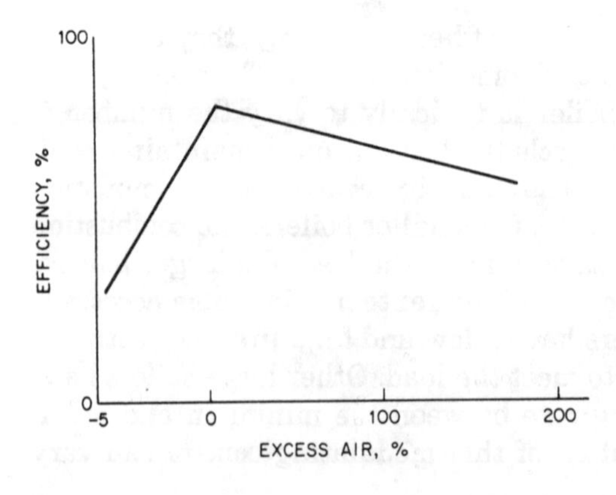

Figure 12.1 Efficiency versus excess air. (*Courtesy of McGraw-Hill,* Handbook of HVAC Design *Fig. 52.22.*)

later. They consist of a stack gas sensor that controls the fuel-to-air ratio to avoid excess air. The sensors measure either oxygen or carbon dioxide in the flue gas as well as stack temperature. The set point of the control loop is the value of oxygen or carbon dioxide that indicates the most efficient combustion. The practical range of excess air control is from 1% to 5%. Figure 12.2 shows the values of O_2 and CO_2 as a function of excess air. As can be seen from Fig. 12.2, the O_2 sensor is more sensitive at low values of excess air and is therefore the preferred sensor. The stack temperature increases with firing rate and with the fouling of boiler tubes. Therefore, an increase in stack temperature for a given firing rate can be used as an indicator of when boiler tubes need cleaning.

Optimizing control for multiple boilers is essentially done by having the number of boilers on line that will carry the load at the least cost. Typically, combustion-type boilers are inefficient at low loads because of the standby costs of fixed minimum thermal losses when the boiler is started. When the loads are over 50%, the stack losses increase with load. Figure 12.3 shows this effect for both one and two boilers with modulated capacity control. It can be seen that the optimum changeover load from one to two boilers is where the unit load costs are equal. This type of curve can be established for specific boilers by measured test or by analysis that uses standby losses and efficiency at minimum and maximum stack temperatures. In this situation, when two boilers are on they should both be modulated to carry equal loads. This provides the characteristic shown for two boilers and is typically more efficient than unbalanced loads. This discussion applies to combustion-type boilers not to electrically heated boilers. In the case

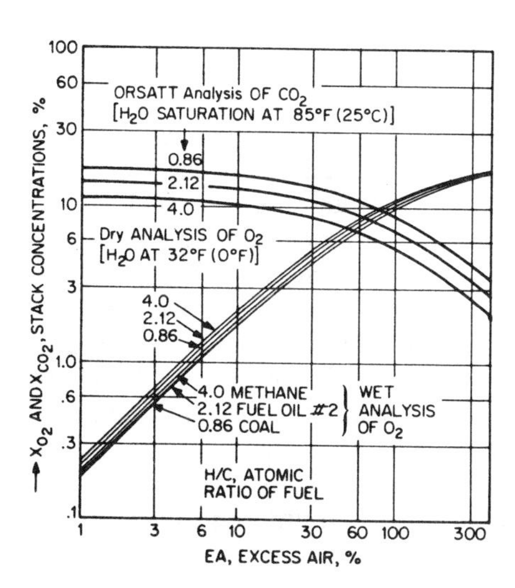

Figure 12.2 Comparison between O_2 and CO_2: Stack-gas analysis versus excess air. (*Courtesy of McGraw-Hill,* Handbook of HVAC Design *Fig. 52.23; originally from Honeywell.*)

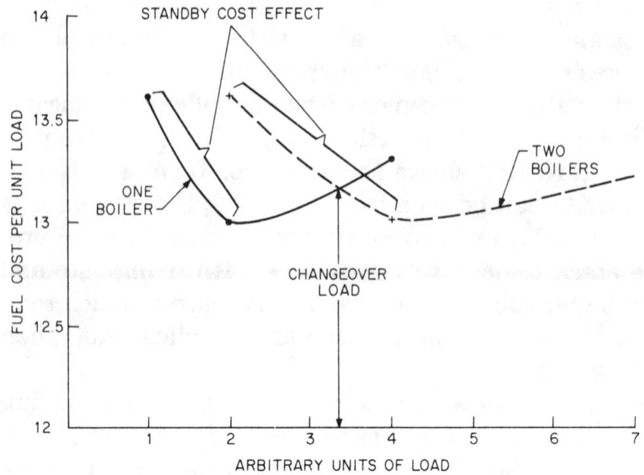

Figure 12.3 Characteristics of one versus two boilers. (*Courtesy of McGraw-Hill,* Handbook of HVAC Design *Fig. 52.51.*)

of multiple electrically heated boilers, each boiler would be independently controlled to maintain both its discharge temperature and its internal high-temperature limit.

Alternate fuel-type boilers have two kinds of burner systems that are controlled individually when that fuel is in use. These controls are applied by the boiler and/or burner manufacturer. The combustion safeguard controls need to be specific to each burner. The dual fuel changeover controls switch from operation of one burner to the other. This switch would be initiated either manually or on the basis of outside air temperature.

Heating converters

Converters as a generic class of equipment convert thermal energy from one temperature level to another. In a practical sense, they are also a means of separating one thermal system from another. The typical uses of heating converters and the method of control are discussed next.

With steam to hot water, the source of steam could be an on-site boiler, district steam, or exhaust steam from another process. The control of a steam to hot water converter is typically a discharge water temperature controller modulating a valve in the steam supply line to the converter. If the supply of steam is limited, the control acts as a high-temperature limit. In this case there would normally be another converter down stream on the water side that has an assured supply of heat to ensure control in the hot water load circuit. The set point on the first controller should be higher than that on the second controller

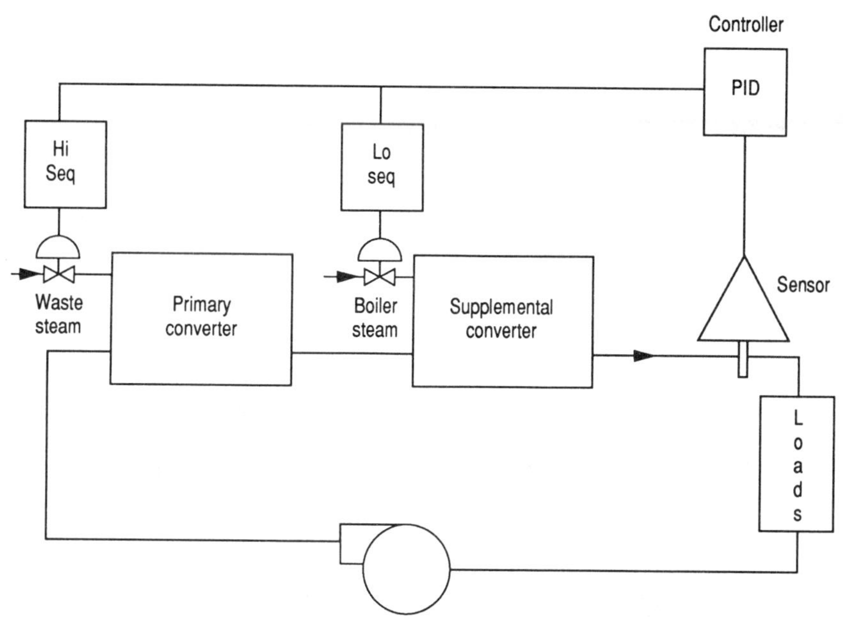

Figure 12.4 Hot water system with waste steam converter and boiler steam converter.

to ensure that all of the limited supply is used before the unlimited (and more costly) supply is used. Figure 12.4 shows a hot water system with a waste steam converter and a boiler steam converter.

Primary hot water to secondary hot water. In some cases the primary hot water is at a high static pressure such as that found on the lower levels of a high-rise building. The secondary hot water can then be at a lower pressure, and all of the equipment in the secondary water system would be less expensive to construct. This use of converters for pressure isolation may not even need to be controlled, unless there is a difference between primary and secondary temperature levels or the secondary water is serving as a supply to a zone of space heating with no further control. In this case the converter is to be controlled as a heating zone with discharge temperature reset by outside air temperature or space temperature. The typical control loop for a water-to-water converter senses secondary discharge water and modulates primary water flow with a valve in the primary water return from the converter.

Chiller–condenser rejected heat

A chiller that can provide rejected heat to a hot water circuit is called a *heat pump cycle chiller*. The primary difference from a normal

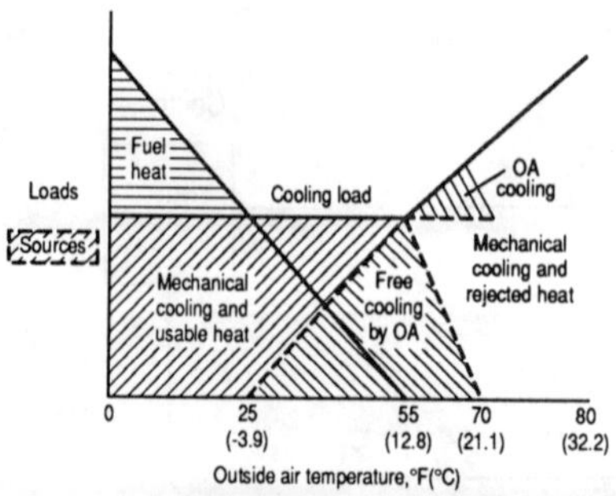

Figure 12.5 Instantaneous load and source relationships. (*Courtesy of McGraw-Hill,* Handbook of HVAC Design *Fig. 52.56.*)

chiller is that it has a double bundle condenser and is applied to operate at condenser temperatures at a heating level normally above 100°F. The practical use of this is to move heat from the interior area of a building with a cooling load to the exterior area of a building with a heating load. In this way the cooling load is a heat source, while in some outside air temperatures, outside air is a cooling source.

Figure 12.5 shows typical heating and cooling (load and source) relationships. It indicates that when the outside air temperature is between 25 and 50°F all heating and cooling loads can be satisfied with a balance of free cooling and mechanical cooling that provides usable heat. Below this range fuel heat must be used; above this range all mechanical cooling heat must be rejected. Although this is only an example, the relative savings magnitude is clear.

The control sequence that implements this balanced use of free-cooling and rejected heat is shown in Fig. 12.6. The heating design water temperature in this case is 110°F. The example of *T*1 hot water control action shows a control action range from 110 down to 98 representing how the sequencing would occur with proportional control action only. This was done for ease of explanation. In practice, a proportional integral control action would be preferred so that the set point of design water temperature could be accurately held under all load conditions. The midrange control function of limiting the amount of outside air used for free cooling can be accomplished as a global control function that becomes a limit in all economizer controls in fan systems. This was covered in Chap. 17 on total system control functions.

Cogeneration heat source

Cogeneration is the generation of both electricity and usable heat from a primary fuel source. By using heat that would otherwise be rejected, cogeneration systems operate at efficiencies greater than systems that generate heat and electricity in separate systems. To justify a

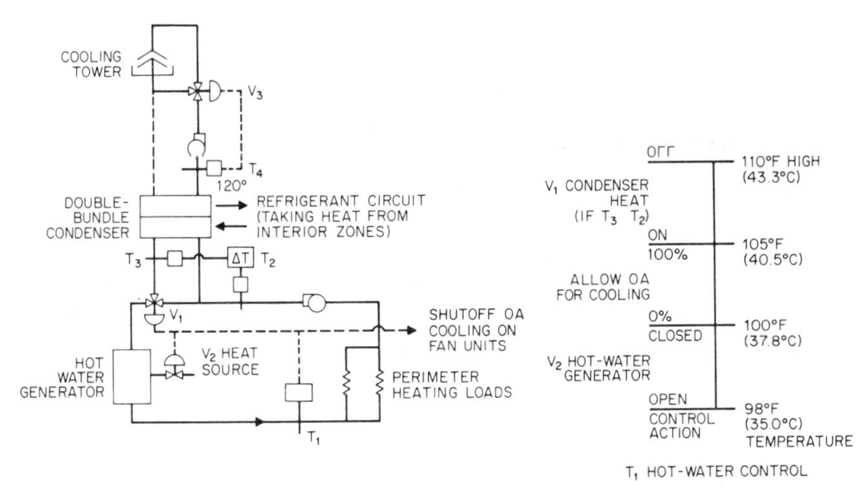

Figure 12.6 Heat pump cycle controls and sequence. (*Courtesy of McGraw-Hill,* Handbook of HVAC Design *Fig. 52.55.*)

cogeneration system, you must examine both electrical and thermal load profiles and the costs of both forms of energy. Generally, relatively full use of the heat throughout the year is needed to pay for cogeneration compared to purchased electricity. Specific load profiles, fuel costs, and specific equipment part load characteristics must be considered in a cogeneration analysis. The choices of fuel and type of prime mover equipment establish the part load characteristics, such as the quantity and temperature of rejected heat. The sizing of equipment depends on the decision to have an isolated system or, if tied into a utility grid, the economics based on load profiles and energy costs. These design alternatives and how they influence control needs will be discussed.

Prime mover choices include reciprocating engines, combustion gas turbines, and steam turbines. Cogeneration plants can be classed as a topping cycle or a bottoming cycle. A cogeneration plant uses a topping cycle when the fuel is used to drive the prime mover to generate electricity, then uses the remaining thermal energy. Both reciprocating engines and combustion gas turbines are used in a topping cycle. A cogeneration plant uses a bottoming cycle when first fuel is used to satisfy thermal loads, then waste process steam is used to drive a steam turbine to generate electricity.

There are two types of steam turbines—back pressure and condensing. The condensing type would typically be used in a bottoming cycle to extract the maximum thermal energy from the waste process steam. The back pressure type would typically be used in the topping cycle as the prime mover of a generator to ensure that its exhaust would provide usable thermal energy. The topping cycle type of cogeneration is usually found in buildings with primarily HVAC thermal loads. The bottoming cycle type of cogeneration is usually found when there are large process loads at high-temperature levels supplied by high-pressure steam.

The temperature level at which heat is rejected is another consideration in matching the load temperature requirements. The most commonly used type of prime mover is the reciprocating engine, which has the lowest temperature level of rejected heat in the range of 250–350°F. The combustion gas turbine has the highest temperature level of rejected heat in the range of 1100°F. In both cases the combustion exhaust is the greatest amount of heat and is at the highest temperature level.

In the case of reciprocating engines, another source of heat is the engine jacket cooling system, which must keep engine coolant in a temperature range between 200 and 235. Figure 12.7 is an example of a typical hot water heat recovery system with provisions for controlling jacket water temperature in the required range. The block marked process loads shows where the heat is used for the HVAC loads without the detail of how the rejected heat is supplemented when necessary. Figure 12.4, which shows a limited source converter in series with a supplementary source converter, is the detail of how to handle the HVAC process load using appropriate converters. There are many different arrangements possible in recovering and using the heat rejected from a cogeneration system, but they all have the same fundamental requirements of providing cooling to the prime mover equipment and providing an assured source of heat to the HVAC process.

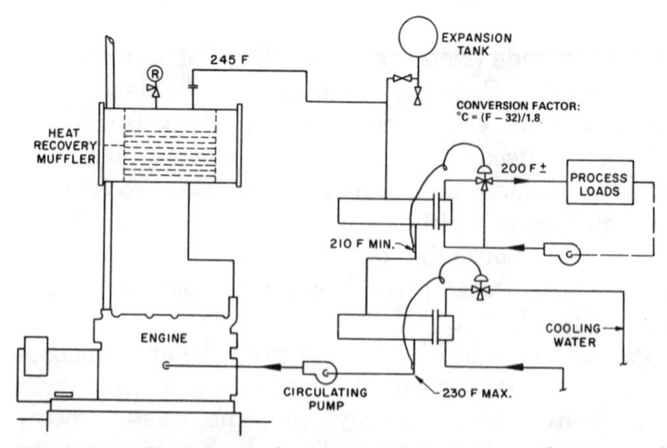

Figure 12.7 Hot water heat recovery system. (*Courtesy of ASHRAE*, 1987 Handbook *Chapter 8, Fig. 5.*)

The overall design and sizing of a cogeneration plant presumed a control strategy that handled electrical and thermal loads in a manner that saved operating costs. The two major choices of control strategy are thermal load following and electrical load following. The implementation of each strategy is to control the capacity of the prime mover to meet the chosen load to follow. The presumptions necessary for each strategy are as follows.

Thermal load following requires interconnection with the electric utility to supply electrical loads greater than thermal loads and to receive excess cogenerated electricity when thermal loads are greater than electrical loads.

Electrical load following requires the generating capacity to handle peak electrical loads and the facility to supplement higher thermal loads and to reject excess thermal supply.

A combination strategy of following the lesser load is necessary if neither of the above presumptions is met. This combination strategy requires that both electrical and thermal supplementary supplies be available to meet the larger load. Control of the prime mover is the implementation of the overall strategy. Besides the thermal control needs that have been discussed, there is need for extensive control of the electrical generating process and the interconnection with the utility. If there are multiple generators, there is further need for control to coordinate generator operation. These generating control systems are specific to and furnished with the cogeneration equipment. They must meet the requirements of the electric utility in protecting the power grid in addition to carrying out their primary functions. The primary functions of these controls are regulation of voltage, frequency, synchronization, and load during all conditions of operation. In addition, there are safety controls to protect both the prime mover and the electrical system in the event of unusual conditions or component failure. These controls also are equipment specific and furnished by the cogeneration equipment manufacturer.

Chilled Water Supply Systems

A chilled water supply system is composed of liquid chiller(s), heat rejection equipment, and chilled water distribution equipment. This section will discuss liquid chillers and heat rejection equipment. The chilled water distribution will be discussed in Chap. 14 with other distribution systems. Three types of liquid chillers are available, defined by the type of refrigeration equipment. They are centrifugal, positive displacement, and absorption. The characteristics of each require different control considerations, and these unique requirements will be discussed for each type of chiller.

Chiller plant optimization

The primary basis of optimizing energy efficiency is common to all three types of chillers. This basis of optimization is the reduction of the refrigerant head. The refrigerant head of a vapor-compression cycle liquid chiller system is the pressure difference that the compressor must create. (In an absorption system the concentration of the solution that carries the refrigerant of water is analogous to the pressure in the vapor compression cycle, and the reduction in refrigerant head reduces the thermal energy required.) The temperature of the liquid being chilled determines the low pressure in the evaporator, and the temperature of the condensing coolant determines the high pressure in the condenser and compressor discharge. The difference between this low evaporator pressure and the high condenser pressure is the true refrigerant head. The temperatures that caused these pressures are sometimes used to measure the indicated refrigerant head. The percentage of load being handled causes variation in the difference between indicated refrigerant head and actual refrigerant head as shown in Fig. 12.8. This is because of the loading effect on differential temperatures necessary to attain heat transfer in the two heat exchangers involved. The indicated head as measured by the difference between entering condenser water and leaving evaporator water is an indicator of the relative refrigerant head without increases from load effect as shown in Fig. 12.8. This characteristic will be used later in dealing with multiple chiller selection. For now the point to recognize is that a decreasing refrigerant head decreases the energy of compression needed to accomplish a given amount of cooling. Therefore, raising chilled water temperature and lowering condenser water temperature decrease the energy needed by a liquid chiller system.

Evaporative cooling towers are the most efficient and prevalent means of rejecting heat from a liquid chiller system. The control of the cooling tower capacity is necessary as the load varies. The conventional way to accomplish this is to measure and control the temperature of the condenser water supply temperature as it comes from the cooling tower. The control actions are to increase the fan capacity to keep the tower discharge temperature from going too high and to bypass water around the tower if the temperature goes too low. Figure 12.9 shows a schematic of this control scheme. Optimizing the cooling tower operation is accomplished by adjusting the set point of this condenser water temperature control loop to as low a value as is safe and attainable with existing outside air wet bulb conditions. The lowest setting is established by the design limitations of the liquid chiller. The design approach temperature of the cooling tower establishes how close the tower can bring the condenser water to the outside air wet

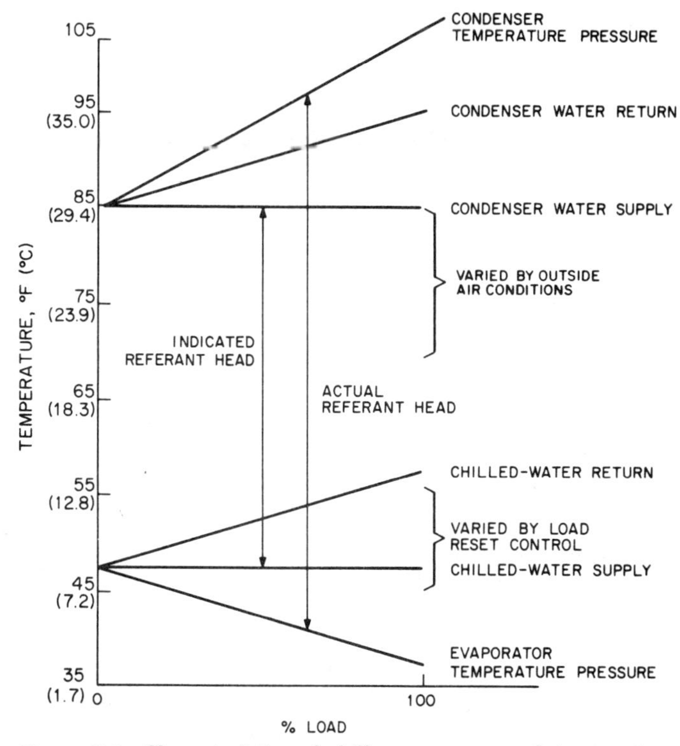

Figure 12.8 Characteristics of chiller pressures and temperatures versus load. (*Courtesy of McGraw-Hill,* Handbook of HVAC Design *Fig. 52.52.*)

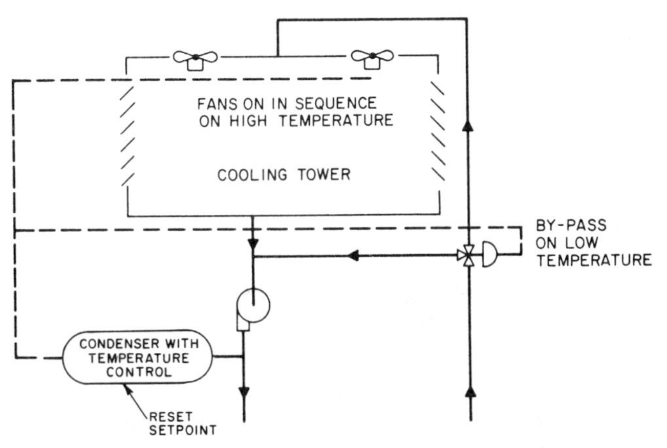

Figure 12.9 Cooling tower control. (*Courtesy of McGraw-Hill,* Handbook of HVAC Design *Fig.52.49.*)

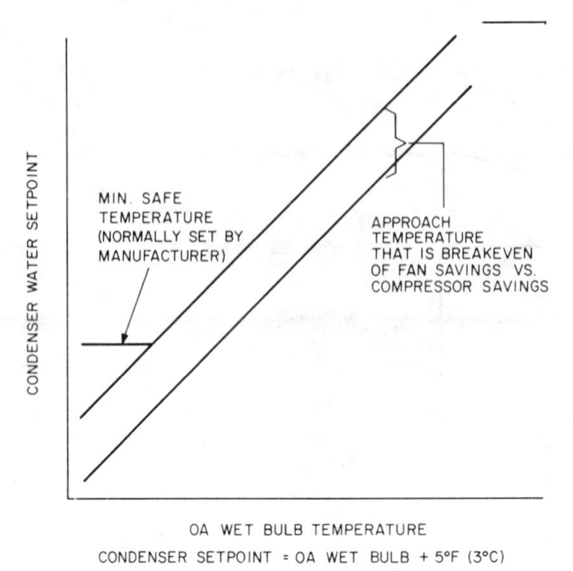

Figure 12.10 Condenser temperature reset by wet bulb temperature. (*Courtesy of McGraw-Hill*, Handbook of HVAC Design *Fig. 52.50.*)

bulb condition. Figure 12.10 shows the reset schedule for this optimizing control of a cooling tower. In addition to reducing the energy used by the compressor, the cooling tower fan energy is not wasted trying to achieve a lower condenser water temperature than is possible.

The other way of reducing liquid chiller energy is to control the chilled water supply at no lower a temperature than is necessary to meet the needs of the load. This is a load reset strategy applied to a chilled water system, similar to the load reset strategy explained in Chap. 10 as it was applied to fan systems.

There are several methods of applying this strategy to a chilled water system. The most comprehensive method is to monitor all significant loads as represented by the cooling coil valves on the fan systems. Any valve approaching wide open means that the supply water temperature is as high as it can go and still carry the load. The reset strategy is to reset the water temperature up and down to keep the most open valve almost open. The other approach to chilled water reset is to use the temperature of the return chilled water as an indicator of load, then either to control to maintain design return water temperature or to reset the discharge water set point from the return water temperature. Neither of these approaches is generally satisfactory. The controlling from return water temperature introduces long cooling load time lags into the control loop that can give poor control results. Both of these return water temperature approaches control to an average load, and a heavy load user may not be satisfied and able to meet its load.

Centrifugal chillers

Centrifugal chillers are generally larger and more efficient than positive displacement chillers. The most common method of controlling capacity is the modulation of inlet vanes. Other means are speed control of the driver or a combination of speed and vane control. Typically, the control is of discharge water temperature subject to a high limit of current draw if electrically driven. Other controls provide safety interlocks for unsafe conditions such as low oil pressure, high pressure, and freezing temperatures. These controls are typically provided by the manufacturer, who includes provision for reset of discharge temperature set point for job site optimizing control.

The control of multiple chillers is a matter of having the right number of chillers on line to meet a load with the least energy. The part load characteristics of centrifugal chillers as well as the existing refrigerant head establish the energy required for a given load. Figure 12.11 shows the part load energy input needs at different refrigerant

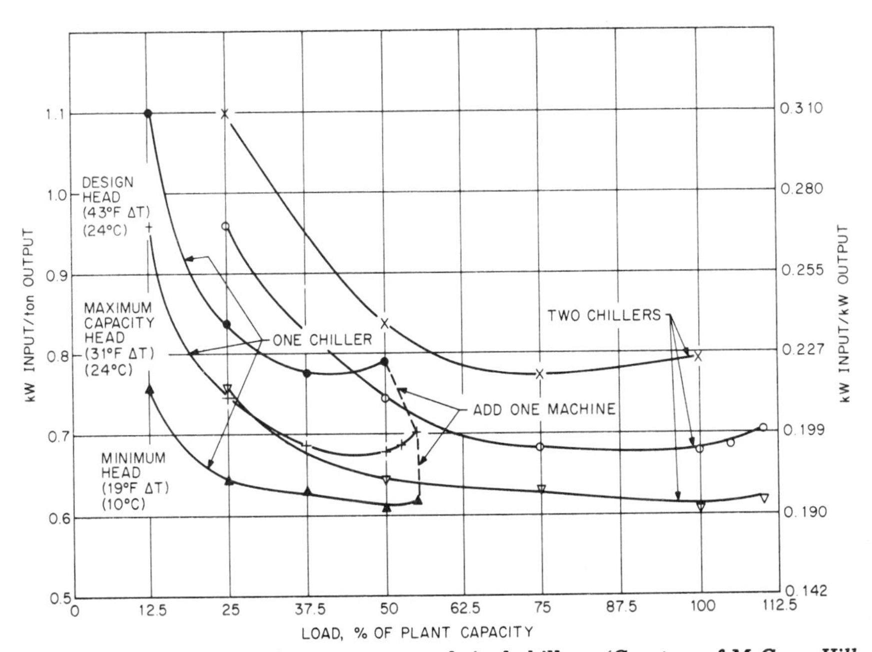

Figure 12.11 Efficiency of one or two equal-sized chillers. (*Courtesy of McGraw-Hill,* Handbook of HVAC Design *Fig. 52.53.*)

heads for vane control of capacity. The higher needs at low loads are primarily due to motor losses, whereas the increased needs at high load and high refrigerant head are due to thermal heat exchange inefficiencies. Note that at lower refrigerant heads the capacity of a chiller increases over the capacity at design refrigerant head. This is because the motor power limit is not reached at the lower work loads prevailing at low refrigerant head conditions. Also note that for the particular machine characterized by these curves, it is always better to use one machine if it can carry the load. This may not be true if the characteristic curve goes up more at high loads or less at low loads. The optimum time to go from one chiller to two chillers or vice versa is a function of both load and refrigerant head.

Figure 12.12 shows a changeover line as a function of load and refrigerant head. It is to be noted that if the indicated refrigerant head as noted in Fig. 12.8 is measured as the difference between condenser supply and chilled water supply temperature, then there is no change with load. This means that all changes in measured power usage are attributed to load changes when representing the curves in Fig. 12.11. This type of analysis can be made for specific machines of the same type that are in used in parallel. The strategy of carrying as much as possible on one machine and then turning on a second machine, works on an increasing load, but can be wrong on a decreasing load if the refrigerant head has changed.

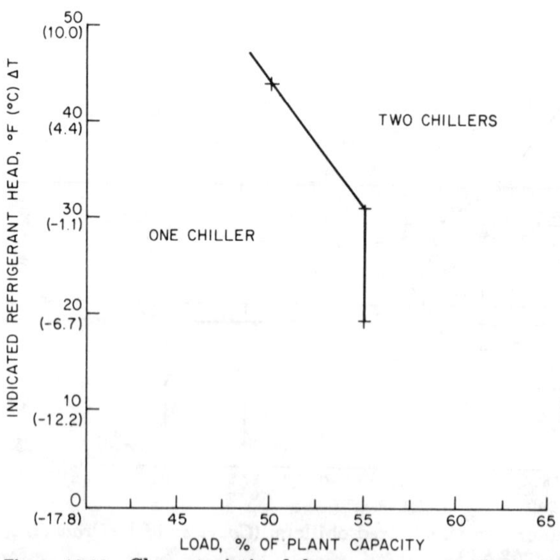

Figure 12.12 Characteristic of changeover reset. (*Courtesy of McGraw-Hill,* Handbook of HVAC Design *Fig. 52.54.*)

Positive displacement chillers

Positive displacement chillers include reciprocating, screw, and scroll types of compressors. Positive displacement chillers with reciprocating or scroll compressors have capacity control in stages of compressor capacity and, in some cases, stages of evaporator coil. The staging of compressor capacity is accomplished by turning multiple compressors on and off. In the case of screw-type compressors, capacity is modulated by a slide valve that varies the active area of the screw providing intake to the compression process. With some reciprocating compressors, cylinders in a machine are loaded or unloaded by holding an inlet suction valve open to unload a cylinder. Whatever the type of compressor, its capacity control is used for discharge water temperature or, in some cases, return water temperature. Other controls provide safety interlock for unsafe conditions such as high head, low suction pressure, low oil pressure, and evaporator low temperature. Typically, all of these controls come with the liquid chiller package from the manufacturer. The capacity of positive displacement compressors is fixed so that control of multiple chillers is by comparison of the load to the rated capacities of chiller sets. The variation in refrigerant head affects the energy used although not the capacity. Therefore, the strategies to reduce refrigerant head apply to positive displacement liquid chillers. The reset of condenser water temperature and of chilled water supply temperature will optimize the operation of positive displacement chillers.

Absorption chillers

Absorption chillers typically have their capacity controlled by varying the heat input to the reconcentration process that provides a refrigerant to the cooling process in the evaporator. The limiting factor in this process is that if the lithium bromide solution is concentrated too much it can crystallize and put the machine out of operation. If a machine has a control system that measures and limits solution concentration, the condenser temperature can be lowered and the energy needs reduced, depending on the solution limit to avoid crystallization. If the machine does not have a concentration measurement and limit control, the normal practice is to control the condenser water temperature near the design level, which prevents crystallization but does not reduce energy needs. The other controls required for an absorption unit are safety controls that guard against loss of cooling water flow, chilled water flow, and low temperature in the evaporator. These controls are typically furnished with the unit by the manufacturer. Normally the discharge water temperature controller set point can be reset. If the unit has a solution concentration limit controller, the condenser water temperature controller set point can also be reset to improve efficiency.

Free cooling cycles

Free cooling cycles are available for central, chilled water supplies when condenser water from evaporative cooling towers is colder than the required chilled water temperature. This is beneficial for systems that cannot use outside air for cooling. The free cooling of the chilled water supply is accomplished by cooling the chilled water with the condenser water without running the refrigeration compressors. This can be done in several ways.

One way in centrifugal chillers is for refrigerant circuits to be provided that bypass the compressor and allow migration of refrigerant from evaporator to condenser and back providing limited cooling capacity. Another way is to bypass the chiller and use condenser water cooling in the chilled water circuit. This can be done by filtering the condenser water and connecting to the chilled water circuit. A third way this can be done is by using a heat exchanger to cool chilled water with a colder condenser water while keeping the two circuits separate. The latter method is preferred because chances of contaminating the chilled water circuit are avoided. In all of these cases, one control consists of providing the proper lineup when conditions will support the free cooling cycle. Another control is to bypass the cold condenser water supply if necessary to avoid subcooling. In these free cooling modes the cooling tower may be used in freezing conditions, and freeze-up protection must be provided per recommendations of the cooling tower manufacturer.

Thermal Storage

Cool storage

Thermal storage is a feasible option of saving operating costs and expanding the capacity of a central chilled water plant. The cost savings result from reduction of demand charges and, in some cases, from the ability to use low-cost energy in off-peak hours. The general approach is to charge cooling storage in off-peak hours and to use stored cooling during on-peak hours to reduce demand. Peak hours are defined in the electric utility contract as those hours when rates are higher for demand, or energy, or both.

Figure 12.13 shows an example of load profile and the optimized use of storage to reduce demand and energy costs. Conventional control is essentially scheduling the charging and discharging of storage by

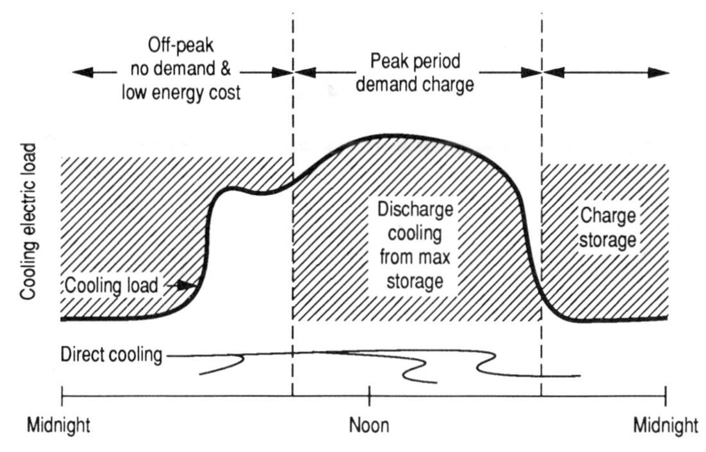

Figure 12.13 Maximum use of storage to reduce demand and energy costs. (*Courtesy of ASHRAE from Symposium CH-89-22-1 Fig. 1.*)

time of day and the need to charge or discharge. Optimized control is controlling the rate of discharge of storage to minimize the electric bill. The optimized control is a complex cost trade-off analysis that uses the storage to the best advantage in reducing the electric bill. The electric utility rate structure and the type of storage influence the basic strategy to be either chiller priority or storage priority. Chiller priority means the stored energy costs more than direct cooling energy and storage is used primarily to reduce demand charges. This is the case when ice is made at night with no energy bargain and the stored energy costs more because of the inefficiency of making ice at very low evaporator temperature. Storage priority means the stored energy costs less than direct cooling during the peak period. This is the case when energy at night costs less causing stored energy to cost less. The other function of optimized control of storage is to use storage at a rate that minimizes and levels the electric demand during the peak hours and, in the case of storage priority, uses all of the storage during the peak hours.

This optimized control was developed as a building automation program for the Electric Power Research Institute and is available for licensing. Figures 12.14 and 12.15 show the hardware and software structure of this system. In addition to this Cool Storage Supervisory Controller (CSSC) program in the central processor of a Building Automation System, there are Direct Digital Control systems required to implement the control of both chiller capacity and rate of discharge of storage to meet cooling load. Figure 12.16 shows the control functions required for this implementation.

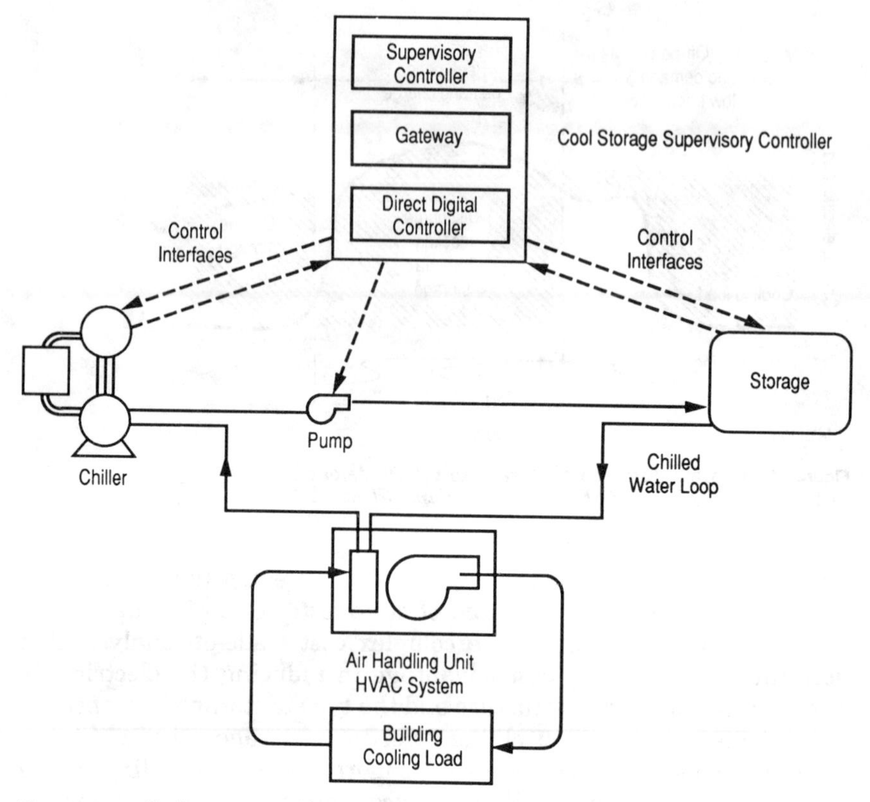

Figure 12.14 Cool storage system and CSSC controller. (*Courtesy of ASHRAE from Symposium CH-89-22-1 Fig. 7.*)

Cool storage has been developed in a number of different forms. These include ice storage and cold water storage. Cold water storage designs are discussed next.

The first design is stratified water storage, whereby cold water is introduced into the bottom of a closed tank and the warm return is taken from the top of the tank during charging. During discharging mode, the cold water is taken from the bottom of the tank and the warm return water is introduced into the top of the tank. The flow of water into the tank is distributed evenly to avoid velocities that would disturb the stratification of the warm water on the top and the cold water on the bottom of the tank.

Segmented water tanks do the same thing with multiple tanks that is done with a stratified tank. In addition to stratification within a single tank, they have adjoining tanks that feed one into the other. The cold tank on one end feeds into the next tank on charge mode and takes from the next tank on discharge. The other end of the row of

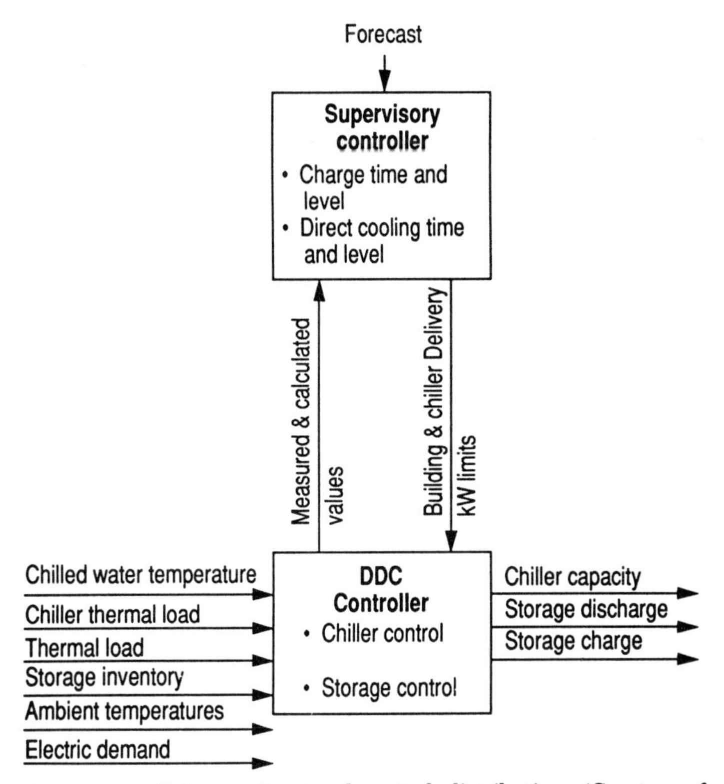

Figure 12.15 Software functional control distribution. (*Courtesy of ASHRAE from Symposium CH-89-22-1 Fig. 8.*)

segmented tanks is the warm end, which feeds through the middle tanks toward the cold end. Figure 12.17 shows the basic circuit of a segmented chilled water storage system.

The empty tank design has multiple tanks with one of them empty. When it is in charging mode, water is taken from a warm tank, cooled, then put into the empty tank until it is full of cold water and the warm tank is drained empty. In the discharge mode, a cold tank is emptied and the previously empty tank is filled with warm water. The switching between tanks must be coordinated to keep a continuous flow and to empty and fill the proper tanks correctly.

Ice on coil storage is a design where a cooling source inside a pipe forms ice on the outside of a coil immersed in a water tank. The water is circulated to provide chilled water to loads and is returned with the heat of the load to melt the ice. The energy needed to make ice is greater than the energy needed to provide an equivalent amount of cooling because making ice must be done at low evaporator temperatures and high refrigerant heads. With the ice on coil design, the more

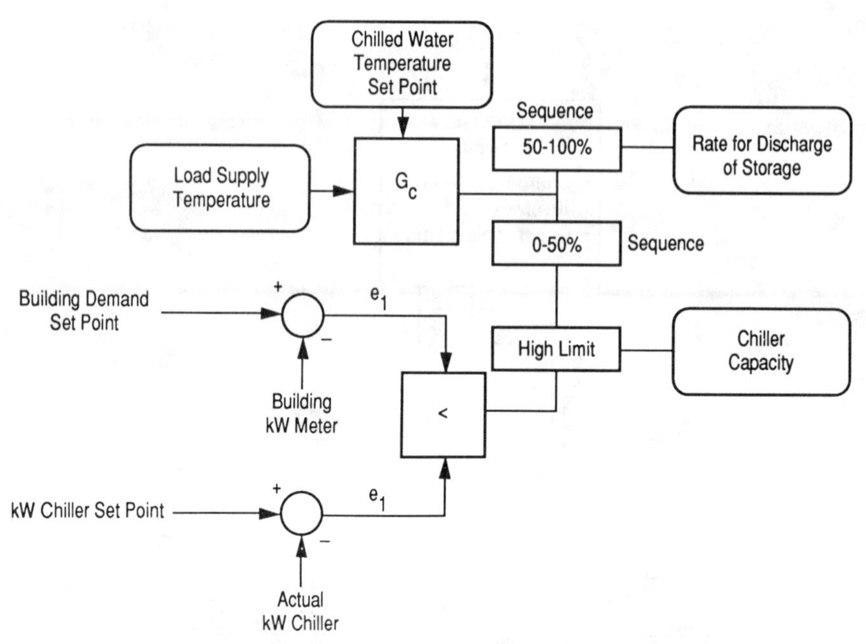

Figure 12.16 Cooling load control implementation. (*Courtesy of ASHRAE from Symposium CH-89-22-1 Fig. 9.*)

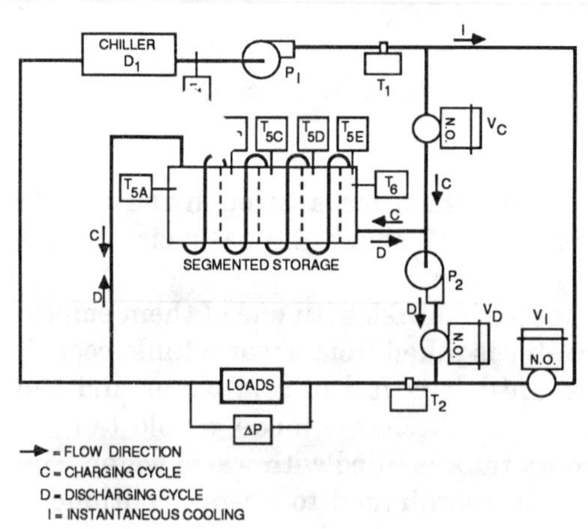

Figure 12.17 Segmented chilled water storage. (*Courtesy of McGraw-Hill,* Handbook of HVAC Design *Fig. 52.57.*)

ice on the coil the lower the evaporator temperature required to make ice and the less efficient the process. Therefore, this system has cost increases proportional to the amount of ice in inventory. The biggest advantage of all ice systems is they take less space than water storage. The ice on coil design can use either low-temperature brine or refrigerant in the coils as the cooling medium.

A brine–ice system uses brine inside a coil immersed in uncirculated water. The circulation of brine in the coil is used to store cooling by freezing the water during charge mode and to melt the ice during the discharge mode. The brine is cooled by a chiller during the charge mode and is warmed by the load during the discharge mode. This design does not lose significant efficiency from inventory level because the ice is made and melted from the inside of the pipe and there is no thick coating of ice during charging.

Ice harvesting designs make the ice then shed it into a bin in which water is circulated to discharge the stored cooling. This system requires the low evaporator temperature to make ice and some heat or other energy in the harvesting process.

Eutectic storage uses a solution that changes state from a liquid to a solid at a cooling supply temperature. This allows the use of conventional chillers as with water storage at comparable efficiencies. It also reduces the space needed for storage. Present materials must, however, be packaged in sealed containers and long-term experience is not available.

Control of storage involves load and stored cooling inventory measurement. The measurement of chilled water cooling inventory is relatively straightforward using volume and temperature. The measurement of ice inventory can be done by thickness measurement, water displacement measurement, or weight of harvested ice measured. The inventory sensing means is typically provided by the manufacturer of packaged ice systems. With water systems, job-applied temperature, level and flow sensors are used to obtain the measurements necessary for calculation of inventory, loads, and thermal energy.

Heat storage

Heat thermal storage is a feasible option when used in conjunction with a source of low-cost heat from another subsystem such as a heat pump cycle chiller or a cogeneration system. The typical implementation of stored heat is in the form of heated water. The design approaches for chilled water tanks can be applied to hot water tanks as well. In general, there are few, if any, winter peaking electric utilities that provide the rate incentives to pay for heat storage, as is the case with summer peaking utilities and cool storage. There have been cases, however, where heat storage has been implemented and presumably justified economically. This justification would be on the basis of heat energy saved so the type of complex optimized control justified for cool storage is not normally appropriate for heat storage. Control for heat storage would then be storing excess heat when available and using it to satisfy heating loads before fueled heat was used. The control sequences would be similar to the use of supplementary heat in the heat pump cycle shown in Fig. 12.6 with use of stored heat coming before the use of fueled heat.

Chapter

13

Heat Pumps and Heat Pump Controls

This chapter will discuss heat pumps, from the smallest unitary types to the large commercial type that are field erected, and controls used with heat pumps. Numerous innovative heat pumps have been designed and used, and some of the control systems used with those unique systems are unique in themselves.

The term *heat pump* is used to describe a refrigeration system that moves heat from one place to another. In point of fact, *all* refrigeration systems are heat pumps in that in all cases, heat is moved from one location (the evaporator) to another (the condenser) and is dissipated into the atmosphere or some other sink such as water. The thing to keep in mind when talking about heat pumps or any refrigeration system or cycle is that the removal of heat from one spot to another, such as the airstream in a duct to the air passing through a condenser, is accomplished with a *pump* such as the compressor that raises the level of the medium being compressed so that the condensing can be done. Without the pump, the medium being used as a refrigerant could only be condensed with a condensing fluid that was colder than the airstream in the duct work. In that case, we might as well use the condensing fluid and not go through the complication of a refrigeration system. Also remember that when we talk about *evaporation and condensing,* we need to think of heating and cooling as relative; for example, 100°F is colder than 200°F, and −100°F is warmer than −350°F.

A thorough study of the refrigeration cycle and a remembrance of some of the facts mentioned above will give you a good insight into the principles of the heat pump, which by the way, are no different than any other refrigeration cycle. The *same* components are used to *heat* a space rather than cool the space. For example, we can take an ordinary window air conditioner and arrange it so that it can be pivoted to blow cold air into the room in the summer and blow hot air into the room in the winter. That is, it would be pivoted around so that the condenser that ordinarily blows outside of the space would blow into the space. The evaporator would then blow cold air outdoors, with the usual problem of trying to cool the outdoor air when it is already very cold. The heat pump then takes the heat energy from the outdoor air and amplifies it with the refrigeration system and compressor, allowing the condenser to blow hot air into the space. Remember that there is heat energy in everything no matter what the temperature is, down to absolute zero. The colder the item, the less the energy, but there is energy and the heat pump extracts that energy with the refrigeration system through the compressor.

Even though the term *heat pump* can be used for all refrigeration systems, for the past 100 plus years engineers have relegated the term to the systems heated for beneficial purposes, where the primary reason for the system is to heat not cool. Note, however, that some units are dual mode and heat and cool in summer and winter. Some systems reclaim heat and provide cooling at the same time. Other units do simultaneous heating and cooling with the refrigeration compression cycle. The main distinction that needs to be remembered is the conversion of electrical or similar energy through the use of a refrigeration compressor *without* additional supplemental heating devices. As an example, the ordinary window unit heat pump heats the space in winter and cools the space in summer without any source of energy other than the electrical power to the unit.

As will be seen, the primary source of driving energy is the electric motor, but many large systems are driven by other means, such as turbines and gas engines. Those systems are, however, the exception.

Several types of heat pump cycles are used in the HVAC industry; they can be described as follows:

1. The closed vapor compression cycle as shown in Fig. 13.1
2. The mechanical vapor recompression cycle with heat exchanger as shown in Fig. 13.2
3. The open vapor recompression cycle as shown in Fig. 13.3
4. The waste heat Rankin cycle as shown in Fig. 13.4

Heat pumps are classified by

1. Heat source and sink
2. Heating and cooling distribution fluid
3. Thermodynamic cycle

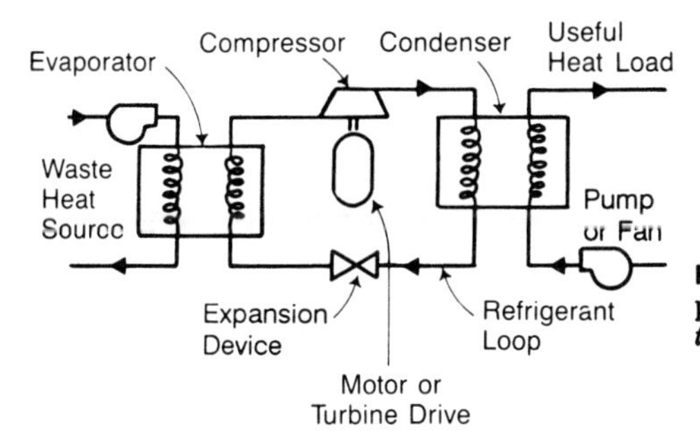

Figure 13.1 Closed vapor compression cycle heat pump. (*Courtesy of ASHRAE.*)

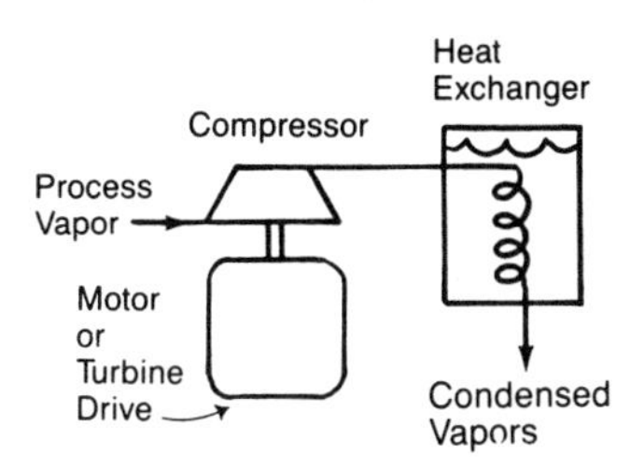

Figure 13.2 Mechanical vapor recompression cycle heat pump. (*Courtesy of ASHRAE.*)

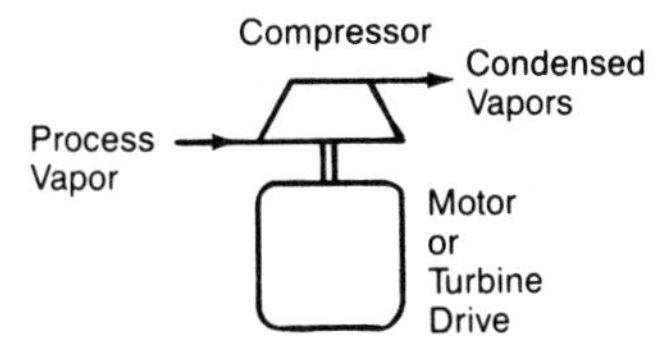

Figure 13.3 Open vapor recompression cycle heat pump. (*Courtesy of ASHRAE.*)

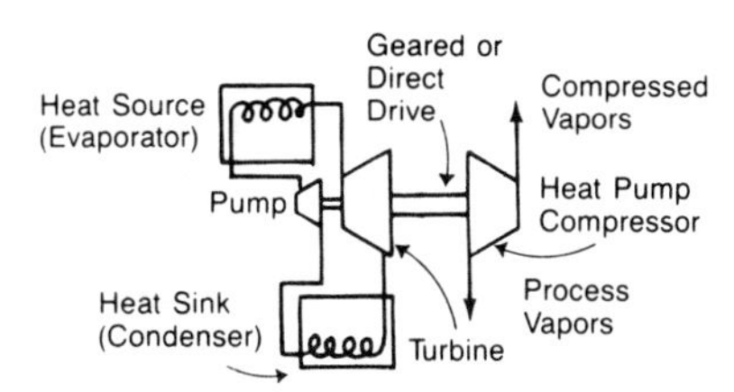

Figure 13.4 Waste heat Rankin cycle heat pump. (*Courtesy of ASHRAE.*)

4. Building structure
5. Size and configuration
6. Limitation of the source and sink

Generally the classifications can be broken down into the following types:

A. Air to air
B. Earth to air
C. Water to air
D. Air to water
E. Water to water
F. Earth to water
G. Internal source
H. Solar heat pumps
I. Waste heat pumps
J. Refrigerant to water

Sometimes there is confusion on the part of HVAC engineers about the term *heat pumps* and when that term needs to be applied. The concept of heat recovery using refrigeration systems has at times been used in conjunction with heat pumps. One section of this chapter is devoted to heat recovery through the use of the refrigeration cycle, but the other sections discuss heat pumps alone.

Air-to-Air Heat Pumps

The air-to-air heat pump is by far the most common heat pump on the market. It lends itself to the easily manufactured unitary type of unit, such as the common plain window unit, with the switch from heating to cooling and vice versa through valves in the refrigerant circuit. In other words, the coil inside the house is an evaporator during one season and a condenser during the other season. The outdoor coil is also reversed from season to season and becomes a condenser in the summer and an evaporator in the winter. In both cases, the heat source and the heat sink are air, and the controls consist of a summer–winter switch and a special refrigerant four-way valve, or a set of two-way valves that allows the refrigerant to reverse its course in summer versus winter as seen in Fig. 13.5. The expansion device is usually a cap tube, and check valves are used to stop the flow of refrigerant in the wrong direction in the different seasons, as can be seen in Fig. 13.5.

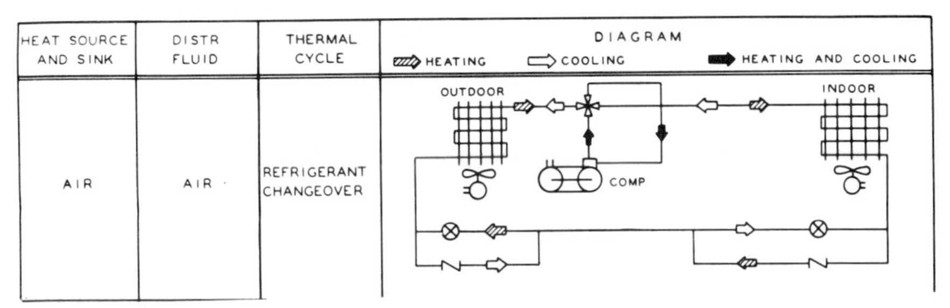

Figure 13.5 Air-to-air heat pump (refrigerant changeover). (*Courtesy of ASHRAE.*)

Other controls, such as the normal window unit, consist of a return air thermostat cycling the compressor on and off. This thermostat must reverse its action between heat and cool modes of operation.

These types of heat pumps are common and are sold in the areas where the cost of power is reasonable and the winters are not too severe. Like most of the air-to-air heat pumps, whenever the ambient temperature is too low in the winter (+20°F and below), the heat pumps have two disadvantages: The amount of heat energy in the outdoor air that can be recovered is reduced. In areas where the humidity is high enough, there is the problem of frost formation on the outdoor coil, which is after all an evaporator during the heating cycle. Units used in those area often have various schemes for defrosting the outdoor coil, including reversing the cycle for a timed period or the use of electric defrosting coils and hot gas injection into the evaporator. In some areas, the codes that apply to heat pumps require that the unit be turned off as a heat pump at a certain outdoor temperature and supplemental electric resistance heating or other means be used to heat the space. When that is done, the advantages of a heat pump are reduced and, in some cases, the owners and engineers revert to conventional systems.

Residential split system heat pumps are also popular as air-to-air systems, especially in areas where winters are not severe and the cost of electrical energy is reasonable. In principle, the units are the same as the window units and the controls are similar. There are switchover valves and thermostats that cycle the compressor, along with in some cases, head pressure controls and low-pressure cutouts.

Some large homes and residences that use what amounts to built-up systems where the evaporator and the condenser coil always act as they do in the conventional systems, with the air redirected at the seasonal change. That is, in winter the room air and/or return air is directed across the condenser and the outdoor air is directed across the evaporator. When the season changes, the reverse is true, and the room or return air is directed across the evaporator and vice versa for

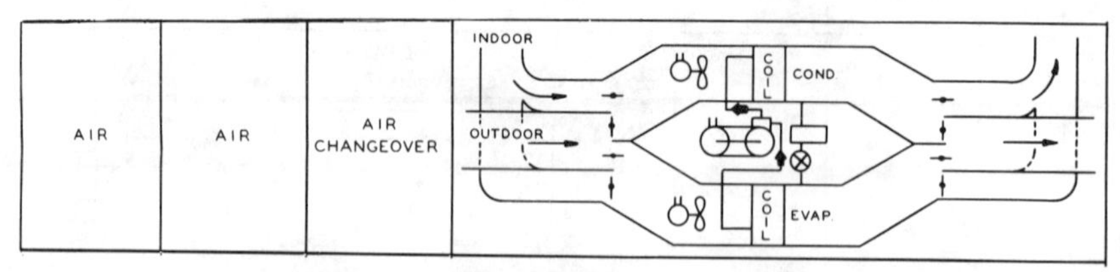

Figure 13.6 Air-to-air heat pump (air changeover). (*Courtesy of ASHRAE.*)

the condenser. This is accomplished through the use of motorized dampers and suitable duct work systems as shown in Fig. 13.6.

Most of the air-to-air systems used for residential heating and cooling are sized on the basis of the heating requirements. They almost always provide sufficient cooling when the selection of both coils and compressor is based upon the heating requirements. Remember, however, as with all selections of components, there are compromises since there is never an infinite number of sizes of compressors or coils, and when a selection is made the next largest size up from the actual size required is used. This means that the selection of the components will always be oversized. This problem exists throughout the HVAC industry, not just with heat pumps.

Earth-to-Air Heat Pump

In recent years the concept of ground-coupled heat pumps for residential buildings has taken hold. Every heat pump requires a heat sink and a heat source. In the case of the air-to-air heat pumps, air is both the heat sink and the heat source. In the case of ground-coupled heat pumps, the earth is both the sink and the source. The earth can be used as a heat sink and a heat source since its temperature swings are not too severe and can be relied upon to be consistent with collected data.

These heat pumps are classified as earth-to-air heat pumps and the coils that are buried in the earth are used to dissipate heat *to* the earth in the cooling season and to acquire heat *from* the earth in the heating season. The coils are usually buried from 4 to 6 ft deep and in sections from 3 to 6 ft apart as serpentine coils. In some installations, the coils are buried vertically and much deeper than in the horizontal systems. The problems with earth coil system heat pumps involve the composition and moisture content of the soil in the area. Since the indoor coil is an evaporator in the cooling season and a condenser in the heating season, the switch-over valves and the other controls needed for an air-to-air system are also needed for an air-to-earth system with buried ground coils.

In the last few years there has been a lot of experimental data about these types of heat pumps, and more and more systems are being tried. The initial cost and excess cost if the buried coils must be serviced due to leaks caused by shifting of the earth have always factors in the decision-making process about these systems. The actual costs of operation have, however, been exceptionally good and over the long haul the data indicate that the systems are cost effective.

The types of pipe used in the buried coils is important, and the engineering of the coils along with the choice of pipe need to be left to the expert designers. Companies that specialize in the installation of these systems can provide all of the engineering needed, as well as the equipment for the complete installation. A rating system has been initiated by Air Conditioning and Refrigeration Institute (ARI) so that users will know whether or not the companies supplying the equipment and engineering are qualified. The certification is known as ARI 330-90, and the manufacturer must allow random testing of their units to use the ARI certification shield on their units.

Water-to-Air and Air-to-Water Heat Pumps

The water-to-air and air-to-water heat pumps are usually relegated to the commercial and industrial uses. In the area of water-to-air and air-to-water heat pumps a unique heat pump system, sometimes called a closed loop heat pump system, has gained popularity in the past 20 years. Basically, the system, which is manufactured by about seven manufacturers, consists mostly of terminal units from ½ to 2 ton capacity with a fan and coil that is an evaporator in one season and a condenser in the other season.

The rest of the system is where the uniqueness is seen. The condenser is a tube-in-tube condenser that is water cooled. The water for the cooling of the refrigerant during the cooling cycle is supplied from a water loop with all of the units in the building fed in a series loop from the same water system. Further, when a unit is switched to heating, the control valves are reversed as in the normal heat pump systems, but in this case the tube-in-tube condenser becomes a chiller and cold water is fed into the loop water system. This system allows units to be on heating or cooling in the same building or zone. From this it can be seen that the BTUs that are removed from one space or zone can in effect be shifted to another space or zone through the loop water system.

These systems work well in the intermediate seasons where some units may be on cooling and others may be on heating. The temperatures of the water loop are also mild, in the range of 90°F maximum and 55°F minimum. In the dead of summer and in the dead of winter when all units are either on cooling or on heating, there is a need for additional cooling or heating of the water loop. In this case, a supplemental boiler is turned on. In the cooling season, a close circuit water cooler is activated to dissipate the heat when all units are on cooling. A closed circuit water cooler is used in lieu of a cooling tower since the tube-in-tube condensers have such small passages that contaminated water from a cooling tower might foul the system. The controls in these systems are again the switch-over valves that allow the refrigerant to be pumped one direction one time and another when the system is switched. The smaller units use a cap tube instead of a set of check valves and an expansion valve. The larger systems use a set of expansion valves and some check valves. These types of systems work well in areas where one occupant wants heating in the summer or one occupant wants cooling in the winter.

The main disadvantage as seen by some designers is the fact that these systems require multiple units to create a flywheel affect and multiplicity of units is a problem to some designers from a maintenance standpoint. The units are manufactured as free-standing units on the perimeter of a building, as ceiling units above a drop ceiling, and as closet units that are vertical with either up-blow or down-blow fan systems. These systems are never used in residential applications, so they are in effect commercial heat pumps.

Water-to-Water Heat Pumps

Water-to-water heat pumps are also common commercial heat pumps since water is an excellent material to use as either a heat sink or a heat source. The water can come from many sources, as will be discussed later, but the important fact to discuss here is the method of switching from heating to cooling and vice versa.

It is possible to switch the refrigerant from one exchanger to another, but the most common method is to keep the refrigerant going to the same exchanger summer and winter and to switch the water circuits as shown (Fig. 13.7). The switching is done through automatic valves that send the return water in one season to the condenser so that it can be heated, with the water supply being used through the

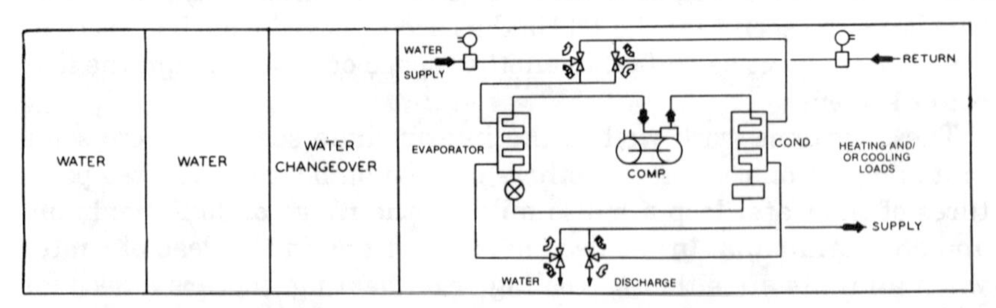

Figure 13.7 Water-to-water heat pump (water changeover). (*Courtesy of ASHRAE.*)

evaporator to cool the refrigerant. The reverse is true in the cooling season, when the return water is sent through the evaporator to be cooled with the supply water being used through the condenser so as to condense the refrigerant. The controls are switch-over valves that are used, in this case, on the water piping. Three-way valves can be used, but usually two-way valves are used as shown in the figures.

Other Commercial Heat Pump Systems

Other commercial systems in use today involve solar-assisted heat pumps and geothermal heat pumps. It needs to be repeated that any medium whereby heat is withdrawn from the medium then pumped up through the use of a refrigeration system is a viable heat pump. As an example, the water in a canopy-covered swimming pool is an excellent source of a heat sink in the winter when the pool is not being used. That same pool can be used as a heat rejection device in the cooling season since the energy is easily dissipated through the large surface area of a swimming pool exposed to the summer temperatures and winds.

Also consider the case in which a commercial heat pump system was designed using the effluent going out of a sewage treatment plant as a heat source. The principle was that the effluent, which looked as clear as potable water, was used as the heat source and was cooled about 10°F. That energy was elevated using a compressor and was condensed, while at the same time making hot water to heat the building. The refrigerant was R-22, so without going to high-pressure vessels, the system made hot water in the range of 120°F to 140°F. The equipment had to be designed to use that low a water temperature. Here is a case where the heat source was excellent and available in large volumes that could be used year around. In the summer, the facility did not use air conditioning except for a couple of window units.

The manufacturer of this system has since sold a number of them and has even given the units a model number. Since the effluent can cause a problem with the tubes in the heat exchangers, the manufacturer was conservative and used stainless-steel tubes in the first units and single-path chillers that could be cleaned. The bottom line here is that this large source of what was considered an unusable medium was used to an advantage with a heat pump.

Heat Pump Components

The components of a heat pump, be it a commercial package or a built-up system, are the same as in any air-conditioning system in almost all respects. Some components of ordinary air-conditioning systems do not, however, lend themselves to heat pumps. Therefore, we will now discuss the various components so that you can see the differences that apply and be able to make the correct decisions as to which components to use.

To start with, remember that the compressor is pumping the refrigerant up to a higher level than in an ordinary air-conditioning system, since we are trying to get the refrigerant to a level where the condenser can make use of the rejected heat to heat the space. For that reason some compressors, such as centrifugal compressors, do not lend themselves to heat pumps.

The high lifts or high compression ratios associated with heat pumps can cause surging at low loads with a centrifugal compressor. A multistage centrifugal compressor is better adapted to heat pump applications, but even here, low-load situations can cause problems. Centrifugals have been used successfully with double-bundled condensers in a heat recovery mode, but those systems are not strictly heat pumps as we have been discussing them this far.

Reciprocating compressors have by far been the most predominant compressor used in the heat pump arena. They are used in heat pump systems ranging from 0.5 to 100 ton capacity. They are the most versatile compressor, having the ability to pump the gases up to the highest condensing temperature and pressure to remove the greatest amount of heat from the systems. In general, the compressor is selected for its heating duty, and the capacity will be more than adequate for the cooling duty provided the proper chillers, and so on, are selected.

Some heat pumps are designed for heating only, so the above precautions are not relevant. The ordinary cooling only compressor has a clearance swept volume of about 0.05, whereas the compressor used in a heat pump should have a clearance volume of 0.025. The compressor used in an air-to-air heat pump can be oversized to get more out of the heat sink at low ambient temperatures, but the disadvantage of this is that at higher ambient temperatures the system must have some sort of capacity reduction on the compressor.

Recently there has been a renewed interest in two-stage heat pump compressor systems. With this system one of the compressors can pump, for example, from −29 to +40°F, whereas the second compressor can pump from 40 to 120°F. With this arrangement, the system can be a two-stage system at low ambient temperatures and a series system during high heat sink temperatures. These two-stage systems can be designed when two compressors are used and or when one compressor is used that has a separation system between multiple pistons built into the compressor. If the loads are properly designed, the later types of compressors are compact and an answer to the problem of low ambient heat sink temperatures.

Rotary vane compressors are not particularly suited to heat pump systems because of the pressure ratios with which they operate. They do move large volumes of gas, but they also have limited capacity reduction systems that are not suitable for heat pumps.

Screw compressors, on the other hand, can be used successfully as heat pump compressors since their ability to pump up to high pressures is inherent. More and more manufacturers are using screw compressors in their heat pump designs. The only disadvantage of the screw compressor is that since large volumes of oil must be injected during the compression cycle to lubricate, cool, and seal the lobes of the compressor, a large oil separation system is required to get the oil out of the refrigerant after it leaves the compressor. The capacity reduction systems used on the screw compressor is excellent and allows the compressor to throttle from 100% to about 15%. A problem, however, is that the horsepower consumption of the compressor is not proportional to the capacity reduction system. That is to say, the compressor uses, for example, more than 50% power when the capacity reduction system has the compressor throttled to 50%.

Absorption systems are used in heat pump applications, but they are the exception not the rule. If anything, they are used with heat recovery systems, not heat pumps.

Vessels and heat exchangers are usually the same types of exchangers used with the normal air-conditioning systems, but the capacities and sizes of tubes and fin spacing might be different because of the service to which they are subjected. Since the pressures and temperatures can be slightly different, it is also possible that the types of materials used in the tubes of the heat exchangers as well as the wall thicknesses of the tubing could be different from the ordinary heat exchangers and vessels used for air-conditioning systems.

The reversing valves used in heat pump systems are designed for refrigeration service and are usually pilot operated to be able to take advantage of the power of the refrigerant moving through the system to operate the valve. Sometimes these valves can be very large and require a lot of power to operate, so the pilot-operated types are the norm.

The expansion devices are the same expansion devices used in normal air-conditioning systems. Remember, when the bulb of a thermostatic expansion valve is placed on the leaving pipe of the evaporator, that sensing area or surface becomes the discharge line of the compressor when the system is on heating, and excessive temperatures and pressures may result. It may be that special pressure and temperature limiting devices will have to be used when the system switches to heating. Also, in some cases it is desirable to use check valves with bypass systems so the expansion devices are in the proper circuits as the system is switched from heating to cooling and vice versa. It is not possible to use, for example, an expansion valve that will work in two directions. In the case of cap tubes used for expansion devices, it is also desirable to provide an accumulator to prevent flooding back to the compressor when the system changes from heating to cooling and vice versa.

A receiver is used in heat pump systems even more than it is used in ordinary air-conditioning systems because the different modes of operation (heating versus cooling) can require different amounts of refrigerant and there has to be a place to store the excess refrigerant when the cycles change.

One of the most important items to discuss when studying heat pumps is the *defrost cycles*. This is particularly true for air-to-air heat pumps. Remember, when the system is on heating in the winter, the normal condensing coil is an evaporator. As such, it tries to cool the outdoor air as it takes the heat energy from that outdoor air and uses it to heat the building with the compressor, which boosts the heat energy up. As the outdoor decreases in temperature, there is less heat energy in that outdoor air, and the coil gets colder and colder to try to get the energy out of that air. If the humidity in the air is high and the temperature drops below the freezing point, ice will form on the coil. The coil must then be defrosted to keep the system working.

One method used to defrost coils is to sense the pressure drop across the coil, which will increase as the coil gathers frost. When that happens the defrost cycle can begin. Another method uses timers to start the defrost cycle, which can be set based upon experience. The defrost cycle can be terminated by a thermostat that measures the temperature of the liquid refrigerant in the outdoor coil. It can also be done with a pressure stat that senses the pressure in the outdoor coil.

A third method is to sense the temperature differential between the outdoor temperature and the temperature of the refrigerant in the outdoor coil. As frost accumulates, the differential between the outdoor air and the coil will increase and the defrost cycle can begin. The actual defrosting is normally done by reversing the cycle for the required time and allowing the system to switch to the cooling mode until the frost has been omitted. This may cause some inconvenience, and some places put a burden on heat pumps for that reason. As an example, some northern states have codes requiring air-to-air heat pumps to switch to resistance heating when the outdoor temperature drops to 20°F, which places a penalty on those types of heat pumps to the point that they are no longer a viable option. Heat pumps that have to operate with resistance heating part of the time and wind up with a coefficient of performance (COP) of 1 and with high electric utility rates wind up at a disadvantage as opposed to the normal HVAC systems.

The COP is the figure used to rate the unit based upon the energy out versus the energy in. That is, if the amount of electrical energy in, for example, is 100 units and the amount of energy out is 100, the system has a COP of 1. Heat pumps can have a COP of 4 or higher. Electric resistance heating typically has a COP of 1, since the amount of energy in is equal to the amount of energy out because all of the electricity is converted to heat in the wire of a resistance heater. Ordinary gas and oil furnaces never have a COP higher than 1, as they are not that efficient. Heat pumps can have a COP higher than 1 because they use the heat of the heat sink along with the heat created by the compressor that is converting electrical energy from a motor to compression energy in the compressor.

It may seem as if a system is getting something for nothing with COP higher than 1 until you remember that a heat sink is contributing part of the energy. This is why so many engineers advocate the use of the heat pump as a viable system for heating and cooling. Also, when the heat pump is in the cooling cycle, the COP is no different than any other conventional refrigeration system that uses the same components.

Heat pumps typically work well with heat storage systems and are becoming increasingly more popular. They are advantageous when the heat source and the heat loads do not occur simultaneously. Heat storage systems are gaining in popularity since the utilities are modifying their rate structure to allow for off-peak rates that are more reasonable. The conventional systems as well as the heat pumps can generate ice and cooling in storage tanks that can be used at a later time for cooling without using the compressor.

Heat Recovery/Heat Pump Systems

Typical operating cycles of heat recovery/heat pump systems are discussed next. As shown in Fig. 13.8, a method of controlling a typical water-to-water heat pump is to provide the control valves that operate with the system and allow the application of either heating or cooling at any time. If valves 2 and 3 are closed and valves 1 and 4 are open, the system is in a heating mode. With the valves in this position, the pump is forcing water through the condenser and making warm water that is supplied to the zone units. The evaporator is taking water from the exchanger, which is using water from the heat sink. The reverse is accomplished when valves 2 and 3 are opened and valves 1 and 4 are closed. When using this system, the differential pressure controls are

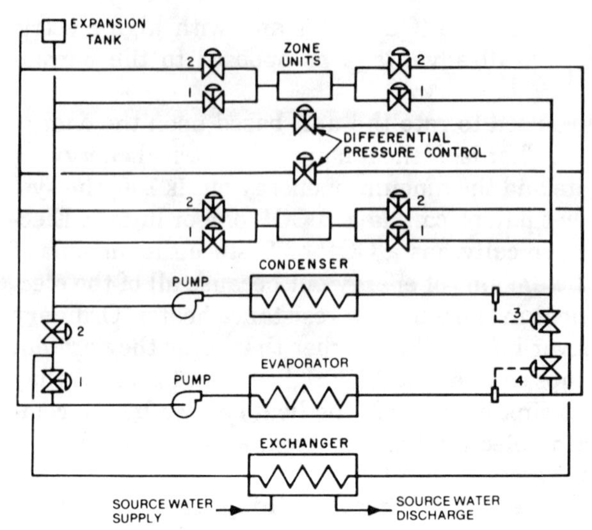

Figure 13.8 Changeover valves in a heat pump and heat recovery system. (*Courtesy of ASHRAE.*)

a good idea to compensate for the differences in the amounts of water pumped into the cooling and heating modes. This is because the chilled water and the hot water are pumped through the same size pipe, but the amount of chilled water needed is much greater and the pipe must be sized based upon the chilled water pumped.

Figure 13.9 shows a typical packaged air-to-water heat pump using a single-stage compressor system. These units are available in up to

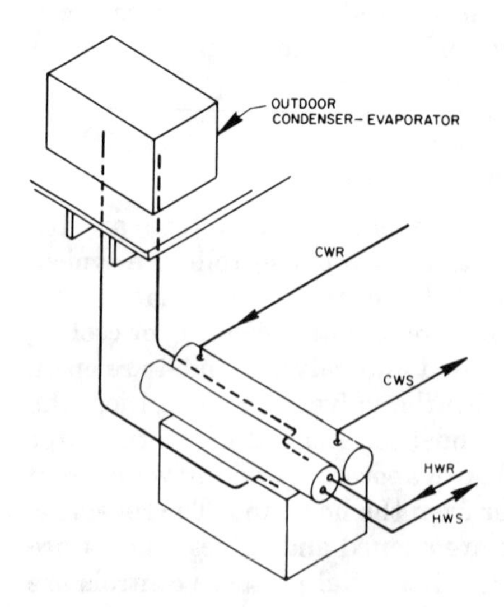

Figure 13.9 Packaged air-to-water heat pump. (*Courtesy of ASHRAE.*)

150 tons of cooling. In this case, the condenser and the evaporator in the package stay as condenser and evaporator and the only operational change is with the outside unit, which is an evaporator at one time and a condenser at the other time. When the system is on a conventional cooling cycle, the chiller is making cool water for the four-pipe system and the condenser outside is rejecting the heat. When the system switches to heating, the inside condenser is making the heat for the four-pipe system and the outside unit is the evaporator with the refrigerant cooling the outdoor air.

The air-to-air system shown in Fig. 13.10 is typical of a system using two coils in an air-handling unit that are supplied with either hot or cool water from the unitary heat pump, depending upon the season. The coils in the air-handling unit are conventional and no different than any typical air-handling unit. The item here that is different is the way that the mediums are supplied to the unit.

Heat Recovery

Some of the principles involved with heat pumps also apply to heat recovery. *Heat recovery* is a system that uses refrigeration in the conventional way it is used for air conditioning and, *at the same time* and through conventional means, recovers the heat that is ordinarily rejected through the conventional condensers, cooling towers, evaporative condensers, and so on. In other words, the heat that is normally rejected is saved and used in the facility. That heat energy is normally

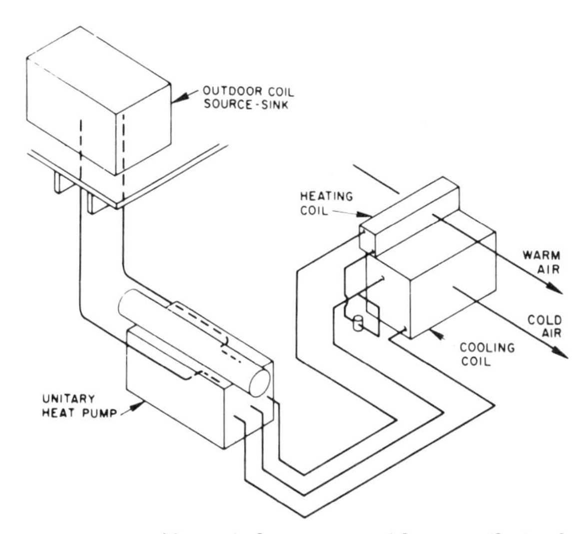

Figure 13.10 Air-to-air heat pump with two coils in the air-handling unit. (*Courtesy of ASHRAE.*)

used simultaneously with cooling in the building. Thus, heat pumps heat when they heat and cool when they cool; heat recovery systems cool and heat at the same time.

The heat generated in a heat recovery system might be used, for example, to heat domestic water in a facility. It could also be used to heat a portion of the building that needs to be heated at the same time as the facility is being cooled. The closed circuit water loop system that was described previously is an example of a heat recovery system where the area being cooled is providing the heat for the area that needs the heat as a result of the heat being rejected.

A common heat recovery system in large tonnages is a system that uses a double-bundle condenser, where part of the rejected heat is fed to a cooling tower and the other part is used to heat, for example, domestic water. This is seen in Fig. 13.11.

Heat recovery systems also work well with the heat storage concept, where heat is not used as fast as it is generated and can be stored for later use. Solar collectors also have a place in some climates with these types of systems. This is illustrated in Fig. 13.12.

Figure 13.13 is an example of two systems that are cascaded hydronically, with one system acting only as a chiller and the other acting only as a heat pump.

Many other configurations of heat recovery systems can be designed and used. The main thing to remember is that the design must use the heat that is normally rejected to heat the space, heat domestic water,

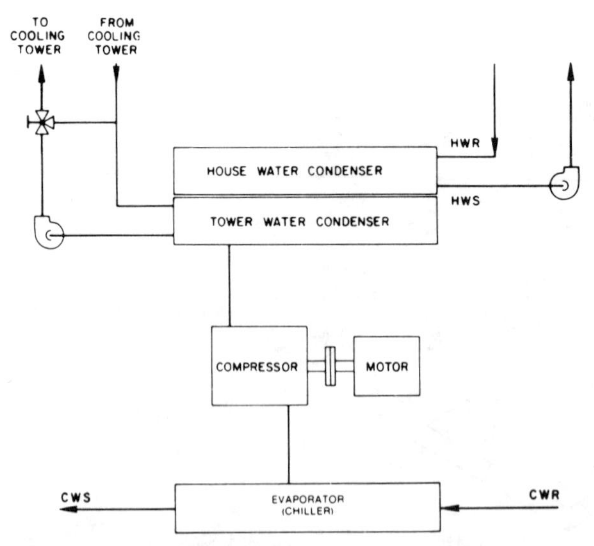

Figure 13.11 Double-bundled condenser in a heat recovery system. (*Courtesy of ASHRAE.*)

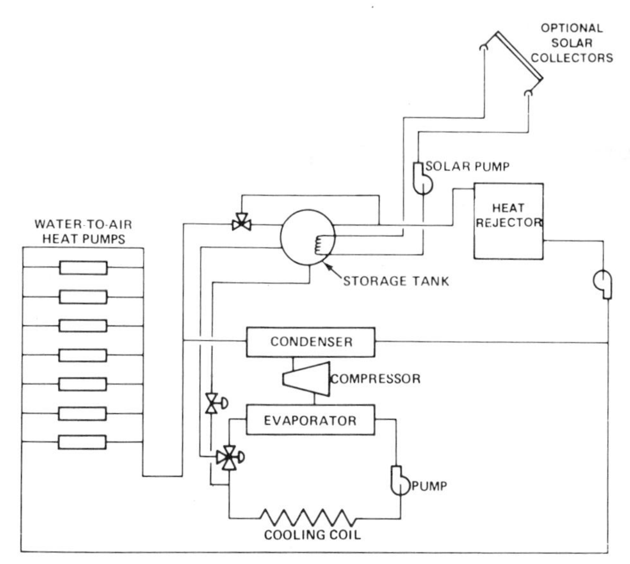

Figure 13.12 Solar-assisted heat pump with thermal storage. (*Courtesy of ASHRAE.*)

or heat something else. The reason heat recovery systems are not used in all systems is that the initial costs and operating costs may exceed the benefits of the energy saved. An analysis of those factors is a must when designing a system. If the system being designed is special and requires equipment that is not in a catalog and is not standard, the saving may not be enough to justify its design and installation.

Supplemental heating equipment mentioned previously is, in some cases, a requirement, and the selection needs to be done carefully so as to not affect the overall COP of the system. Realistic outdoor temperatures need to be used, and lower nighttime temperatures need to be taken into account when designing a heat pump system. If the designer just uses the lower nighttime temperatures, the system might be overdesigned and the resulting overall COP could suffer. Other items that can be designed into the system include proper insulation, dependable heat sources within the building, closing the outdoor air intake when on the night cycle, and adjusting the ventilation rates to the exact amounts required by the codes.

Also remember that the water temperatures generated by heat pump systems may not be as high as those generated by conventional boilers and heat exchangers. This means that the equipment used to heat the spaces, such as the coils, radiators, and convectors, must be sized to use the lower water temperatures and still heat the space. If, however, the refrigerant selected (such as R502 or R504) is a high-temperature refrigerant and the designer and operating personnel do

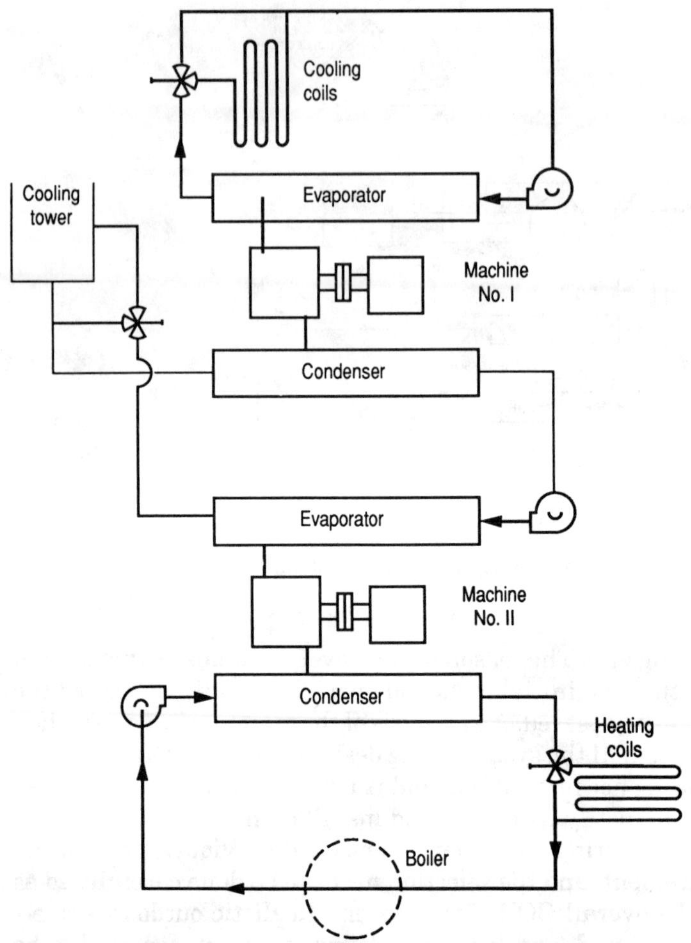

Figure 13.13 Two systems cascaded with chiller and heat pump. (*Courtesy of ASHRAE.*)

not mind the use of high-pressure vessels and compressors, the systems can generate conventional water temperatures. This, however, is the exception and not the rule. Usually any vessel above 400 psi requires special welding and certification and is not normally found in the HVAC field.

As far as the controls that are used is concerned, they are still the same conventional controls used in any other system. They are just used differently. There are thermostats, control valves (both water and refrigerant), as well as all the other refrigerant devices.

In summary, this chapter familiarized you with heat pumps, which are not much different than conventional systems as they use the same components in a slightly different way. Think of the heat pump as a window air conditioner that is on a pivot and can be rotated in summer and winter to use the condenser to heat the space in the winter and the evaporator to cool the space in the summer. Like anything else in the HVAC field, engineers and designers are constantly making improvements and modifications to heat pumps. The heat pump portion of the HVAC field, in fact, offers the greatest challenge in terms of innovations and improvements.

Chapter

14

Distribution Systems of All Types

HVAC distribution systems include hot and chilled water and steam piping systems. Water piping systems are typically closed systems (not open to the atmosphere) that include the pumping systems that circulate the water. The control aspects of distribution involve pressure and flow into the system to transport thermal energy from a central supply subsystem to the using subsystems. The quantity of flow required in a system is determined by the control valves associated with each using subsystem to meet its load. The requirements of the distribution system are to deliver the necessary water or steam at a pressure and temperature that is controllable by each load's control valve. The sizing of a piping system must be designed to delivery the maximum required flow to all parts of the distribution system. The selection of the pumping in a water system must provide design flow at the maximum pumping head required. The control of the distribution system must maintain controllable pressure differentials in all parts of the system under all conditions of load and flow. The range of flow variation and the system configuration influence how best to control a distribution system. Therefore, this chapter will be organized by system type and configuration.

Water Distribution Systems

Two-pipe systems

A two-pipe system, the most basic system with a supply pipe and a return pipe, is used for both hot and chilled water circuits. The arrangement of these pipes can be either a direct return as shown in Fig. 14.1 or a reverse return as shown in Fig. 14.2. The direct-return piping arrangement is appropriate for a constant flow system because balancing valves in the branch circuits will allow the same pressure

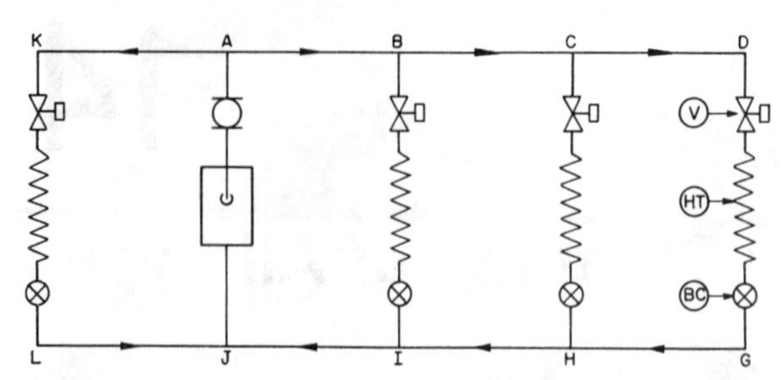

Figure 14.1 Typical direct-return closed-loop piping circuit. (*Courtesy of ASHRAE*, 1987 Handbook *Chap. 13, Fig. 10.*)

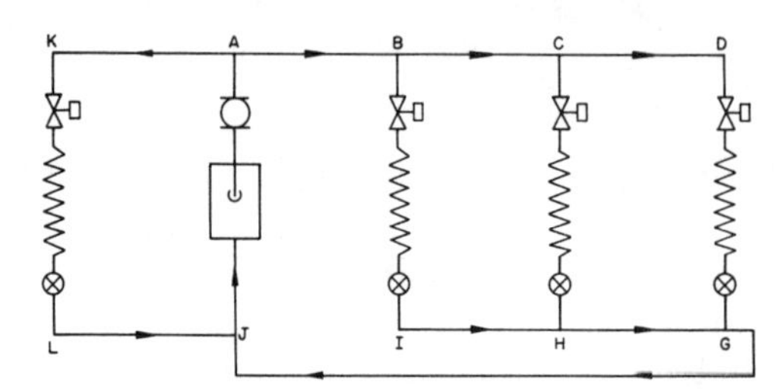

Figure 14.2 Typical reverse-return closed-loop piping circuit. (*Courtesy of ASHRAE*, 1987 Handbook *Chap. 13, Fig. 11.*)

differential to be set across all loads and control valves. The reverse-return piping system is appropriate for a variable flow system because all branch circuits see the same pressure differential as flow changes. This is because each load circuit sees the same length of total supply and return piping. With variable flow, balancing valves do not give a constant pressure drop and therefore cannot balance between different load circuits with different flows.

A system becomes a constant flow system by the use of three-way bypass valves for control of loads. Figure 14.3 shows examples of a three-way valve control circuit and a two-way valve control circuit. The circuit controlled by the three-way valve has constant flow. Constant flow circuits have the advantage of being able to be balanced with load circuit balancing valves but have the disadvantage of al-

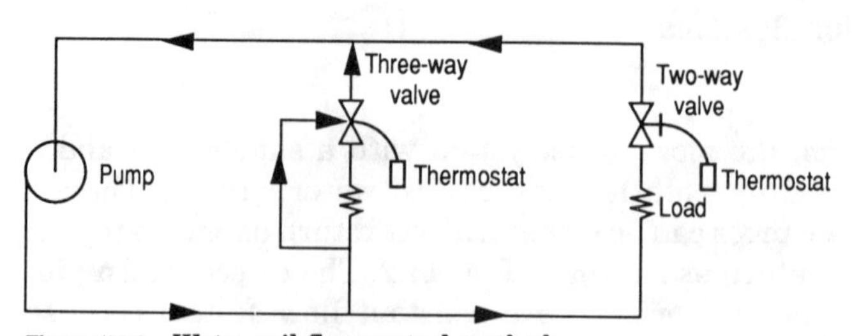

Figure 14.3 Water coil flow control methods.

ways using maximum pump energy even at low loads. In a constant flow system, variation in load causes variation in temperature differential between supply and return.

In variable flow systems, some variation in pressure differential across branch circuits and their loads and control valves will still provide good control. When pressure differential across a branch circuit is more than two or three times the design differential on which the control valve was sized, however, there is likely to be poor control. This poor control could be hunting (cyclic overcorrection) or large offset from control point with changes in load. Therefore, system regulation and branch circuit balance should limit the variation in differential pressure at branch circuits. Variable flow systems with reverse-return piping are automatically balanced between different branch circuits but may require overall pressure regulation as total flow varies. Variable flow systems with direct-return piping arrangements are a design compromise but are frequently used because the direct return entails less costly piping and variable flow gives lower pumping costs. Low load conditions can, however, cause large imbalances in pressure differential between branch circuits that can create control problems. Figure 14.4 shows how branch pressure differential changes in a direct return system when the flow is reduced. The method of differential pressure control can alleviate some of this and will be covered later under the subject of pumping control and differential pressure regulation.

Three-pipe systems

Three-pipe systems are a heating supply and a cooling supply with a common return. The control valve or valves on a coil use either heating or cooling but not both. This is accomplished with a special design of a three-pipe valve that sequences control of heating and cooling or with two valves. This system allows one unit to be on cooling and another unit to be on heating.

Three-pipe systems were popular when first costs were the primary consideration. The energy efficiency of the system was greatly influenced by the temperature of heating return water mixed with cooling return water. If terminal unit coils had enough heat exchange surface that they could produce return water temperatures close to the entering airstream temperature, the units could be reasonably efficient in spite of mixing heating and cooling returns from different units. This system is not as popular as it once was but is explained to help you understand existing systems.

A weakness of the three-pipe system is that all of the design piping loss in the return piping system can cause unbalanced conditions.

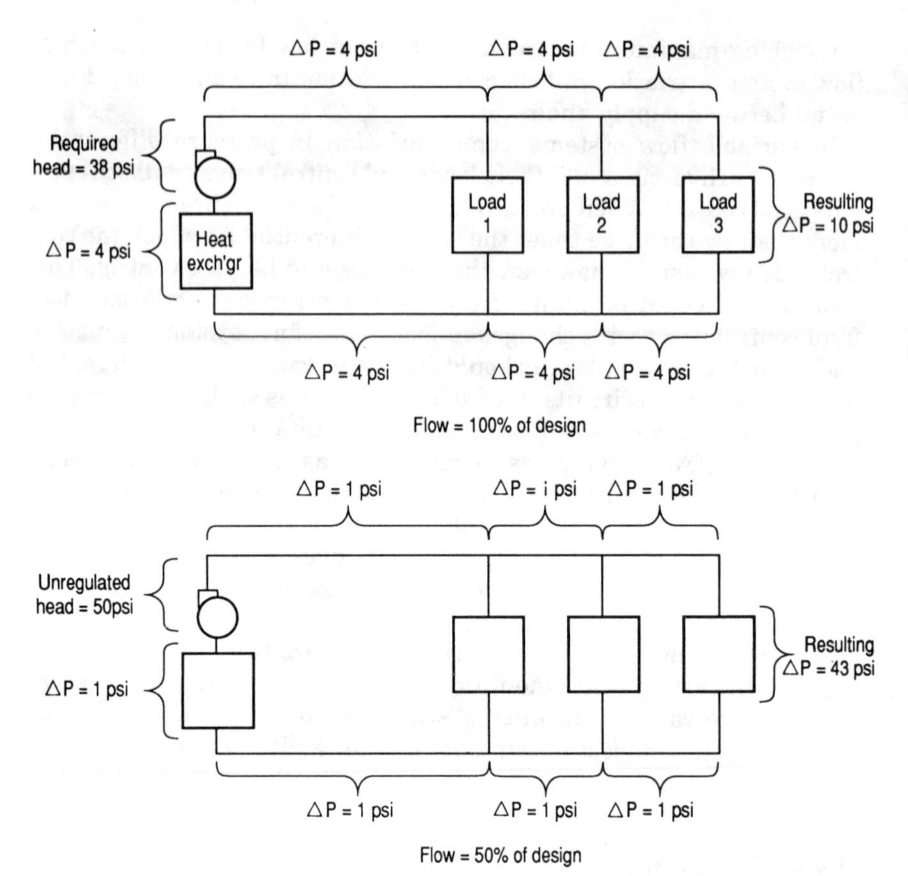

Figure 14.4 Direct-return system pressure drop change with flow variation.

When the design piping loss is greater than the design drop through the control valves, there can be either high differential pressure across control valves or reverse pressure differential. This effect is shown in Fig. 14.5 for the two cases of differential pressure sensing at either end of the reverse-return piping system.

The results would be even more unsatisfactory with a direct-return piping system. The result of sensing at the end of the main gives reverse pressure differential on the less loaded hot water supply. This has caused reverse flow from the return into the off-season supply. Some systems were fixed with check valves to prevent reverse flow, but this still results in no flow available from the off-season supply. The location of the differential pressure sensing to the beginning of the supply piping corrects reverse pressure but causes very high-pressure differential across the off-season supply that may unseat the close off to that port. The design solution is to provide large size piping

3 PIPE DISTRIBUTION—REVERSE RETURN

Large piping system—Return pipe	ΔP = 15 psi
unit valve	h = 4 psi
unit valve	h = 1 psi

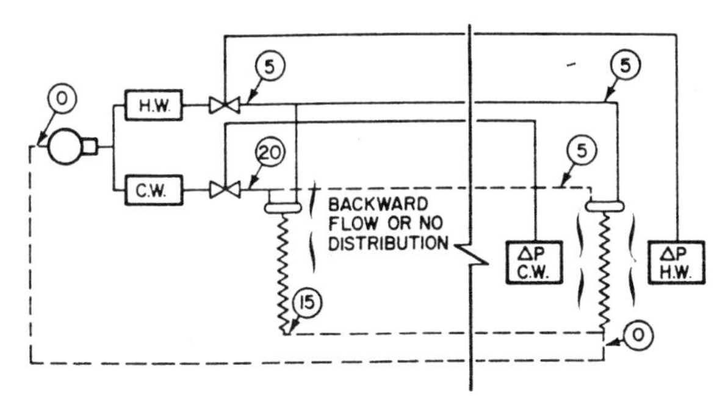

MAXIMUM C.W.–MINIMUM H.W. FLOW ΔP CONTROL AT LAST UNIT
UNSATISFACTORY

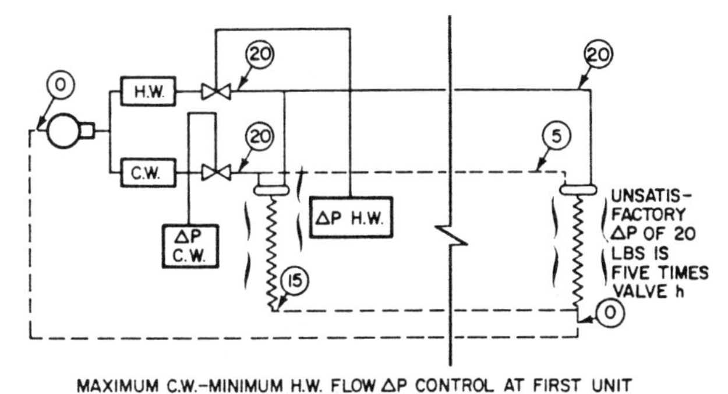

MAXIMUM C.W.–MINIMUM H.W. FLOW ΔP CONTROL AT FIRST UNIT
UNSATISFACTORY CONTROL

Figure 14.5 Three-pipe distribution with reverse return. (*Courtesy of Honeywell, 77-5101 Fig. 15b.*)

to reduce piping pressure drop. Given an existing system, finding a compromise location for sensing differential pressure in the middle of the piping system is a corrective approach that may reduce problems.

Four-pipe systems

Four-pipe systems are essentially two two-pipe systems tied to a common load coil through two sets of control valves that operate in sequence. Figure 14.6 shows the difference between schematic circuits for two-, three-, and four-pipe systems as they connect to one terminal unit. In actual usage there are numerous terminal units so the use of four control valves per unit becomes expensive. The use of separate heating and cooling coils

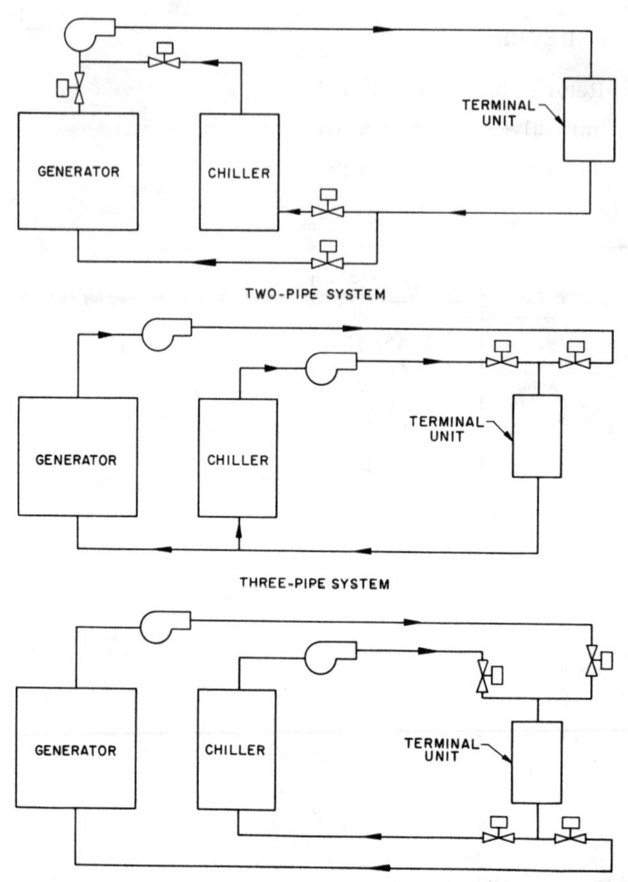

Figure 14.6 Schematic diagrams of two-, three-, and four-pipe systems. (*Courtesy of ASHRAE*, 1987 Handbook *Chap. 13, Fig. 9.*)

needs only two valves so the cost of an added coil is balanced against the cost of two added valves. The control of the distribution of a four-pipe system is the same as the control of two two-pipe systems. The only distinction is that a four-pipe system should have an equalizing line between heating and cooling expansion tanks so there is no pressure difference between the hot and cold water systems that would unseat control valves at the common coil locations.

Pumping Control and Differential Pressure Regulation

Pumps applied in HVAC distribution systems are typically centrifugal pumps. The flow and pressure head created by a centrifugal pump is the balance point where a pump characteristic crosses a system re-

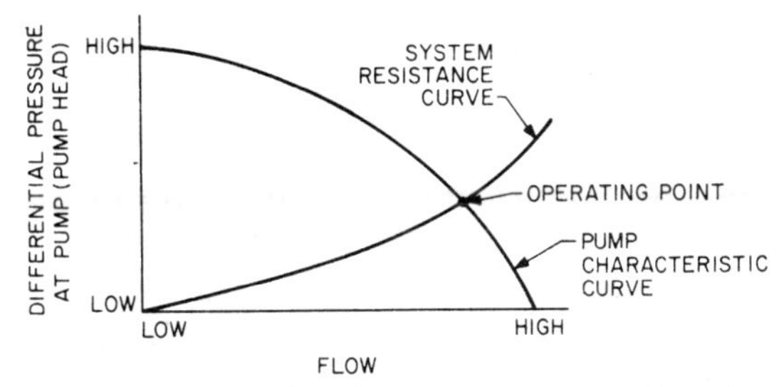

Figure 14.7 Water distribution system operating point.

sistance curve as shown in Fig. 14.7. The pump characteristic curve represents the fact that at the left end of the curve all energy is expended to create maximum pressure. Then as pressure need is reduced, more flow results until at the minimum pressure the maximum flow results. The system resistance curve represents the fact that for a fixed resistance of piping, the pressure loss increases by the flow squared. This represents the resistance in the piping system with the control valves wide open. The control valves, however, throttle close to reduce flow and the summation of this action is represented by the movement of the system resistance as shown in Fig. 14.8.

This buildup in pump pressure at low flow is added to the decrease in piping loss at low flow that was shown in Fig. 14.4. If this buildup in pressure is more than two or three times the design pressure drop across control valves, a means of pressure regulation is needed in the system.

Pressure regulation can be achieved by controlling pumping capacity or by bypassing water flow at some point in the system. Pumping capacity can be varied by throttling the pump discharge with an added valve, by changing the pump speed, or by changing the number

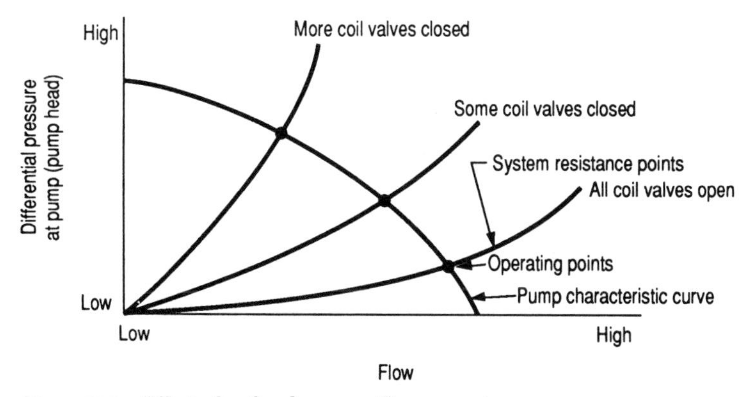

Figure 14.8 Effect of coil valve operation on system curve.

of constant speed pumps online. Pressure regulation by varying pumping capacity saves pumping energy, while pressure regulation by bypassing water flow does not save energy. The specifics of each means of control are as follows.

Throttling pump discharge is accomplished by adding a valve in the pump discharge and controlling it from a differential pressure controller. Figure 14.9 presents a schematic for this control and an explanation of how this affects total system resistance while limiting the pressure drop across control valves. The added throttling valve must be a design suitable for the high-pressure drops it will create. Typically, double-seated construction and metal-to-metal seating are appropri-

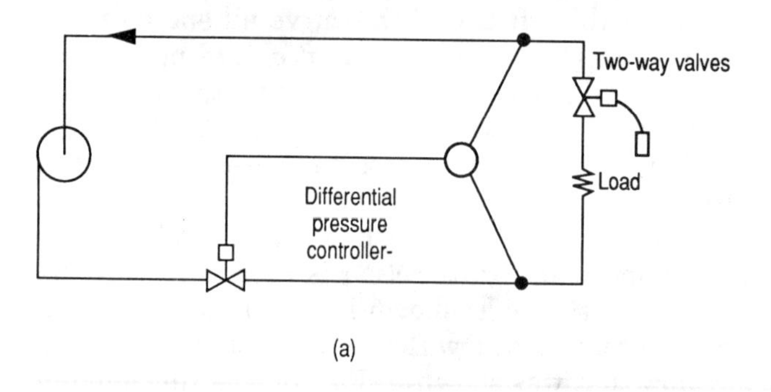

(a)

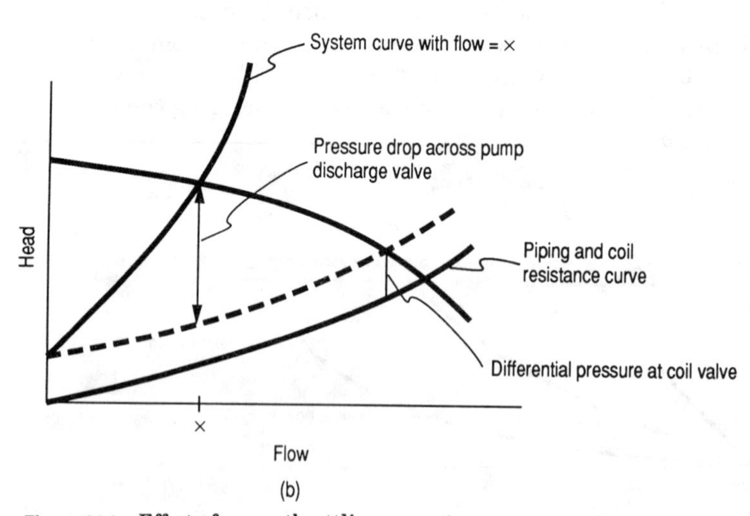

(b)

Figure 14.9 Effect of pump throttling on system curve.

ate. This method of control does reduce pumping and saves energy, although the pump is less efficient when throttled.

Variable speed pumping gives the greatest savings in energy under low loads. This control approach is shown in Fig. 14.10. The figure also shows how the pump curve is shifted as speed is reduced. Although variable speed control is a more costly implementation, the cost is frequently paid back in a short time because of the improved energy savings. When there are multiple pumps in parallel, they should all be run at the same speed whenever they are on. If they are not, the lower speed pump does no pumping until its head equals the higher speed pump. The correct way to control multiple pumps with variable speed control is to start and stop pumps as capacity needs require and to control the speed of all pumps on line from the same modulating control signal.

Multiple constant speed pumps can provide stages of control that keep differential pressure within reasonable limits as flow requirements change. Figure 14.11 shows this control scheme and the way pump characteristics change when two pumps are operated in parallel. The measurement of flow is the most reliable way of ensuring that an adequate number of pumps is on line. An analysis of the pressures

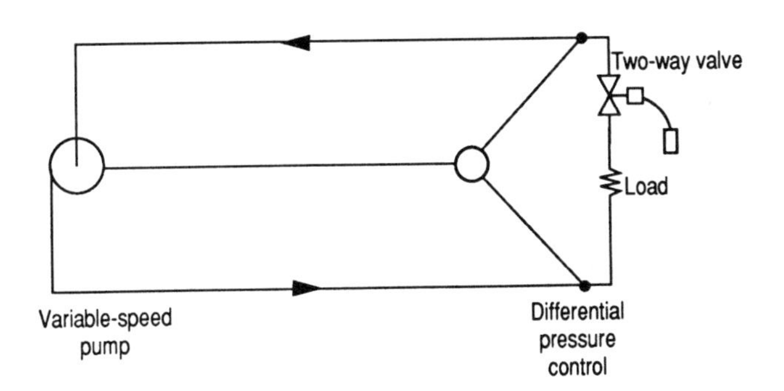

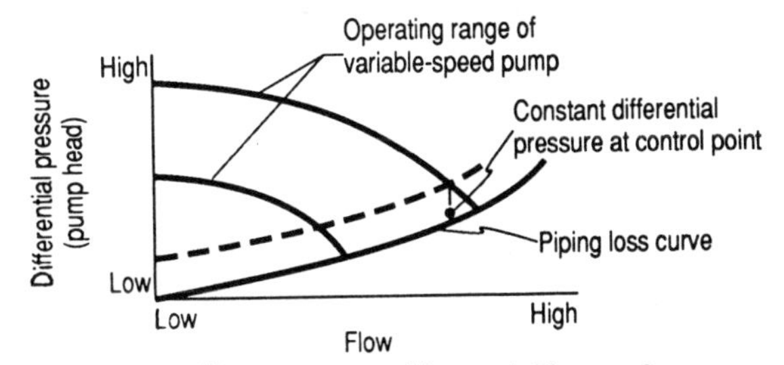

Figure 14.10 Pressure control by variable speed pump.

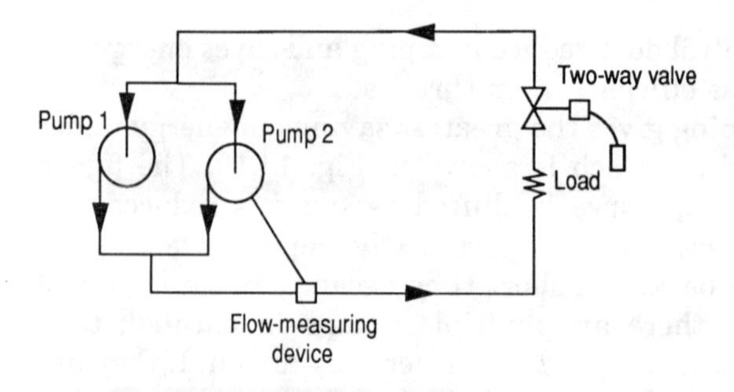

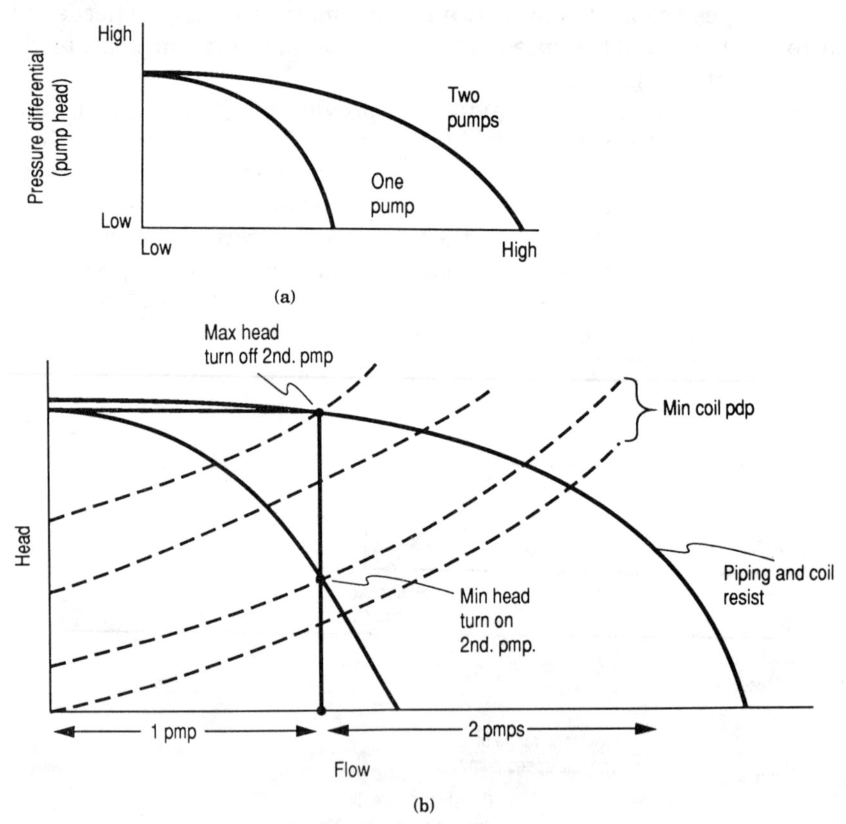

Figure 14.11 Pressure control by multiple pumps.

that prevail when changing from one pump to two pumps is necessary and is marked on the pump curves. This method could also be used to establish differential pressure settings if that sensing was used instead of flow sensing. This method saves significant energy because most of the time only one pump is used.

Bypass control of system pressure provides some limitation of differential pressure seen by control valves but does not save energy. Pump bypass is the simplest implementation because the bypass is all done at the pump location. However, it only limits the rise in pressure up the pump curve by causing the pump always to operate near design flow. This is shown in Fig. 14.12, which also shows the increase in differential pressure because of decrease in system piping loss at low flows. This method accomplishes the most for pumps with steep curves. It may be satisfactory if the piping loss is relatively small so differential pressures limits are not exceeded by the decreased piping loss at low flows.

System bypass control where water is bypassed out in the system is illustrated in Fig. 14.13. This implementation gives better control be-

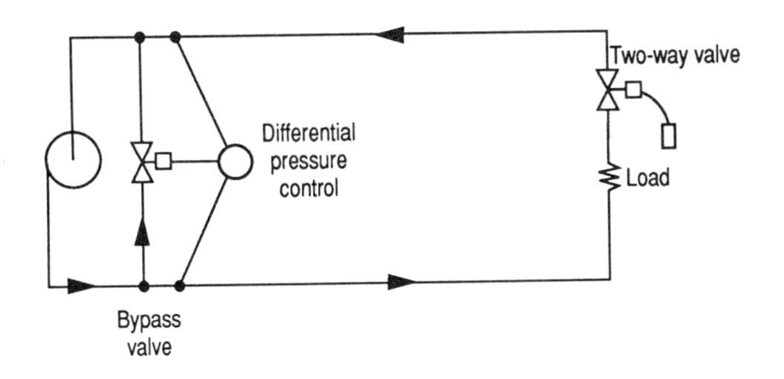

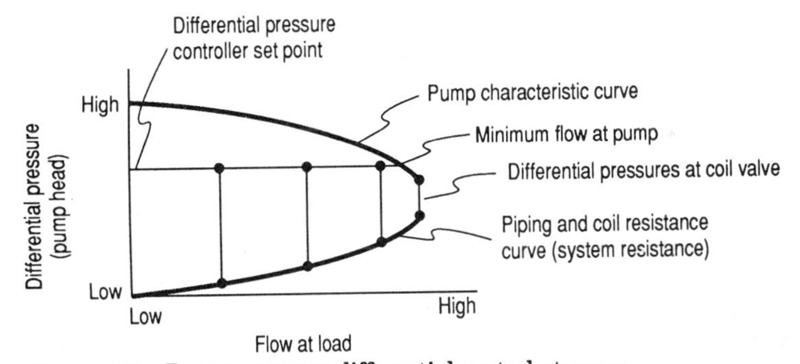

Figure 14.12 Bypass pressure differential control at pump.

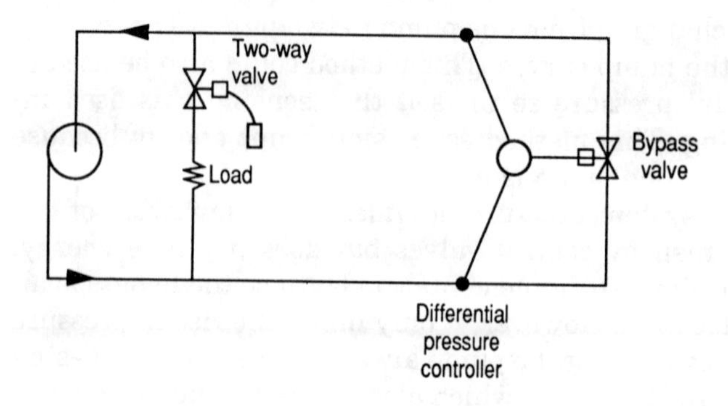

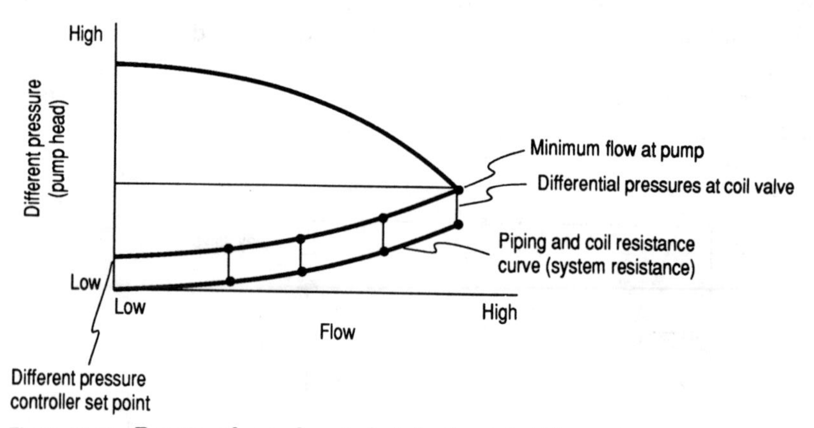

Figure 14.13 Bypass valve and control at the far end of the system.

cause it compensates for piping loss changes. The only complication is that multiple bypasses must be used if there are multiple main branches in the piping system.

Primary–Secondary Pumping Variations

The purpose of primary–secondary pumping systems in HVAC systems is to isolate the secondary pumping system from the primary pumping system. In this way the primary system can be a constant flow system making a primary supply available to multiple secondary pumping systems that supply different loads. There are several methods in which secondary systems tie to primary systems and control flow from the primary to the secondary. Three of these control methods are shown at three different zones in Fig. 14.14. The top two zones in Fig. 14.14 use modulating valves to control how much of the pri-

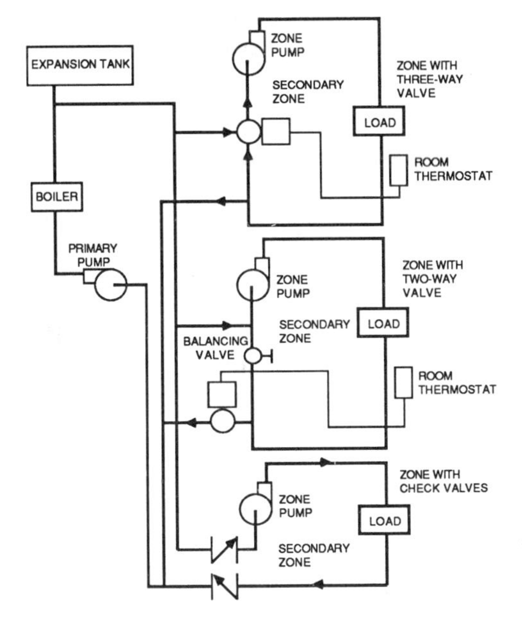

Figure 14.14 Primary–secondary zone pumping control methods. (*Courtesy of Honeywell, 77-1132 Fig. 66.*)

mary supply is taken into the secondary zone. In both of these zone cases, the zone pumping is presumed to be a constant flow, and the primary flow into the zone mixes with return zone water to vary the temperature of the constant water flow to the single load. These examples are for variable primary flow and constant secondary flow with variable temperature of secondary flow.

If a secondary pumping circuit has multiple load zones, they could be controlled by three-way coil bypass valves to maintain constant secondary flow. The multiple secondary zones could also be controlled by two-way valves causing variable secondary flow. In this case, all of the differential pressure control considerations previously explained need to be considered in the control of the secondary pumping system.

When there is variable flow in the secondary or primary circuit, care must be taken in how the primary pumping pressure differential could be reflected into the secondary pumping circuit. If the primary piping is arranged as a reverse-return system as shown in Fig. 14.14, then all three secondary pumping zones see the same primary pumping effect. The secondary zone closest to the primary supply has the longest return path to the primary return. If primary flow varies, regulation of primary pumping can be accomplished to minimize pressure differential from the primary pumps causing upsets in the secondary zones. If the primary piping arrangement is direct return, the nearest secondary zone has short interconnection runs to both the primary supply and the primary return and we would therefore see a large pumping differential from the primary pumping. In this situation, a common design practice is to place a throttling valve in the interconnection between the primary and secondary circuits and to control it to reduce the pressure differential seen from the primary system. Figure 14.15 shows an example of this type of primary-to-secondary interconnection.

Primary piping arranged as a one-pipe system connects both supply and return interconnections to a single primary pipe, this avoiding any primary pressure affect in the secondary system. This approach does, however, moderate the temperature of the primary water as it feeds successive zones of secondary pumping. This would mean that downstream pumping zones would have different supply temperatures and may need different design temperatures for using equipment.

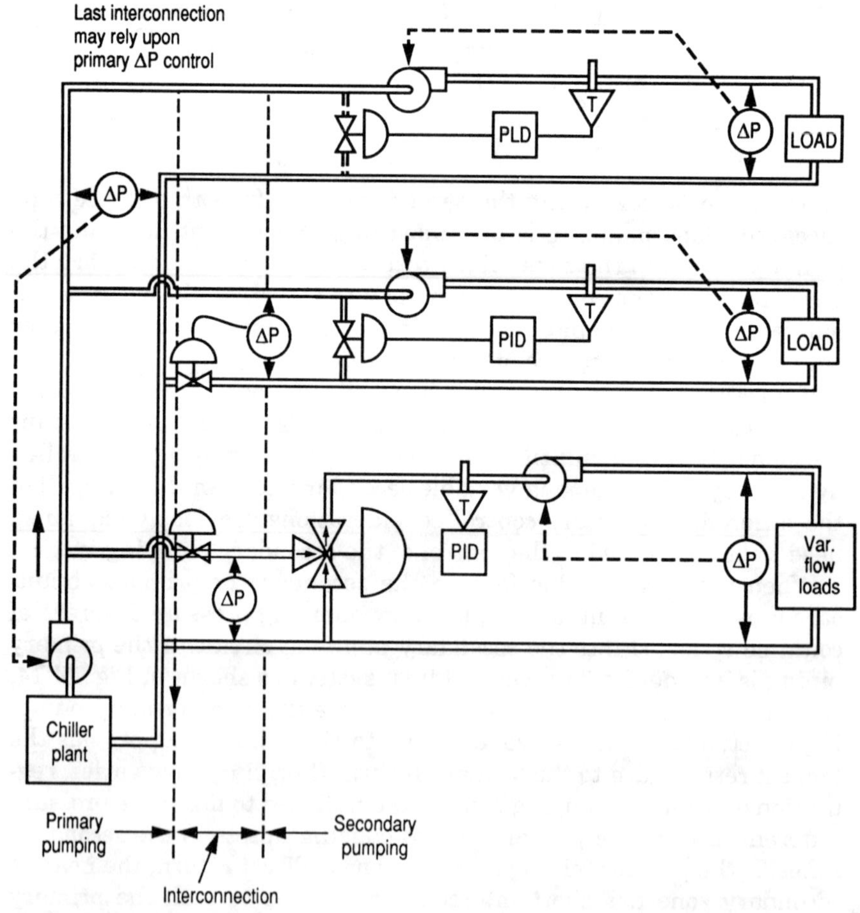

Figure 14.15 Pressure differential control of primary-to-secondary interconnection.

Steam Distribution Systems

Steam distribution systems must provide piping to deliver steam to using subsystems and must provide the equipment and piping to return the condensate. The pressure and temperature of the steam being distributed establish some of the equipment and control requirements. Therefore, high-pressure systems and low-pressure systems will be examined separately. Other variations in system design and control relate to how the condensate is returned. Therefore, single-pipe, two-pipe gravity return and two-pipe vacuum return are different system designs that will be examined separately.

High-pressure steam systems are involved with HVAC systems when the high-pressure (and high-temperature) steam was generated for some other purpose, such as use in a high-temperature process or in driving a turbine prime mover. Because HVAC processes ultimately use steam at condensing conditions, the control requirements specific to high-pressure steam supplies relate to the limitations of control valve application. The limitations of a steam control valve are the temperature limits of the materials and the limitations of maximum pressure drop to be less than half the absolute pressure of the supply. (Absolute pressure equals gauge pressure plus 14.4 psi.) This limit of half the absolute pressure of the supply is called the *critical pressure drop limit* and applies to all steam valves. With high-pressure steam supplies, one or more pressure-reducing valves in series with the final control valve may be necessary to abide by the critical pressure drop limit on each valve. Generally, a supply of 125 psig requires two stages of pressure reduction, whereas a supply of 50 psig requires one stage of pressure reduction. Figure 14.16 shows the arrangement of a PRV station. Whenever saturated steam has its pressure reduced, the resulting steam has a temperature equivalent to the original pressure until some heat is removed. This must be taken into account in the system design by desuperheaters or by using equipment with superheat temperature ratings.

One-pipe steam distribution systems have a single pipe that supplies steam in one direction and returns condensate in the other direc-

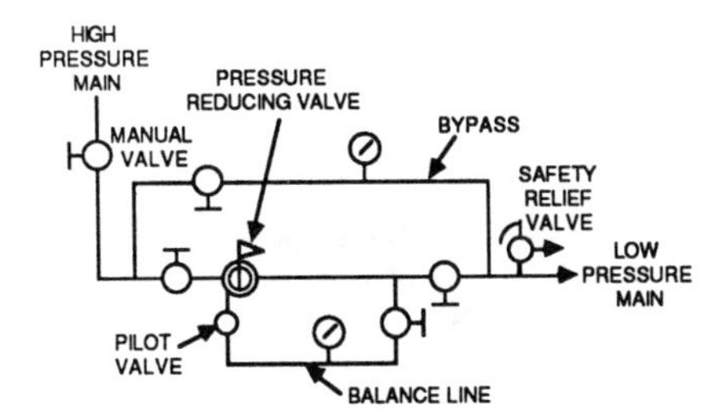

Figure 14.16 Steam pressure reducing valve station. (*Courtesy of Honeywell, 77-1132 Fig. 95.*)

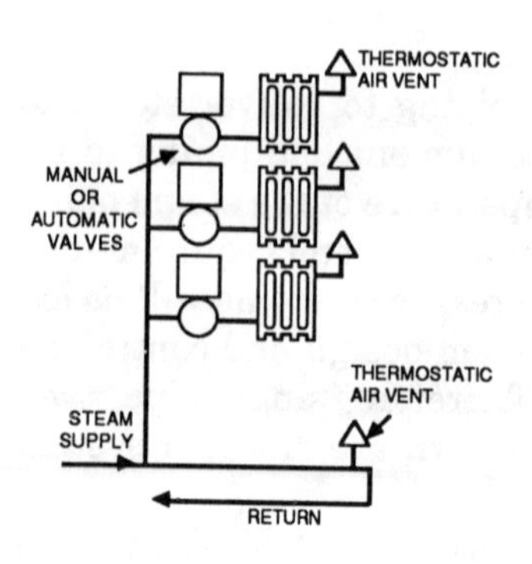

Figure 14.17 One-pipe steam system. (*Courtesy of Honeywell, 77-1132 Fig. 96.*)

tion. Figure 14.17 shows an example of this type of system supplying radiators. Note that no traps are needed but there are air vents. The condensate is returned to the boiler by gravity so the piping must be graded. The run-out piping and each radiator valve must be large sized, and the valve must be opened in a two-position manner to allow steam to flow to the radiator at the same time condensate flows backward through the same pipe and valve. This system is old fashioned and is only suitable for small systems.

Two-pipe gravity return systems have steam supply lines separated from condensate return lines by traps that keep steam and condensate separated. Figure 14.18 shows this type of system. The automatic valves may be modulating and sized for the steam load. The loads in a two-pipe system may be radiator, coils, or steam-to-water converters. Because of the gravity return nature of this system, the return lines need to be graded to either a receiver or the boiler. A receiver is used on an open return system where the low point of the gravity return is below the water line of the boiler. In this case, there must be a condensate return pump that pumps condensate back into the boiler. This is normally operated by a float switch in the receiver tank.

A two-pipe vacuum system has a vacuum pump added to the system to give added pressure differential to help in the control of steam dis-

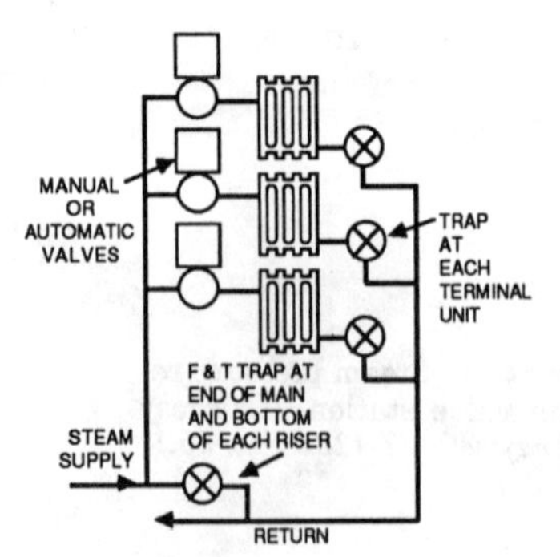

Figure 14.18 Two-pipe gravity return system. (*Courtesy of Honeywell, 77-1132 Fig. 97.*)

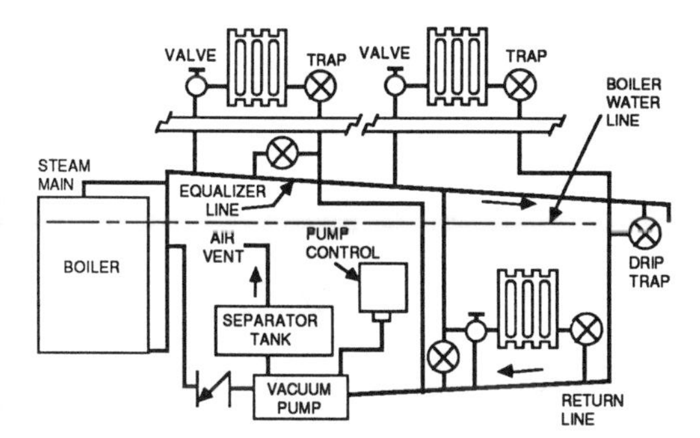

Figure 14.19 Two-pipe vacuum return system. (*Courtesy of Honeywell, 77-1132 Fig. 99.*)

tribution. The amount of vacuum can also be controlled on some systems to vary the temperature of the steam when the system pressure goes into the vacuum range. Figure 14.19 shows a vacuum system. If the pump control set point is set to more vacuum at lower loads, the system is a variable vacuum system. In this case, the entire system is normally controlled as a single zone by resetting the amount of vacuum according to load. When there is final control of each zone by individual valves, the system is typically not operated at vacuum pressures in the supply side of the system. This is because normal control valves are not rated to operate under vacuum conditions and vacuum leaks would deteriorate system performance.

Chapter

15

Supervisory Control and Total System Optimization

Definition and Historical Background

Supervisory control is the total system monitoring and overall control of the local subsystems. Overall control includes functions such as manual overrides, optimizing modification of discharge local loop set points, optimizing start–stop of subsystems, and controlled interaction between subsystems. Total system monitoring includes functions such as alarm reporting, energy measurement and calculation, logs, and trend reports.

Supervisory control has been given different names at different times, depending upon the situation and the functions to be accomplished. It has been called a building automation system (BAS) when the primary focus was on automating as much as possible to save labor. It has been called an energy monitoring and control system (EMCS) when the focus was on saving energy by both automatic control and manual control with the aid of energy monitoring. It has been called an energy management system (EMS) when the focus was on saving energy by specific automatic control programs. It has been called a facility management system (FMS) when the scope of control went beyond HVAC control and/or beyond a single building, such as including fire, security, or manufacturing systems. Since this book focuses on HVAC control, the term *building automation system* will be the one most frequently used.

Supervisory control started in the 1950s with hard-wired centralization of start–stop switches, sensor readouts, and remote controller set points. These were usually accompanied by graphic representations of the systems being controlled and were called supervisory data centers. The next step in the early 1960s was the multiplexing of the wiring from remote devices back to the central panel. This multiplexing time-shared the function wires for start–stop control, sensor measurement, and control point adjustments to connect to one system at a time. The switching of the common function wires was from a decimal matrix of wires that controlled the switching relays connecting the function wires. The graphic for a connected system was displayed from a photographic slide projector that selected the picture of the connected system. This development reduced the number of wires needed and the space needed to store and display graphic representations of systems.

The next major step as we went from the 1960s to the 1970s was to use solid-state electronics to provide serial data communications over two wires and the digital logic to decode these communications and provide control logic. During the 1970s rapid development of solid-state digital electronic components led to continued enhancement of these systems, which were essentially custom-designed central processors. Also during the 1970s energy management programs were developed and added to these systems to fill the needs for energy conservation brought about by the oil embargo and energy crises. During this period, minicomputers were coming into use on very large projects with many sensing and control points and the need for energy management programs and better operator interface and management report capabilities.

During the 1980s the development of microprocessors led to the development of smart remote panels. These remote microprocessors were used to gather data as well as for direct digital control (see Chapter 9). A short time later, the personal computer and its third-party software capabilities came into use as the central processor, with emphasis on enhancing the friendliness and flexibility of the operator interface and graphic displays. At the same time that smart remotes were given DDC capability, many of the energy management programs previously implemented in the central processor were implemented in the remote panels. This distribution of intelligence led to the use of peer-to-peer communications, where one controller could talk to another without going through a central processor. In addition, the use of a hierarchal architecture on large projects came into use so that a single processor or controller could communicate on a separate bus with a number of subordinate controllers. The most recent developments have been of small-size zone controllers that are a whole new bottom level of subordinate controllers in this hierarchal architecture of a BAS.

This history of the development of building automation systems is based largely on the actions of the major HVAC control companies who have sold complete systems on an installed basis. At the same time, there have been niche products that provided some of this functionality and were sold on other than an installed basis. There has also been a move among major HVAC system manufacturers to provide their systems with controls, including versions of building automation systems.

System Configurations and Communications

The almost universal use of distributed intelligence in the form of smart remote panels has led to a hierarchical structure for large-size systems. Figure 15.1 shows a hierarchy of function and different size processors to accomplish these functions. The lower end of the system presumes two sizes of microprocessor controllers, called system level controllers and below that zone level controllers. As the names imply, the application fits the HVAC system hierarchy of a fan system feeding a number of zones of airflow control. This fits both the physical and functional relationships between the fan system controller and the zone controllers.

All of the zones supplied by the fan system can communicate back to the fan system controller most directly and simply by a master–slave, poll–response protocol. The information regarding all zone loads is then available to support dynamic load reset strategies and to provide complete load information for the fan system.

The communication between all system level controllers is best done by peer-to-peer communications, which support a variety of func-

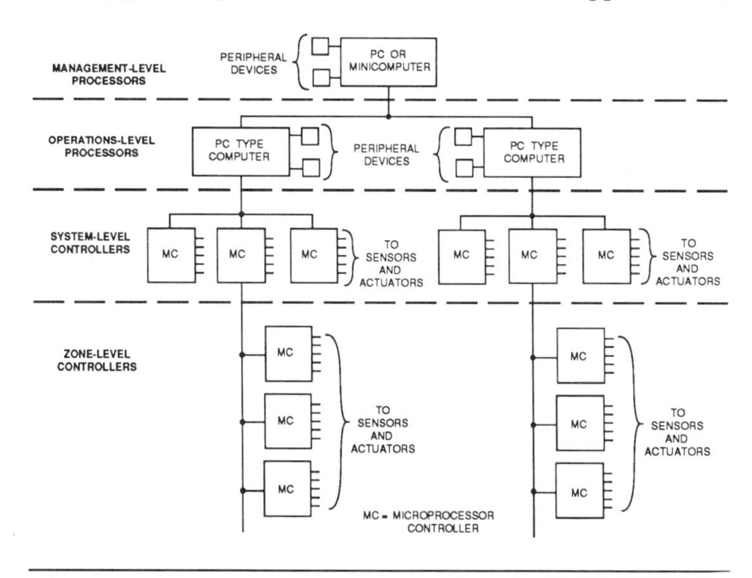

Figure 15.1 Hierarchical configuration of a building automation system. (*Courtesy of Honeywell, 77-1124 Fig. 4 c1855.*)

tions by different types of HVAC subsystems. The functional distinctions at the upper end of the system are primarily in how information is manipulated. The operational level deals with real time information in the short term to guide ongoing operations, while the management level deals with summaries of information for different time periods and compares them to an historical base. The levels shown represent a distinction in function that might be done by different processors on a large job or by a common processor on a smaller job. This means that a hierarchy of function does not necessarily require a hierarchy of hardware. The flexibility of distributed processing provides this capability, but its practical use depends upon software to fill the needs of a particular system. These needs can be represented by the required control, the data, and the data manipulation and report formats.

The system configuration chosen should support the provision of closed-loop control with sensors and actuators in the same processor panel. Supervisory control programs should have direct communication with the subsystems they are controlling, and operational control centers should have frequent updates on data by communications with speeds appropriate to the application. That is, critical alarms should be reported fast enough to avoid problems. The operational control processors should have the summary data on all controlled subsystems. That is the set point, actual value, and controlled actuator position of each control loop. In addition, trend logs should be available to give periodic measurement of controlled results. Cumulative measurements of energy used should be available when energy management is an operational responsibility.

The highest level of the functional hierarchy is the management level that includes summaries of the operational information over periods of time. The requirements of the upper part of the system, which is the operator interface, include not only prompt reporting of measured conditions but also historical summaries and averages over extended periods of time. Therefore, the system configuration and communications must provide adequate point data, speed of communication, and historical storage space to support the project. The software must provide the data manipulation, operational displays, and management report formats required. Benchmarks to be determined are speed of response for critical alarms, point capacity for current conditions, and number of historical values to be saved for specific time periods.

Note that the complete hierarchy is not necessary for smaller sized projects. Any level or combination of levels can be implemented to fit a given situation. Also, growth over time can add to both size and number of levels.

Proprietary Versus Open Systems

The fact that systems developed by the major manufacturers are proprietary in nature and sold as complete systems has led to a perceived difficulty in expanding a facility over the years. This difficulty is both in a technical sense and in a business sense. In a technical sense, the challenge of using new technology as it becomes available and bridging to existing systems has led to the addition of translators between old and new systems at high costs. In a business sense, the incompatibility among proprietary systems has led to a perception by some owners of high add-on costs because of a noncompetitive situation. Other owners have avoided these situations by requiring that original bids include unit pricing of add-ons.

The perception of high add-on costs plus the proliferation of manufacturers and generations of system designs have led to an industry effort to move toward standards for an open system design. The ultimate objective of an open system design is to have interoperability of equipment from different vendors. The movement was started by the formation of an ASHRAE Standards Project Committee SPC-135 in 1987. The name of the SPC is EMCS Communications Messaging; the name of the proposed standard for public review in late 1991 is BACnet: A Data Communication Protocol for Building Automation and Control Networks. When approved, this standard will serve as a design guide to equipment designers and suppliers so as to enable different devices in a system to communicate with each other.

BACnet defines EMCS specific data structures, called *objects,* as well as the operations that can be performed on the objects. These operations are described as *services* and result in messages that are sent between devices. The BACnet standard allows four different ways to communicate between devices. These ways to communicate are combinations of data link protocols and physical layer requirements that are existing or proposed standards that cover a range of communication performance and cost. The BACnet standard also provides for the use of repeaters, bridges, and routers to support a variety of topology and mix of types of networks.

A standard for an open communications protocol only enables interoperability, it does not provide it. The remaining steps for interoperability are consistent, functional use of the information and operations transmitted via the messages between devices and consistency in which functions are carried out in what type of devices. This constitutes the design of the system.

The system functions are considered application functions and are outside the scope of the BACnet standard. The broad classification of application functions can be further identified by what they do in making the BAS work or providing specific types of control such as

EMS or DDC. BACnet allows a variety of system designs as long as the messages between devices conform to the BACnet standard.

The general pros and cons for each type system are as follows:

Proprietary system pros

- Systems have been in use and have a track record
- Single source responsibility
- System engineering done primarily by the vendor
- As new technology becomes available, BAS manufacturers have the freedom to implement without being tied to a standard

Proprietary system cons

- Add-ons require support from the original system vendor

Open system pros

- Flexibility of equipment selection
- Flexibility of add-ons
- Open environment sets the stage for a system design that provides interoperability of BACnet devices

Open system cons

- Requires more complete system engineering that is custom to each project rather than a manufacturer's standard design
- If mixed vendors, responsibility for total system rests with owner or system designer and prime contractor
- Performance degradation due to standard support
- Open systems do not yet exist or have little track record

The evaluation of an open system versus a proprietary system goes beyond the evaluation of one proprietary system versus another. The evaluation involves more complete system engineering, which defines not only what functions are to be done but also where and how they are done. This is especially true if a multivendor system is being planned. Ultimately, the evaluation becomes a cost–benefit comparison for the situation. The costs and benefits of the open system's flexibility are subjective judgments until such time as consulting engineers can specify open systems and manufacturers have BACnet product designs to bid with.

The first step in the specification task is defining the open system in detail for all types of communicating devices on the local area network (LAN). The next step is identifying the BAS functions required in each type of device. This includes both monitoring functions and HVAC control functions. For example, you must specify which devices provide operator interface function and which devices provide sensing, alarm monitoring, DDC control, or EMS control. Are there global or multidevice strategies such as electrical demand limit control or chilled water load reset? And what is the required interaction between devices? These steps by the consulting engineer constitute the open system design and presume knowledge of the overall BAS system design as well as the functional capability of each BACnet device in the system.

The BACnet standard does not define devices but defines 9 functional groups commonly used in providing building automation function and 20 objects that are standard data structures used in providing function. In addition, it defines the services that are the rules by which objects are created and used in providing function. It also defines 6 conformance classes that are the levels of services provided in a device. The conformance class of each required BACnet device in the system must be specified by the system designer to ensure comparable bids.

The first step for a manufacturer designing a BACnet device is identifying the BACnet objects and services required in each type of device to provide system functionality in a standard manner. A BACnet *object* is a standard data structure defined by a set of properties and data types. Examples are points, devices, tables, loops, groups, directory, calendar, command, schedules, and programs. The BACnet services required for each type of device are the programmed actions that create and use the data objects to support the required functionality of the type of device. This presumes knowledge of the device function in the total system. The BACnet standard defines the requirements of services by six separate conformance classifications. Each successive level adds functionality and has the functional services of the lower level. A specific BACnet device design would be represented by a protocol implementation conformance statement (PICS) prepared by the manufacturer to identify the BACnet options implemented in the device.

A proposed open system should be examined by the system designer-specifier as to how all control applications and all monitoring applications are accomplished. The same questions asked about where data files and specific programs are accomplished in proprietary systems should be asked for open systems. In addition, the question of how control and monitoring applications are accomplished may determine the services needed in a device.

The primary areas of coordination are in the distribution of data files and system functionality between devices in the system. Examples are where alarm monitoring is done, where command processing is done to avoid duplicate or missing functions, the location of EMS programs and DDC programs, and any custom time event programs. Some of the coordination involves information passed between these programs or duplication of commands. Each of these programs can command the same devices, and there must be a workable way of resolving contention when different programs send different commands to the same device. The BACnet way of accomplishing this requires both command priority coordination and relinquishing of commands by all programs that can command any point. The important thing in this system engineering task is that the coordination be assured to give proper total system functionality.

Possible Supervisory Functions

Supervisory functions include both operator interface functions and automatically scheduled or software program initiated commands. The supervisory nature of a function does not necessarily mean it is a centralized implementation. The automatically scheduled commands are typically implemented by a standard program that is applied by entering data files that identify specific information as to the command that is sent, where it is sent, and when it is sent. These types of programs can be identified as time event programs (TEPs), and functions done by these programs are discussed more fully later.

Other software programs that provide supervisory control commands are standard energy management system programs. The implementation of these programs is also typically accomplished by the entry of data files identifying the specifics of the application. Although some standard EMS programs are generic in their general nature, they are typically specific implementations by each manufacturer. The economics of using standard programs usually justify using a manufacturer's standard, although there may be some minor proprietary differences from one manufacturer to another. The specifics of standard EMS programs will be discussed later.

Customized Global Control Functions

Besides the standard program functions mentioned above, there are some generalized types of functions that can be classed as supervisory control. Global control functions are a class of functions that deal with multiple processor locations to accomplish a systemwide function or control strategy. An example is electric demand limit control, where electrical loads throughout a building or facility are shed to maintain a limit on the total demand. Another example is the load reset of a central chilled water supply temperature just to satisfy the greatest demand fan system. This requires monitoring all fan system chilled water valve positions and is therefore global in nature.

A third example of a global control function is the use of rejected heat from a central chiller to supply heating to the fan systems while an outside air economizer cycle acts as a source of free cooling. The global strategy is to limit the amount of outside air as necessary to have enough rejected heat to maintain the heating load.

The strategy of load reset came into common use in the 1970s when energy conservation became an important consideration. It was most beneficial on fan systems providing simultaneous heating and cooling, such as reheat systems and hot and cold deck fan systems. There is the potential to use a load reset strategy whenever the energy put into a supply system can be reduced and still meet load needs. This is true not only for supply temperature levels but also for supply pressure levels in variable volume fan and pumping systems. These strategies that respond to changes in load have also been called *dynamic control* because they deal in the dynamics of load changes. Whatever they are called, they require close study of the building load dynamics and the part load characteristics of all energy-using equipment to meet the changing loads.

For instance, on a VAV fan system reset of discharge temperature may use more fan energy than is saved at the chiller and therefore waste energy rather than save it. It may be more beneficial to reset the fan supply pressure just to satisfy the required static pressure at the end of the duct runs. Whatever the load reset or dynamic control strategy, it is probably global in nature and is implemented by a custom control program that uses information from a number of remote points. Therefore, custom programming capability should be looked upon as another means of providing supervisory control.

Standard Energy Management Functions

Optimum start and stop

Optimum start and stop is one of the most commonly used standard EMS programs. Here, the outdoor air and inside space temperatures are used to calculate the lead time necessary to condition a space to achieve comfort conditions by the time of occupancy. There are several variations of the optimum start program. The simplest is the use of summer and winter constants that are manually tuned until the lead time is accurate. Another variation is a form of adaptive program that uses the measured experiences of start-up to adjust the summer and winter multiplier constants automatically. The same principles are used for the optimum stop program, except the lead time on stopping should let the space temperature drift to the end of the comfort zone at the time occupancy in the building ends. Typical data files for these standard programs include design of outdoor and inside temperatures for summer and winter, times of the start and end of occupancy, multipliers for summer and winter, and addresses of the sensors that measure inside and outdoor temperatures.

Demand control

Demand control is a program that measures total building demand and the end of each demand period used by the electric utility. It projects the electrical demand in each demand period and calculates how much must be shed to maintain the set point of the demand limit. The demand program commands loads off and on to shed the net amount necessary while observing the maximum off times where specified for some loads. The loads to be controlled are identified in data files, which include the command addresses, the size of load that is controlled, maximum and minimum off times, and in some cases the relative priority.

The implementations of demand control by different manufacturers can include some proprietary features. They may have several methods of projecting the usage to the end of the period. Some utilities record a series of instantaneous demand readings, then use a 15-min time frame to scan the recording for the highest usage in any contiguous 15-min interval. This is called a *sliding window* method of metering; there is a calculation method by the same name that keeps projecting out 15 min to calculate what must be shed.

A more typical type of utility metering is using a fixed time period, such as 15 or 30 min, and measuring the energy used in that time period as the measurement of demand. In this situation, the typical demand control program uses a measured pulse sent from the utility meter at the end of each period to synchronize the demand control program with the utility meter. Then the projected usage is specific to each period in real time.

Whatever method of metering is used, the demand calculation must match it. The usual way to accomplish this is by using a data file that specifies the selection of the correct algorithm, the addresses of the utility meter counter, and, if appropriate, the address of the end of the period pulse relay and the length of the demand period. There can be times when all loads that can be shed are shed and the present demand set point cannot be maintained. At these times, the demand control program should send an alarm message to alert an operator so he or she can take manual action if so desired. Although demand control is by its nature a global program, the use of distributed intelligence has led to some schemes of demand control that are of a distributed type and pass shed values from controller to controller. In this way, each individual DDC controller sheds what it can and passes on a value of what remains to be shed.

Duty cycle was one of the first EMS programs to be widely implemented in the 1970s when the energy crisis hit. At that time it was relatively easy to implement from existing centralized start–stop capabilities. Since then it has decreased in usage as better EMS strategies have evolved. The increased use of VAV systems is not good for duty cycling of fans because the fans go to higher volumes when they are turned back on. Also, load reset became a better strategy on reheat and double-duct systems. There are, however, still valid applications for duty cycling, such as constant volume, single-zone fan systems. Duty cycling was originally done at central processors, but now with distributed intelligence, it is more typically done at remote panels. As a standard EMS program, it is implemented by a data file that identifies points to be cycled, the on–off periods, start–stop times, and any temperature compensation sensors and ranges.

Load reset and zero energy band

Load reset and zero energy band are two programs that have some common functionality. The load reset program is simply resetting a supply temperature to a more moderate value that just satisfies the greatest demand zone. The most typical applications are on cold air supply temperature in fan systems and on chilled water supply temperatures from chillers. The energy savings on fan systems are from decreased use of heat and cooling, where heating and cooling oppose each other in reheat or double-duct systems. The energy savings on chilled water systems is from the decreased refrigerant head on the compressor and from the decreased energy needed to provide a ton of cooling. The zero energy band (ZEB) program is used on fan systems that provide both heated air and cooled air. The ZEB refers to a range of space temperature where neither heating nor cooling is required. The load reset of cooling supply is from the highest space temperature related to the upper end of the ZEB range. At the same time, the load reset of the heated air is from the coolest space related to the lower end of the ZEB range. The data files that implement this program identify the addresses of zone sensors and hot and cold deck sensors and the ZEB range of temperature limits.

Night cycle and night purge

Night cycle and night purge are programs to maintain safe conditions during the unoccupied period and, in the cooling season, to use outdoor air at night when it is a valid source of cooling. The night cycle program cycles the fan on as necessary to maintain a low limit in the winter and a high limit in the summer. This program does not allow the use of outdoor air, as there is no need for ventilation and the heating or cooling of outside air. The night purge program compares space conditions of temperature and moisture content to outside air conditions and uses outside air to do free cooling when necessary and possible.

Enthalpy control

Enthalpy control calculates enthalpy (total heat content) of both outside air and return air and, when in a cooling mode, uses the airstream with the least enthalpy to minimize cooling costs. The instrumentation to measure enthalpy directly is a wet-bulb sensor, which is relatively expensive and requires frequent maintenance. Therefore, a common way to measure enthalpy is to measure the dry-bulb temperature and relative humidity and to do the calculations or use lookup tables to convert the values to enthalpy. This conversion is typically done in a standard EMS program, although the resulting control action is implemented in the local loop control of the fan system's dampers. With DDC, the enthalpy override is a command to a relay function in the DDC control loop. This override function closes the outside air down to the minimum amount needed for required ventilation. When not overridden, the local loop control uses outside air as a source of free cooling to maintain space temperature or discharge air temperature.

Interface to Local Loop Control

Interface to local loop control is a frequent application when dealing with DDC and other programs such as EMS or smoke control programs. The nature of DDC control is that the program is reexecuted frequently and its commands are updated each time. Therefore, the most natural interface from other programs is to have their commands input to the DDC program, which then implements the overriding logic. Another approach for handling command contention is to assign a priority for each command. There are different priority schemes, some of which are proprietary in nature.

Different commanding programs can have different needs. The demand control program may need to know if a shed command has been acted upon in order to determine how much more to shed. The fan start command of a smoke-purging program would need to override the stop command of an optimum stop program, then stop when the purge cycle was ended. Then the optimum start program would need to be able to start the fan, but duty cycle or demand control could stop it and restart it. The point is that all sources of command to a point have to be considered from both a start command and a stop command requirement. Whether DDC logic or a command priority scheme is used, all command contention must be resolved.

Other interface considerations of DDC and other programs include the set points for PID loops in the DDC program that come from other programs. These are software points commanded by the other programs and used as inputs to the DDC programs. Another type of interface is when a DDC program calculates a value that becomes an input to another program. Again, the output of a software point from one program becomes the input to another program. The use of descriptive names for these software points of interface makes it easy to keep track of the functions being implemented. The interface to the operator can then show the present values and status of all parts of the programs.

Operator Interface Functions

There are two general types of operator interface functions: modifying the data files that define the system configuration and accessing the measured data contained in the system. Accessing measured data is accomplished by the following functions:

Display of logical group point lists with descriptions and current values and conditions. Commandable points are identified and commands are selectable, usually via a keyboard. See Fig. 15.2 as an

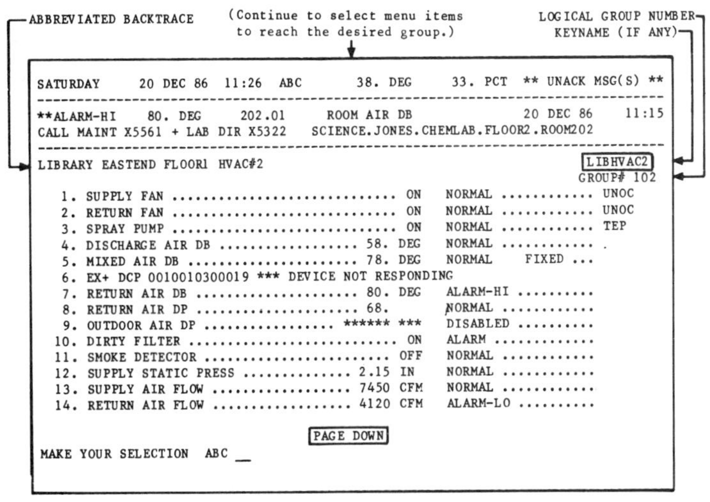

Figure 15.2 Example of logical group display. (*Courtesy of Honeywell, 74-2557 page 13.*)

example of a logical group display. Displays are presented when requested by the operator.

Display of a graphic picture of the system, with points and current values shown. Commandable points are identified and commands are selectable, usually via a keyboard or mouse device.

Alarm displays are automatically put on the cathode ray tube (CRT) screen, one at a time until each is acknowledged. Alarms and acknowledgments can also be logged onto a printer with a time stamp.

Reports print out on a time schedule or by operator request. Reports include system reports, all point logs of value and condition, alarm logs, trend logs of selected point values, and custom-defined logs.

Modification of the data files that define the system configuration includes the following operator interface functions:

1. System configuration by assignment of devices to the system and enabling of devices to the communication bus. This includes all devices that are a part of the communication network.

2. Peripheral assignments to assign the functional roles played by operator consoles and printers. These functional assignments segregate different consoles to different uses by different-type operators, such as HVAC operators and security operators.

3. Operator assignments that define operators, passwords, and access levels for each operator.

4. Access provisions that are a means of constructing site-specific menus. Typically, these menus go through several logical levels of location identity to get to each specific subsystem that is identified as a logical group of points; Fig. 15.3 shows an example. Note that the abbreviated backtrace on the logical group display in Fig. 15.2 is the path through the site menu in Fig. 15.3. Direct access may also be arranged by the assignment of key names to each logical group.

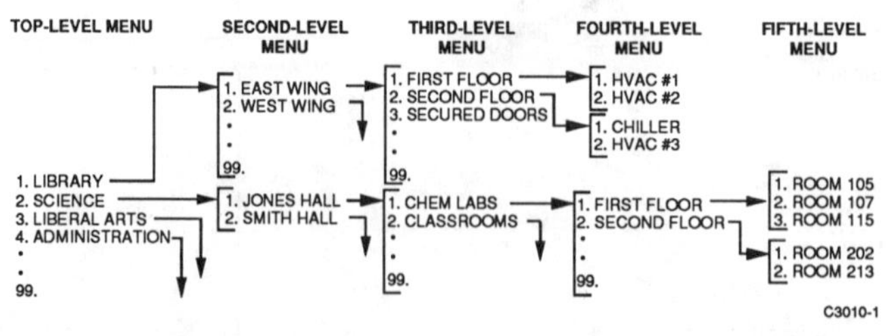

Figure 15.3 Site-specific menu penetration. (*Courtesy of Honeywell, 74-2557 page 12.*)

5. EMS, DDC, and TEP program enable status and parameter values. There are templates that are used to enable and disable programs and to change parameter values.

Custom Programming and Time Event Programming

Time event programs are standard programs with limited logic and scheduling capabilities. Typically, a TEP can send commands based on a time test or logic test of up to two conditions. When these capabilities will satisfy a need, a TEP is implemented by filling out a data file. This is simpler than implementing a custom DDC program.

The capabilities of TEPs, DDC programs, standard EMS programs, and report writing programs may not satisfy all programming needs of some projects. In this case, custom programming in a high-level language would be the solution. Such custom programming requires the software utilities to write, edit, compile, and link the program, then to generate the media on which it is stored or to load it into the processor that will execute it.

Application Engineering Functions and Tools

The application engineering functions associated with a BAS project include the following:

1. Summarizing hardware points required by location and type.

2. Sizing and locating panels to minimize costs of hardware and installation by considering trade-offs of panel costs versus home run wiring costs.

3. Providing stable control by ensuring that the sensors and actuators for closed loop control are in the same panel. If a remote measurement is to control a process, consider making that remote measurement a master that resets a submaster controller with a sensor in the same panel that has the actuator. In this way, fast stable control is achieved in the submaster control loop in the panel near the process being controlled, while a remote condition is satisfied by resetting the set point of the submaster loop. This is especially appropriate for controlling fans or pumps that must satisfy minimum pressures at the end of distribution runs but are controlled by a discharge sensor in the submaster control loop.

4. Defining locations of DDC, EMS, and TEP programs; establishing the software point requirements that associate with these pro-

grams and providing for the points that need to communicate between panels; providing for the points that are necessary to carry information to the operator interface.

5. Establishing input–output (I/O) device data files that provide the hardware information needed for microprocessor panels to process the input and output signals.

6. Establishing the point data files for all points that identify required functional information. This includes scales, engineering units, descriptions, and special functions such as alarm limits and global communications.

7. Establishing program data files such as EMS and TEP data files.

8. Writing and testing custom programs and establishing whatever data files are required for these programs.

9. Defining and establishing operator interface data files that provide system displays and graphics as required.

10. Preparing documentation that represents a submission of plans for approval. This includes point lists, sequences of operation, display groups and their point content, system display examples, and equipment specification sheets.

Specifying Building Automation Needs

The specification of requirements for a BAS should include the following items:

1. Schedules or lists of subsystems and definitions of what is to be monitored, controlled by DDC, or local hardware control commanded by supervisory commands. Sequences of control to be specified for each type of subsystem and for any EMS control specified. Examples of subsystems are fan systems, chilled water systems, and hot water systems.
2. Lists of hardware input points grouped by subsystems and any special functions noted, such as analog alarm limits, normal digital state, and global communication.
3. Lists of hardware output points grouped by subsystems and type of control action defined; for instance, analog or two position, normal position with no power, and what is controlled such as valve, damper, motor starter, or variable speed controller.
4. Definition of what types of points are to be displayed in the operator interface for each subsystem and which are to be alarmed or

commandable. These points should be defined by their function in the subsystem, such as discharge air temperature with alarm limits or discharge control set point to be commandable.

5. The required number and types of logs and reports should be defined. The required historical files should be defined by point lists, frequency of logging, and length of time to be retained.

Check Out Commissioning and Acceptance

The check out of a BAS is a matter of viewing all system displays that are supposed to be in the operator interface and confirming that the point content is correct and that the alarm and command functions are operational for the intended points. An operational check should also be made of all programs. DDC programs can be checked by simulated execution, with fixed inputs and noted outputs for proper values. EMS programs and TEP scheduled programs can be checked by logging real time execution inputs and output results. Also, an operational check of system performance should be made. This can be done by observing the speed of alarm reporting when the system has low activity, then high activity, as well as observing the time to bring up graphics and reports.

Documentation

To a large extent the modern software-based BAS is self-documenting in that the data files within the computer can be viewed. The primary hard copy documentation needed is the explanation of the data file's contents plus the listing of what system identities are used in identifying the point data files. Figure 15.4 gives an example of the data file types and their contents that are needed in a BAS. Data file formats will vary between manufacturer's systems so specific information defining data files and their contents should be a deliverable with the specific system. The operators manual for a system should have a table of contents similar to that in Fig. 15.5. The custom programming facility should have a language manual with contents similar to that in Fig. 15.6.

Data File Types

Name	Description
System	Hardware characteristics and software packages included
Processors	Processor (device) type, address, and characteristics
Engineering Units	Engineering unit names and types
Point Segregation	Segregation categories
Point Seg. Desc.	Segregation category groups
Interpolation Tables	Sensed vs. normalized (linear) values for nonlinear sensors
Ana. Input Device, Ana. Output Device, Dig. Input Device, Dig. Output Device	Nonstandard sensor and actuator devices
Matrix Boards	Matrix board parameters
Ana. Input Points, Ana. Output Points, Dig. Input Points, Dig. Output Points	R7044 point definitions (analog/digital, input/output, address, and characteristics)
Totalizer Points	R7044 totalizer point definitions
CNAP Global	Global and originator point definitions for passing information from R7044s to Control Network Automation Protocol ("C-NAP") Controllers
CNAP Peer Global	Global and originator point definitions for passing information from C-NAP Controllers to R7044s
Program Time Sched.	Occupancy and program activation schedules (start/stop times for each day of the week)
Excep. Schedules	Exception times for occupancy schedules
Time/Event Programs	TEP definitions
Time/Event Ana. Pts., Time/Event Dig. Pts.	TEP points
Time Initiators	Time initiator definition
Event Initiators	Event initiator definition
HVAC Sys. Data	Fan or Heating, Ventilating, and Air Conditioning (HVAC) system definition
HVAC Sys. Actuators	HVAC actuator list
HVAC Sys. Sensors	HVAC sensor list
Duty Cycling	Duty Cycling parameters (e.g., cycle period, maximum OFF time)
Unoccupied Period	Unoccupied Period Programs and their parameters
Optimum Stop	Optimum Stop parameters (e.g., zones and zone sensors)
Enthalpy Control	Enthalpy Control parameters (e.g., latent gain, minimum outdoor air enthalpy)
Zero Energy Band	ZEB/Load Reset parameters (e.g., band or subregion boundaries, temperature ranges)
Power Demand Control	Power Demand Control parameters (e.g., demand limit, interval)
Power Demand Loads	Power Demand loads to shed
Power Demand Total.	Power Demand totalizer list
Spare Data	Amount of space to reserve in the device for future use
Point Desc.	Point descriptors
Alarm Messages	System alarm message list
Trouble Messages	System trouble message list
Run-Time Messages	System run-time message list
Access Level Desc.	Access level descriptor list
FS90 Board	FS90 board definition
FS90 I/O Assoc.	FS90 input/output point associations
FS90 Point	FS90 point definition
FS90 Sub-device	FS90 subdevice definition, slot, and board type

Figure 15.4 Date file types. (*Courtesy of Honeywell, 74-3538 pages 6 and 7.*)

TABLE OF CONTENTS

INTRODUCTION 1
Organization of This Manual 1
System Architecture 2
Micro Central Architecture 2
Networked Architecture 2
Sample Configurations 3
Applicable Literature 5

GENERAL INFORMATION 6
Screen 6
Keyboards 7
Standard Keyboard 7
F1-F10 Keyboard 8
ADM 12 Keyboard 9
Error Messages 10

ACCESS 12
Sign-On/Sign-Off 12
Window Switching 13
Data Access Keys 13
Site Menus 14
Keynames 14

POINT DATA (DATA ACCESS Key) 14
Point Groups 14
Direct Access to Point Groups 17
Commands 18
Override Commands 19
Fixed Points 19
Alarms 20
Alarm Display 20
Multiple Alarm Conditions 20
Alarm Acknowledge 20
Alarm Classes 21
Alarms in Dialup Systems 22
Graphics 22

TEMPLATE OPERATIONS 23
Template Format 23
Automatic Data Display 25
Menu Path Indicator 25
Template Data Entry Example 25

REPORTS (REPORTS Key) 27
Report Requests 27
Report Samples 28
Scheduling Reports 34
Trend Reports 34
Standard Reports 36
Trend Setup/Enable 37
Trend Setup 37
Trend Enable/Disable 38

MODIFY SYSTEM DATA (MODIFY Key) 39
Operator Assignments 40
Operator Passwords 40
Operator Access Levels 41
Operator Auto Signoff 42
Operator Segregation 42
Peripheral Assignments 44
Console Assignment 45

iii 74-2557

Figure 15.5 Contents of an operator's manual. (*Courtesy of Honeywell, 74-2557 pages iii, iv, and v.*)

Console Backup Assignment 47
Console Segregation 48
Console Header Points 49
Printer Assignment 49
Printer Backup Assignment 52
Printer Segregation 52
Default Printer Assignment 53
Log of Operator Changes 54
System Configuration/Diagnostics 55
Device Enable/Disable 55
Device Assignment 56
Device Diagnostics/Init. 57
Application Command Trace 59
Point Command Trace 60
Portable Terminal Override 60
Board/Port Configuration 61
Comm. Bus Configuration 64
System Text 65
Point Descriptors 66
Alarm Messages 66
Run Time Messages 67
Trouble Messages 67
Segregation Descriptors 68
Access Level Descriptors 68
Time/Schedule Control 69
Time Date Set 69
Time/Occupancy Schedules 69
Exception/Holiday Schedules 71
Daylight Savings Schedule 71
Point Monitoring/Control 72
System Point Enable/Disable 72
Analog Alarm/Warning Limits 73
Analog Point Calibration 74
Digital Point Runtimes 75
Runtime Enable/Disable 76
Point Alarm Lockout 76
Point Segregation Table 77
Physical Address Display 79
Fixed Mode Set/Reset 79
Time/Event Program Control 80
TEP Enable/Disable 80
TEP Initiation 81
Central TEP List Assignment 81
Central TEP Mode Parameters 82
Central TEP Analog Point Lists 85
Central TEP Digital Point Lists 85
Central TEP Time Initiators 86
Central TEP Event Initiators 87
Applications Control 89
Distributed EMS/DDC Parameters 89
Distrib. EMS/DDC Enable/Disable 91
Comfort Limit Setup 92

MISCELLANEOUS OPERATIONS 93
Collect Historical Data 93
Save/Restore Data Files 93
Compile DeltaNet Pascal Programs 93

DEFINE KEYNAMES 94
Keyname Concepts 94
Keyname Processor 94
Access to Keyname Processor 94
Keyname Command Format 94
Keywords 95

Figure 15.5 (*Continued*)

Predefined Keynames 96
Keyname/Keyword Reports 97
Keyname Names 97
Keyname Help 97
Escape from Keyname Command 97
Keyname Commands 98
Add/Insert 98
ADD/INS KEYN 98
ADD/INS MENU 100
ADD/INS GRP 102
ADD/INS PNT 103
Change 104
CHG KEYN 104
CHG MENU 104
CHG GRP 104
CHG PNT 104
Delete 105
DEL KEYN 105
DEL MENU 105
DEL GRP 105
DEL PNT 105
Rename 106
RENA KEYN 106
RENA KEYW 106
RENA MENU 106
RENA GRP 106
Issuing Prompts 108
Mail 111
Keyname Miscellaneous Notes 112
ENTY Command 112
Keyname Main Help Message 112
Specifying Two Choices in a Prompt 112
TEP-Activated Keynames 112

APPENDIX A: REMOTE DIALIN CAPABILITIES 113
Dialin to Micro Central PC 113
PC Dialin Hardware Configuration 113
PC Dialin Operation Overview 115
Set Up Hayes Modem 116
Set Up Ports 116
Micro Central Dialin Port Setup 116
Remote Dialin Port Setup 117
Enter Terminal Emulation Mode on PC 117
Dial into Micro Central (with Concurrent DOS) 118
Sign Off and Disconnect (with Concurrent DOS) 118
Dialin to Gateway 119
Gateway Dialin Hardware Configuration 119
Gateway Dialin Software Configuration 120
Operation—CRT or PC with Concurrent DOS 120
Step 1: Configure Gateway Modems 121
Step 2: Connect Cables 121
Step 3: Edit DMTECOM.DAT 121
Step 4: Set Up PC/CRT Modem 122
Step 5: Dial Gateway 122
Step 6: Perform Desired Commands 123
Step 7: Disconnect 123
PHONE Command Batch Files 123
PC with MS-DOS and Mirror II 124
Install Mirror II 124
Dial into Micro Central (with Mirror II) 125
Sign Off and Disconnect (with Mirror II) 125

APPENDIX B: POINT INFORMATION REPORT DATA 126

v 74-2557

Figure 15.5 *(Continued)*

TABLE OF CONTENTS

INTRODUCTION 1
Applicable Literature 1
Programming Steps 1
Definitions 2

DDC PROGRAM DESIGN AND OPERATION 3
General 3
Typical Control Loop 3
Analysis of Application Control Requirements 5
Program Design 6
Overview of Pascal Statements 9
Program Structure 9
Statement 1: Include File 10
Statements 2 and 3: Comment Lines 10
Statement 4: VAR Work Area 11
Statement 5: Main Procedure 11
Statements 6 and 7: Procedure Declaration 11
Statements 8 through 10: Source Code 12
Statement 11 CPROG Statement 12
Timing Guidelines 12

LANGUAGE STATEMENTS 13
Notation Conventions 13
Naming Conventions 13
Indention 14
INCLUDE Statement 14
Comment Lines { } 14
Constants TRUE/FALSE 15
Data Types 15
BOOLEAN 15
SHORT_INTEGER, WORD_INTEGER, INTEGER 16
REAL 16
RECORD 16
Point Records 17
Status and Flag Bits 18
Residual Priority 18
Totalizer Reset 19
Output Point States 19
Miscellaneous Information 19
Pseudopoints 19
Data Declaration Statements 20
CONST 20
TYPE 21
VAR 22
Control Statements 23
PROCEDURE 23
FUNCTION 25
BEGIN, END 27
Assignment (:=) 27
IF-THEN-ELSE 28
CPROG 29
Operators 30
Built-In Functions 31
DDC Operators 31

iii 74-5566

Figure 15.6 Contents of a language manual. (*Courtesy of Honeywell, 74-5566 pages iii and iv.*)

TABLE OF CONTENTS (Continued)

Analog-Controlled Analog Output—ACAO 32
Analog-Controlled Digital Output—ACDO 34
Adaptive Control—ADP 37
Send a Command to a Point—COMMAND 39
Get Date—DATE 41
Digital-Controlled Analog Output—DCAO 41
Set Delay Time—DELAY 44
Find Maximum—MAX2 - MAX7 46
Find Minimum—MIN2 - MIN7 47
Proportional-Integral-Derivative Control—PID 48
Translate Input from One Scale to Another—RATIO 52
Reversing—REV 54
Sequence—SEQ 55
Test Point State—STATE 57
Get Time of Day—TIMEDAY 59
Control a Counter—TIMER 61

DDC SIGNAL PROGRESSION 62
General 62
PID Design Guidelines 65
DDC Command and Residual Priorities 68

STANDARD PROGRAMS 68
BONES—Standard Skeleton Program 68
STD1—Discharge Control Heating Coil 68
STD2—Discharge Control Cooling Coil with Damper Sequencing 69
STD3—Discharge Control of Serial Heating and Cooling Coils with Damper Sequencing 69
STD4—Mixed Air Control of Dampers 69
STD5—Mixed Air Control of Dampers with Economizer 70
STD6—Discharge Control of Hot Water Converter with Outdoor Air Reset 70
STD7—Return Airflow Control Fan Tracking 70
STD8—Supply Air Static Pressure Control 70
STD9—Ventilating Airflow Control Constant Volume 71
STD10—Space Control of Serial Heating and Cooling Coils with Room Humidity Override Plus Reheat 71
STD11—Space Control of Heating with Discharge Air 71
STD12—Discharge Control Heating Coil Using ADP 71
STD13—Discharge Control Cooling Coil Using ADP 72
STD14—Mixed Air Control of Dampers Using ADP 72
STD15—Mixed Air Control of Dampers with Economizer Using ADP 72
STD16—Discharge Control Hot Water Converter with Outdoor Air Reset Using ADP 73
STD17—Power Demand Logs 73
Sample Program 74

COMPILER MESSAGES 74

QUICK REFERENCE 74
Formats 74
Reserved Words 77

Figure 15.6 *(Continued)*

Chapter

16

Operating and Maintaining Control and HVAC Systems

This chapter will delve into the operations and maintenance of all types of HVAC systems. It will include the systems as well as the controls involved with the systems. Some of the systems mentioned are not in vogue today. But since many of them are in the field, it is imperative that their owners and operators be aware of the problems that can develop if they are not operated correctly or maintained properly.

The discussions of the operation of systems will center on energy conservation and safe operation and maintenance. The systems discussed will be both primary and terminal systems and, in some cases, will be combined in one area. The chapter will also include the state of the art as far as supervisory control systems is concerned, both from and operation and maintenance viewpoints.

The maintenance of all types of control systems will be a major portion of the chapter. The discussions will involve all types of controls installed today as well as those that have been installed over the past years. Finally, troubleshooting will be covered, concentrating on the proper and fastest methods.

Operating and Maintaining Primary Systems and Controls

Hot water and chilled water systems are in use in many areas of the country and, in many cases, are replacing outdated steam and DX systems. The days of the gravity hot water system, for example, are gone, and all hot water and chilled water systems now use pumps. The original hot water systems that involved radiation and boilers were gravity types. That is to say, the hot water circulated only because it rose in the pipes and the cold water fell back to the boilers. In today's modern systems, the pumps are the heart of the systems. With the boilers and chillers or heat exchangers and piping, the pumps present one of the largest possibility for problems in hot water and chilled water systems. Pumps come in many configurations from plain in-line circulators to the split-cases and special nonclog types. In all cases, they are operated by electric motors or other standard drivers. If the pump is connected to a driver that is not shipped with the pump from the manufacturer, the r/min of the pump is important and must be taken into account when operating the pump in the system.

Other factors that need to be addressed when selecting and operating pumps in a hot water or chilled water system include the following:

1. Maximum and minimum flow
2. Net pump suction head
3. Use of intermittent or continuous flow
4. Operating pressures and temperatures
5. Ambient temperatures
6. Number of pumps needed (whether series or parallel configuration)
7. Voltage
8. Water chemistry
9. Expansion tank location and removal of entrained air

When operating water pumps in water systems, one of the most important issues involve the changes in head pressures that are the result of the devices that are opened or closed as the system is operated. As an example, if after the system is started it only sees three-way and not shut off valves, the pressures and pressure drops encountered would be stable and could be accounted for. If the valves and other devices are of the shut-off types, the pressures and pressure drops in the system will vary and the pumps and systems must be designed to allow for the varying pressures and pressure drops. A study of the pump curves in the *ASHRAE Equipment Handbook* will show that as the pressures in the system rise, the system curve will rise up on the pump curve and the power requirements will change drastically.

Systems that have only three-way valves for control and a fixed head pressure and pressure drop do not need to be concerned with multiple pumps in parallel, multiple pumps in series, multispeed pumps, or variable speed pumps. The economics of the systems that are involved will determine the type of pump to use. The piping arrangements for series or parallel pumps can be obtained from the manufacturer's literature or the ASHRAE handbooks. The methods of controlling pumps in those situations again depend upon the type of system involved. Basically, however, differential and positive pressure controllers that sense the change in system pressure operate the variable speed or two-speed systems, as well as the two-pump setups. The important item to remember when operating pumps is that they use energy. Therefore, if they are pumping at a constant rate when they are not needed, when the loads are reduced, energy is wasted.

One other item that needs attention here is the fact that sometimes the selected pumps are not the correct ones for the medium being pumped. An example is a pump with impellers made of a material that will not stand up to the chemical composition of the fluid being circulated. Another example is the situation whereby chemicals are added to treat the water without proper information as to the composition of the pump impeller materials. The operation of pumps in any hydronic system involves careful study as to the intent of the system and the reason for circulating the mediums (water). A study of the flow diagrams to understand the system is a must. Without that information, the probability of loss of control or damage to the system and/or the pump is likely.

Pumps

As far as maintenance of pumps is concerned, remember that in many cases the pumps have seals that prevent the fluid from leaking around the shaft of the pump in the center of the impeller. Those seals need attention and need to be changed or repacked in accordance with the manufacturer's recommendations. In many cases, the pump impeller, due to unusual low-suction pressures or flow problems, can cavitate, which in turn can cause damage to the pump impeller. When that happens, there is usually a significant change in noise level. Damage to the impeller can result; the pump needs to be opened up to check for damage, and the cause of the cavitation must be corrected. This is usually because the net pump suction head was not maintained, or the system was "air-bound." The bearings of the pump need to be packed with grease in some cases or oiled in accordance with the manufacturer's recommendations. The maintenance of the bearings of the prime mover (motor) should also be of concern to the operator, and a regular schedule to service those items needs to be set up.

Boilers

The operation and maintenance of boilers and chillers will be discussed here, as they are reasons we have piping systems and pumps that move the water in the piping. The discussion of boilers will be limited to hot water boilers, since the operation of steam boilers in a hot water system is merely to provide steam for a heat exchanger. Steam boilers will be discussed in the section on steam distribution systems.

All boilers shipped today for heating hot water in a hydronic system are either a water-tube or a fire-tube type. For water-tube boilers, the water is in tubes with the fire around the set of tubes; for fire-tube boilers, the opposite is true. At normal pressures that do not generally exceed atmospheric pressures, hot water boilers provide hot water at temperatures from 100°F to about 200°F. Some systems use high-pressure–high-temperature hot water (up to 400°F), but they are the exception.

Hot water boilers are shipped from the manufacturer with all the control and *safety* devices approved by the various agencies, such as the NFPA and American Society of Mechanical Engineers (ASME), and they should *never* be compromised. For example, if the controls and safety features are tapered with when natural gas is in the fuel for a boiler, the operation of the system is akin to playing with a bomb. When in doubt, a qualified, certified mechanic or technician should be called in to analyze a problem before allowing the boiler to continue to operate in an unsafe manner. Boilers are fired with all types of fuels; but, again, they are designed for certain fuels, which must be used to operate the boiler safely. Changing fuels without proper consultation with the manufacturer can be disastrous. All of the safety devices should be tested periodically in accordance with the manufacturer's recommendations, and records of the tests should be kept.

In some cases, a manufacturer may recommend that the boiler control system not be the type that controls the boiler in accordance with outside temperature. That is to say, they may not recommend a lower water temperature for the boiler when the outside temperature is mild. In that case, the controls may have to fire the boiler at a normal high temperature and some other device such as a three-way mixing valve be used to vary the water temperature for the system in accordance with outside temperature. Furthermore, the use of those systems of control may cause the lowering of the flue gas temperatures in the boiler, which can cause acidic deposits that are damaging to the internal surfaces and to the flue. The bottom line is that boilers should never be operated differently from the manufacturer's or insurer's (IRM Factor Mutual, etc.) recommendations.

Maintenance items for boilers include oiling the motors used in induced and forced draft fans, paying attention to fans and liners, which may need to be replaced as the boilers get older, and checking the inside of a boiler in the summer when it is normally shut down for unusual conditions that may cause a shutdown at the height of the heating season. If a central monitoring system is not available, a visual check of the temperatures and pressures at the boiler should be a daily task.

Chillers

Chillers that provide chilled water for a hydronic system for air conditioning, although not dangerous like boilers, do have items that need to be checked to ensure proper operation and not cause damage to the chiller or the system.

Furthermore, when chillers break down as a result of improper operation or maintenance, many building occupants may get very angry at the loss of air conditioning. Chillers come in many sizes and styles, but in the modern buildings of today, they are usually one of four types. Centrifugal, Reciprocating, Absorption, or Helical Screw.

The term *chiller* applies to the package that generally consists of the compressor, the driver for the compressor, the chiller vessel that cools the water, and the condenser that cools the refrigerant in the system. Although some engineers refer to the chiller as only the vessel in the package, for our purposes, the chiller will be the entire package.

In large tonnages, such as those systems above about 150 tons of refrigeration, the tendency has been to use centrifugal machines. Reciprocating machines use refrigerants designed for positive pressures instead of operating in a vacuum like a centrifugal. Centrifugal chillers tend to operate smoother than reciprocating packages, but they have some operational problems. Since they operate with a vacuum, they need purge systems that can remove the noncondensable gases such as air that tend to get into the system through the gaskets, bolt holes, and so on. The purge system must be kept in good working order if the chiller is to provide the capacity intended in the design. Some larger units have special hermetic motors that cannot be recycled quickly, and thus they have built-in timers. These timers that require up to 30 min of off time before they can be restarted. This fact can cause problems if the differential on the chilled water side is too close, asking the machine to have frequent starts. This is particularly true if there are several machines on a system that are operated in parallel, with inlet vane control systems that typically cannot go below about 20% of capacity. This means that as the load decreases and one machine shuts off at 20% capacity, the other cannot pick up the slight increase that is going to take place when it is already at 100% capacity. The bottom line from a control standpoint is that thought must be given to the way the machines are programmed to lead lag each other.

Centrifugal chillers have a tendency to surge at low capacities; if that persists, the problems must be corrected or damage can result.

The surging is usually caused by the compressor operating to the left of the surge envelope.

The capacity control system in a centrifugal compressor consists of inlet vanes that impart a swirling motion to the gas as it enters the impeller to change the capacity of the machine. By operating those vanes, the capacity can be reduced to a level of about 20%. Those inlet vanes are controlled on start-up and while running by the inrush current to the driver. On start-up the unit can run away with itself and blow out the starter and other electrical devices, so the inrush current must be limited. This system must not be bypassed, or again serious damage will result. The sequence involved in the start-up of a centrifugal packaged chiller, such as having the chilled water pump running and having the condenser water pump running with proven flow, must be followed to the letter or damage will result. The operation of this type of package must be left to qualified personnel, who have had the appropriate training and who have read the manufacturer's instructions.

The lubrication instructions from the manufacturer also need to be followed; they will depend upon whether or not the unit is a hermetic or an open machine. The manufacturer will provide a recommended schedule of maintenance for centrifugal chillers, even to the periods suggested for complete tear down. These schedules should be followed, and the tubes of the chiller and condenser vessels should be routed and cleaned on an annual or biannual basis as recommended. In addition, the megging procedure for the windings of the hermetic motor on an electric-driven chiller should be followed.

There is nothing better than a good preventative maintenance (PM) schedule, where centrifugal chillers are concerned. I have seen 500-ton chillers that have been on a good PM schedule in operation and running like a Swiss watch after 30 years. Pages of text have been written about the proper maintenance of centrifugal compressors by organizations such as Building Owners and Managers Institute (BOMI) and ASHRAE. To list all that information here would be redundant. Suffice it to say that the best place to get the proper information is from the manufacturer's literature provided with the chiller.

As indicated earlier, the smaller chiller packages tend to be reciprocating types that have one or more compressors to cool the chilled water. These units use positive displacement compressors, and, as such, they use a different refrigerant than the centrifugals. There is no need for a purge system as the gas tends to leak out, not in. Like all chiller packages that have compressors connected to them, however, they must be operated in accordance with standard procedures.

First, these machines have a crankcase that is charged with oil and that is the most important part of these units. Without proper attention to the oil pressures in the system, the compressor will be scrambled. All units come with controls that will operate to shut the unit down if the oil is not being provided in sufficient pressure and quantity to the parts that need the lubrication. Reciprocating compressors also need the proper devices and controls to see that they do not have to pump *liquid,* since they are designed only to compress gas. Again, if these devices are not present or are compromised, damage can result. Many piston heads on reciprocating compressors have been severely damaged when a slug of liquid entered the suction valves. The sound that develops at that point is unmistakable, and action needs to be taken immediately. The important point to remember here again is that the controls must not be bypassed in the operation of the chillers. The high- as well as low-pressure switches are there for a reason and need to be checked and to be in good operating condition at all times.

Most systems also have ASME and other agency approved blowout plugs that will blow at a very high pressure to prevent damage should the need arise. If they are needed and actually operate, the cause must be investigated and corrected and the plugs replaced. The maintenance of these types of chillers follows the same pattern as that for centrifugal units, except here the proper operation depends on the oil in the crankcase and the PM schedule recommended by the manufacturer needs to be followed to the letter. The tear down of the units needs to take place as recommended, and all parts such as piston rings, and bearings, need to be replaced when they wear out. Often it is better to hire qualified people from the manufacturer's service group to do that kind of maintenance.

Helical screw compressor chiller packages are becoming more prevalent in today's HVAC market, as they are positive displacement machines that use less space per ton than centrifugal machines and they have as good a system of capacity control as centrifugal systems. They too, however, must be operated with all of the safety devices in place and in accordance with the manufacturer's recommendations.

Oil is a critical part of the screw compressor, as it is injected during the compression cycle to seal, cool, and lubricate the helical rotors. The amounts injected are large and must be removed from the refrigerant before oil enters the system and coats all the surfaces. For that reason, the oil removal and capturing system must be operated at all times the unit is running.

The safety controls are also part of the package, just as with the reciprocating and centrifugal packages. Devices such as high- and low-temperature cutouts and high-pressure cutouts are all part and parcel of the chiller package. Maintenance is a similar problem as with the reciprocating and centrifugal units, and the manufacturer's recommendations should be followed. The return of the oil to the compressor as a result of the oil injection system needs proper maintenance of the controls that assure that return.

Absorption chillers are a breed unto themselves. They use a salt such as lithium–bromide that has an affinity for moisture to create the chilled water used in the system. The moisture-ladened salt is heated to drive off the moisture, and the process is repeated. Although this is an oversimplified explanation of the units, the important point to remember is that there is no compressor and the units require a source of heat such as steam, hot water, or a gas or oil flame. The controls furnished with the units are designed to see that the salts do not crystallize and prevent the unit from operating, so they must be in place and operating satisfactorily. The capacity controls consist of items that vary the heat energy input to the unit, such as a steam valve controlled by the temperature of the chilled water created by the unit. Other items are the pumps that move the mediums around the system in the package in accordance with the manufacturer's recommendations. The pumps must be operating to prevent damage to the system. They must be lubricated periodically as recommended, and the controls need to be checked to see that they are operating within the range set up by the manufacturer. Absorption units typically have few moving parts, except for the pumps and the controls that furnish the heat to the chiller.

The distribution systems involved with hot water and chilled water systems, once installed and working properly, generally do not require any particular expertise to operate them. If the operators are to get most out of the systems, however, they need to know as much as possible about the flows and capacities of the systems and the piping. They should, for example, know about the pump capacities, the temperatures of the fluids in the pipe, and the pressure drops and heads of the system. If, for instance, there are no thermometers to measure the water temperature in the system, or if the gpm flows of the pumps are not known, or if the pressure drops are not indicated by gauges installed in the piping, the operators will never be able to operate the systems at the highest efficiency.

A discussion of the various types of hot water and chilled water piping systems is not in order at this time since you can learn about those systems from other well-qualified texts. The most important things to remember here are how the controls operate the systems and the fact that compromising the *safety* controls of any of the system can cause damage to the equipment and maybe even injury to the operators. As far as maintenance of the piping systems is concerned, there is little to do once the systems are in and working. It is when they cause problems due to a lack of instrumentation that maintenance is a factor.

Steam distribution systems

Steam distribution systems, although not as common in the HVAC arena as they were a few years back, need to be mentioned since their piping and other parts are different from those in hot water systems. Remember that steam is used to give up its heat content when it condenses and changes to a liquid, called condensate. In a way it is like a refrigerant that *absorbs* heat as it changes from a liquid to a gas, except that the medium is going in the opposite direction—it changes from a gas (steam) to a liquid. Steam is easier than water to circulate in a high-rise building as there is less weight involved than with water and the height of the column of steam is not a factor as it is with water in, for example, a 40-story office building.

The pipes that carry the steam do not need to be pitched, but the pipes that carry the condensate back to the steam boiler do need to be pitched to make sure the condensate returns to be reheated and turned again into steam. In some cases, a condensate pump is used to ensure that the condensate gets back to the boiler, and it should be operated anytime the system is in operation. After a few years the condensate lines need to be checked to see if there are leaks in the fittings that can affect the ability of the condensate pump to return the condensate to the boiler, not to mention the damage that can be caused by the leaking hot condensate.

All steam systems use condensate traps. The main reason for traps is to hold the steam in the heat exchanger, such as a coil or radiator, until the heat is removed from the steam and it is turned into condensate. That is, the system is arranged so only condensate, not live steam, leaves the coil or radiator. The system should never be operated without the necessary traps.

In most cases, the largest maintenance item in any steam system is the traps. They cause 90% of the problems in old steam distribution systems, and are given the least attention by maintenance personnel. To be affective, a steam distribution system *must* have all the traps in good working order.

Primary air systems

Primary air systems consist of central station air-handling units that are generally located in major equipment rooms of commercial buildings or, in some cases, on the roofs of commercial buildings as weatherproof units that supply conditioned air to the spaces below the roof. In general, there are two classes of these units: constant volume and variable volume. They can be further classified as single-zone units, multizone units, induction air-handling units, double-duct air-handling units, and variable volume units, which can be further classified as single-duct and double-duct variable volume units. A further classification is possible on some of the units; they can be low-pressure, medium-pressure, or high-pressure units. The low-pressure units operate from ¼ to 1 in. static pressure; the medium pressure units operate from 1½ to 3 in. static pressures; the high-pressure units have been known to operate at up to 8-in. static pressure.

The operating characteristics of all the above types of units are generally the same. The units more or less consist of outdoor air, return air automatic dampers, and, in some cases, automatic exhaust dampers operated by the control system to perform a specific routine or algorithm. The other items, which depend upon the type of air-handling unit, are filters, heating and/or cooling coils, a fan to move the air through the duct work, and in severe climates that tend to be dry in the interiors a humidifier to add moisture in the winter.

The heart of the air-handling unit is the controls supplied with the unit; they are either field mounted and installed or factory mounted and shipped. The available cycles are as indicated in other chapters and should be operated in accordance with the data furnished by the controls or air-handling unit manufacturer. The starting, stopping, and operating of the units need to take into account the size of the unit, the current supplied to the motor on the fan, and the safety devices installed to protect the equipment. Items such as the freeze stat and fire stat, for example, should never be bypassed, since serious consequences can result. If a unit shuts off as a result of a freeze stat or a fire stat, the reason *must* be investigated before the unit is restarted.

The automatic control systems need to be checked by the operator from time to time. Filters need to be changed when necessary; if the systems are automatic, filters need to be investigated when they go into an alarm condition. Other safety items such as disconnect switches near the fan motors need to be installed and should never be bypassed. Guards on fan belts are there for a reason, and the fans should *never* be run with them off. If there are automatic clock systems that start and stop the fans on a daily basis, they should be installed in conjunction with hand-off-auto switches so that test positions are available and the auto position can be bypassed. Some areas in which very large fans are used will have only push-button switches so that an operator *has to be present* when the fan is started.

Several items must be considered concerning the maintenance of air-handling units. First are the controls, which will be covered in a separate section of this chapter. There is also the oiling of the electric motors on the unit. Further, the filters, along with the automatic dampers used with the unit, are a large part of any maintenance program. Sometimes pumps are used with things such as sprayed coils, and they too need attention at times. The coils themselves can act like a filter and become clogged with dirt and dust; therefore, they need maintenance if they are to perform as designed. If the units are used in conjunction with sprayed coils that have pumps and sumps, they need special maintenance because of the problems involving salts and other things in the water that cause lime deposits on all surfaces of the system. Without deionized water, the problems can be enormous to the point where some operators have elected to remove the sprayed coil systems and install dry coils. The bottom line again is that common sense as well as a good PM program as recommended by the manufacturer or the contractor and engineers who designed and installed the unit need to be followed.

Operating and Maintaining Terminal Equipment

The terminal equipment covered in this section includes items such as VAV boxes, fan-coil units, double-duct boxes, unit ventilators, through-the-wall units, closed-loop heat pumps, radiators, convectors, and induction units. Items whose characteristics are similar will be discussed as a group.

VAV and double-duct boxes are similar in that they generally have damper actuators that are controlled from the space sensor or room thermostat. Once these units are calibrated and set up their operation and maintenance consist of seeing to it that the dampers are freely operating and not binding. The ongoing maintenance consists of checking the units from time to time to see that the actuators have not slipped on the damper shafts or that the air lines in the case of pneumatic controls have not been broken or eaten away by rodents. In the case of pneumatic as well as electric controls, vibration can sometimes play a role in the controls on the boxes. Sometimes the lack of perfect filtration can cause problems at the boxes because dirt and dust can be trapped in the air passages. For that reason, the boxes should be inspected periodically to see if they need to be cleaned.

Fan-coil units, unit ventilators, and induction units are similar devices that are generally mounted on the outside wall under a window in an office building or a school. With the exception of induction units, they have fans and filters, as well as coils that may or may not have hot water or chilled water flowing in them at the appropriate time. Unit ventilators also have outside-air and return-air dampers to perform the ASHRAE Cycle I, II, or III. Their controls need to be studied and understood for proper operation of the units.

Fan-coil units usually do not have outside air supplied to them, and the controls consist of a thermostat either in the room or in the return air of the unit that starts and stops the unit fan on demand. Some have valves that are controlled by a controller. Others have two coils in them, one for chilled water and one for hot water. Some have two pipes running to them for hot water in the heating season and chilled water in the cooling season. Some have separate chilled water and hot water lines with a common return, whereas others have two supply lines and two return lines.

Unit ventilators are similar except that they can take in outside air as part of the control cycle so as to prevent overheating in classrooms. They have been used predominantly in classrooms where there is no air conditioning.

Induction units are similar to the fan-coil units, but the movement of the air in the space through the unit is a result of high- or medium-high-pressure primary air that is piped to the units through a nozzle in the unit that induces the room air to pass through the unit and its coil. The control cycle in the unit is similar to the control cycle in a fan-coil unit that uses valve control not fan start–stop control. The proper operation of the induction units depends not just on the units themselves but also on the PAD unit as described in other chapters.

Through-the-wall and closed-loop heat pump units contain a complete refrigeration system, including the compressor. Through-the-wall units are nothing more than a glorified window air conditioner. Closed-loop heat pumps have all the items in a window air conditioner, except for the method of condensing the refrigerant; this is where the similarity ends. The refrigerant in each unit in a building is condensed in a tube-in-tube condenser that is supplied with water from a system; all the units are piped together so one common water loop supplies all units. This effectively allows a unit that is cooling in one room to supply warm water to the unit in the next room to heat that room.

As far as maintenance of these terminal units is concerned, proper care for the filters is necessary; in many cases, this is the one item that is neglected. In many motels that use fan-coil or through-the-wall units, for example, 90% of the complaints from room occupants will be about poor heating or cooling as a result of filters that have not been changed or cleaned. The lubrication of the small motors that are typically a part of these terminal devices is another area that needs attention. In classroom unit ventilators used in some schools, a problem can result if the damper actuators are not checked and serviced periodically. They can, after a time, slip on the damper shafts and thus get out of adjustment. A good PM schedule that includes all of the above as well as the items unique to the terminal unit involved is a *must.*

Radiators and convectors

About the only items that need to be checked when it comes to radiators and convectors, other than the controls themselves, are the steam traps used with them.

Operating and Maintaining Controls

The operation and maintenance of building control systems is the most important topic in this chapter, as the controls and control systems are the heart of any HVAC system or component. Too often, operator's problems have stemmed from improper operation and maintenance of the controls, not the equipment being controlled. In almost all cases, the problem is a result of a lack of understanding and knowledge about the controls themselves. It is therefore imperative that operators be *trained* to operate and maintain the control systems installed in modern commercial buildings.

In this section most comments on the operation and maintenance of controls are generic and involve pneumatic, electric, and DDC controls. Some items, however, are unique to one system or another and the generic concepts will not apply. The chapter ends with a section on the operation and maintenance of building automation systems.

Pneumatic controls

In the case of pneumatic controls and systems, the manufacturer's instructions concerning operation and maintenance should be followed as closely as possible on all things, especially the air compressor in a pneumatic control system. The air compressor is the heart of a pneumatic system, and without it the control system is worthless. The two biggest enemies of a pneumatic control supply air system are oil and water in the control lines. The procedures used to prevent those items from entering the air tubing have been discussed and should be adhered to religiously. The condition of the supply control air to the system *must* be checked periodically, even as frequently as daily. The air supplied should be dry and free of all traces of oil.

The controllers in the spaces or attached to the duct work need to be installed and operated in an area that is free of excessive vibration and is not subject to unusually dirty or wet conditions. Duct-mounted instruments should be at an accessible height and installed with a thermometer with the same style of capillary as the controller; in the case of a sensor or transmitter, an indicating receiver gauge should be used.

Maintenance of pneumatic controllers should be limited to calibrating and checking the operational characteristics of the devices. (Calibration will be covered in another section in this chapter.) Room sensors (transmitters) can get out of kilter at times due to unauthorized tampering. If that happens and the unit will not operate, the unit needs to be sent to a qualified technician at a control company for readjustment of the pivots, linkages, and so on. If some signals used to control the cycles seem to be way off from what they should be, it is possible that other things are wrong with the system, such as capillaries not sensing the correct temperatures in the duct work due to the type and style of the capillary or its location in the duct work. Often, stratification of the air in the duct work will give a false reading to the controller or sensor, and the only way to correct that is to move the capillary, use a different capillary, or eliminate the stratification.

Pneumatic controls installed in panels are usually put there so they can be adjusted by operators, not occupants of the building. Set point, range, authority, and so on can be adjusted by a trained operator to suit the requirements of the building. Maintenance of these panels is no different than maintenance of devices that are not panel mounted. It generally consists of calibrating the set points, ranges, or spans, as well as defining the authority one controller has over another. Sometimes clocks and switches are mounted in or on the panels; they may or may not need periodic adjustment as the system changes from heating to cooling and vice versa or from one time cycle to another (e.g., daylight savings to standard time). The electric–pneumatic devices such as EP and PE switches that are sometimes mounted in or on these panels can fail, and in that case about the only solution is to replace the offending item. There is a term used in the control industry, tweaking the controls. What that means is that after calibrating and setting the control points, the operator needs to let the system operate for a time and settle in, then perhaps fine tune it from the first rough tune-up settings. This concept is used extensively in pneumatic controls to ensure that they operate as designed.

Auxiliary items in the control scheme seldom if ever need adjustment. Devices such as the highest pressure selectors usually operate without moving parts other than diaphragms. These items are throwaway devices, so replacement is the only solution when they fail. Other auxiliary items such as the clocks, PE switches, and EP switches (also called solenoid air valves) may require some adjustment but very little maintenance. One exception might be the transducers that convert a pneumatic signal to an electrical one or an electrical signal to a pneumatic one. They have operating point adjustments as well as span adjustments to suit the control cycles.

The devices controlled in any control system consist of the valve and damper actuators, and they need to be maintained and operated by the controllers in order for the system to function properly. To start with, the valves in a pneumatic system have moving parts consisting of the stem that moves up and down and the spring with an airtight diaphragm that is part of the actuator. To be sure the valve is controlling properly, the operator must be able to ascertain whether or not the valve moves its stem up and down at the air pressures from the controller that correspond to the spring range of the valve. To do this, the operator needs to disconnect the air line to the actuator and connect a hand-operated squeeze bulb (the same kind as that used by a doctor to check blood pressure) with an accurate gauge attached so he or she knows where the valve is operating. As an example, a valve with a 4–8-lb spring may actually operate at some other span such as 5–9 psi. This span is due to all the external forces such as pressure drop through the valve and the friction in the stuffing box or packing. Once that is known, the operator is in a better position to calibrate the controllers controlling the valve.

As far as maintenance is concerned, the biggest item involved with valves is the packing around the stem of the valve, which should be replaced as necessary and adjusted so that it does not leak, yet allows the stem to move freely. The normal control valve involved in low-pressure water and steam systems has a composition disk that can wear out and needs to be replaced from time to time. The diaphragm can also spring a leak if it is subjected to very high temperatures and may have to be replaced from time to time. Valves should never be installed where the body of the valve with the stem and actuator are upside down. It should never be more than 90 deg from vertical.

Damper actuators need to be checked to see that they operate within the spring range stamped on the unit. Some minor adjustments may be needed in terms of the starting and stopping points of the actuator. The biggest problem with damper actuators is the linkage that is driven by the actuator. The linkage can slip on the damper shafts or be installed backward so the damper rotates opposite to what was intended. The maintenance of damper actuators is not as important as the maintenance of the dampers themselves. They must always be maintained so they actually rotate with little or no effort without the actuator attached and the fan running. If they cannot do that and have a tendency to bind, the binding must be corrected to allow the actuator to operate them easily. Most dampers come with bearings that are either factory sealed and lubricated or that never require lubrication, such as nylon and other composition materials. *Damper bearings that are in the airstream should never be lubricated, since all that does is attract dust that will cause the dampers to bind.* Dampers that are not visible to the operator should be checked periodically to see that all the blades are connected to the appropriate linkage and no blades are slipping. Dampers with special seals on the blade edges and sides need to be checked for wear on those seals, and the seals need to be replaced as necessary.

Electric controls

Electric controls and systems also have operation and maintenance needs. The controls discussed here are the typical *electric controls not the electronic controls.*

Some commercial control items on the market for many years are often referred to as three-wire, 135-Ω pot devices, since that is what is used for the controllers and controlled devices in the systems. The operation of these items constitute a modulating control system with electric controls. These systems use, for example, a 24-V ac electrical current that is supplied to the controllers and the controlled devices.

The operation of the actuators, which are generally 24-V shaded-pole reversible electric motors, drives the valve stems through gears to control the steam or hot water in the system. The same is true of the actuator that operates the damper through a suitable linkage. The motors (and they are true motors as opposed to the so called pneumatic motors) are immersed in oil to keep them cool, and the power they generate is proportional to the size of the electric motors. Maintenance of them, like that of the pneumatic operators, consists of checking the packings in the case of a valve and checking and adjusting the linkage in the case of a damper actuator. Added to that is checking for the potential wearing out of the gears in the motor and/or the breaking or wearing out of electrical parts, such as the balance relay used in the actuators. Care must always be exercised when opening up the motor for service because of the potential of burns from the hot oil that encases the motor winding. Running an actuator through its paces can be accomplished, as with the pneumatic actor and the squeeze bulb, through the use of a 135-Ω handheld rheostat. Some auxiliary devices used with pneumatic systems are also used with electric systems and, like pneumatic systems, generally do not require much maintenance since they are switches and relay-type devices that should be replaced when they fail.

Many other two-position electric controls are used in conjunction with the modulating devices just discussed. Some are low-voltage and some are line-voltage items that are used, for example, to start and stop the fan motor on a fan-coil unit to maintain room temperature. In the case of line-voltage electric controls, the items that need to be checked and maintained are the contacts of the controller, as they will typically become pitted if the amperage of the device being controlled is too high. The same thing is true of the contacts of relays and starters in the system.

Calibrating Controls

The calibration of controls and systems is an area that is often confusing to the novice because of the many types of controls and sensing items with which to contend. As an example, you need to think about the calibration of thermostats, humidistats, static pressure stats, and other pressure stats. Also, the control outputs can be pneumatic, electric, or electronic, and the signals put out from the controller can be modulating or two position. The procedure explained here will apply to all those options, regardless of the item involved.

To begin, the operator must have accurate devices to read the conditions in the room, duct work, or pipe. That is, to calibrate a room thermostat, you must have an *accurate thermometer*. The same applies to the calibration of the other controllers mentioned. The principles are simple:

> Turn the dial to where you are. (If the space temperature is 70°F, set the dial to 70°F.) Then turn the calibrating screw (all instruments have one) so the output of the controller is in the midrange of the spring in the device being controlled. An example is a pneumatic valve with a #4 to 8 spring. In that case, set the output of the instrument to 6 lb. Now, stop and remove the screwdriver from the calibrating screw and turn the dial of the controller to what you want. An example is setting it to 75°F.

In the case of multiple actuator ranges, calibrate to 7½ psi, which is the middle of the 0–15 psi range. This method will work for all controllers, regardless of the type and style. In the case of two-position electrical devices, since they have a differential from on to off and vice versa, turn the calibrating screw to turn the electrical output on. When the temperature being controlled gets to the set point on the dial (75°F in the example). This method will work with all controllers if the following items are known: the value of the medium being controlled, the differential of the input of the controller, and the output of the control signal at the device being controlled.

One of the items we mentioned is a device, such as a thermometer, used to check the condition of the mediums being controlled. One of the greatest sins a designer can commit during the design phase of a project is to skimp when specifying the diagnostic device necessary to operate and maintain the system. For an operator to be able to calibrate, operate, and troubleshoot a modern HVAC system in a commercial building, *he or she must have adequate diagnostic capabilities at hand.* The equipment can be portable or permanent; that is not as important as whether or not it is available.

In terms of troubleshooting, the first thing to consider is that the operator *must understand the system.* After that there is a method for troubleshooting, sometimes called the cascade method, that will work if it is followed logically. This method is particularly suited to troubleshooting controls and control systems. The cascade method can best be described in Fig. 16.1, where the problem is described in the box at the top and the various solutions are shown in the boxes below. This is the same type of logic diagram often used when there are several options to the solution of a problem. It is also the language of many computer control diagrams in which gates are used to determine the path of the electronic signal involving the complicated algorithms of software. If the concept is used in all control troubleshooting, chances are very good that the problems can be solved, remembering again that a

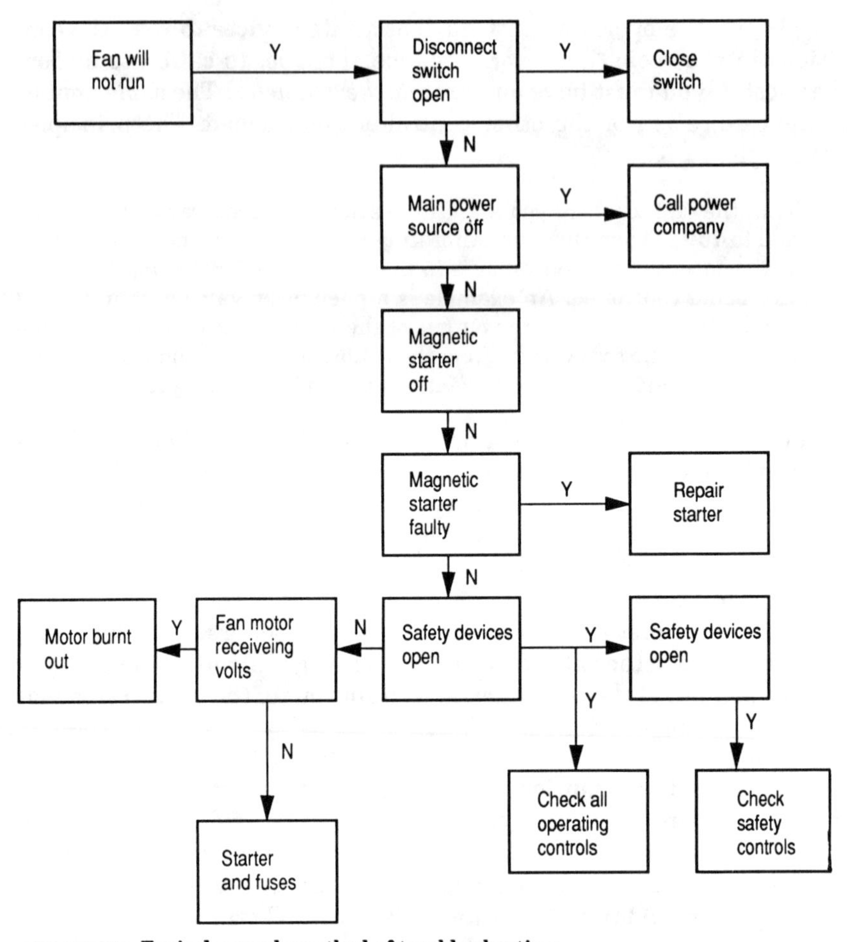

Figure 16.1 Typical cascade method of trouble-shooting.

very large part of the time the fault lies with the system and not the controls. In fact, 40% to 50% of the time, a control problem is the result of a system that is *too large*. All the controls in the world will not solve the problems of a poorly designed system.

Operating and Maintaining Building Automation Systems

The operation of a building automation system involves its monitoring, logging, and command capabilities to help control the systems in a building or facility. The use of a particular BAS requires a detailed knowledge of the system capabilities and procedures. Proper training and the use of a users manual as a reference are essential.

The typical BAS monitors itself and reports out-of-range values and no-response conditions when they occur at any point. These indications of trouble should be followed up to resolve the problems. In addition, there are times when sensor conditions can be compared; for example, when a heating or cooling process is turned off, return an discharge sensors should read the same. These are opportunities to cross-check calibration accuracy (unless the process valve is leaking through). Logs are another excellent tool to check for consistency and reasonableness of different measurements. The maintenance of the hardware of a BAS is essentially the same as the maintenance of any electronic equipment. The majority of equipment is self-checking, and a malfunction is shown by a message with an error code or by a change in operation of an indicator light on an electronic board.

The maintenance of software in a BAS is a matter of keeping backup copies of data files and using them if a file is corrupted or is changed in error. The version number of a DDC program should be made a parameter so it can be displayed to show what version is loaded and working in the controller. From time to time, updates to the software that drives a BAS or DDC controller may be produced by the control manufacturer. These may contain mandatory corrections or enhancements. If any updates are installed, the compatibility of the new software with the rest of the system should be known and checked by the manufacturer's representative.

Summary

In this chapter we have told you about some of the items of concern when it comes to the maintenance and operation of HVAC systems and the controls that are associated with them.

The chapter dealt with the primary systems and the terminal systems of all types. Those primary systems included the boiler systems, including steam and hot water systems, chilled water systems, and the related ancillary equipment. The pumps of both the hot water and the chilled water systems were explained.

Some of the chapter was devoted to the safety devices that need to be maintained. Primary equipment, such as the chillers, was discussed as well as the maintenance of those devices.

Controls were discussed from the standpoint of the maintenance required. Information on terminal equipment was also discussed. Air-handling units of all types, with the work that needs to be done to keep them in good working order we've explained in this chapter. The chapter also included the very important term *calibration* and the knowledge of how to implement a system of generic troubleshooting.

Chapter

17

Total Facility Approach to Planning Controls

Planning what controls are needed depends upon the design of the mechanical system and the philosophy of the organization managing the facility. The design of the mechanical system was necessarily influenced by the philosophy of the organization that planned and purchased the facility. Therefore, if the organization consisted of knowledgeable owner-occupants, the design probably included life-cycle costs. If the facility were of a speculative venture, the design criteria was probably first cost, with only the need to meet minimum building code requirements.

This chapter will cover the control selections associated with total mechanical systems. These selections will be examined from the perspective of the type of organization that will be occupying and running the facility. Example situations and desirable control system features, for those situations will be defined. Then, the content of specifications for the plan and spec bidding process will be given. Finally, criteria for negotiated scope contracts will be suggested.

Building Usage and Zoning

Building and system zoning are important criteria in the actual operation of a building. The schedules of usage of different areas and the needs for heating and cooling each area are what establish the HVAC system loads and the diversity of those loads. Good HVAC system zoning to meet needs while not conditioning spaces unnecessarily is a major contribution to energy efficiency. The zoning of final space control with terminal units for each room and the use of a fan system for each floor or common load area are typical practices. The common load area may have been defined on the basis of time schedule of use and/or similar loads because of exposure. The heating zone and the cooling zone should be the same area to accommodate the changeover from heating load to cooling load.

The range of control features that could be used in different situations include the following:

1. Optimum start–stop to turn fan systems on and off per an occupancy schedule. The unoccupied mode maintains safe temperatures without any ventilation. If central supply of heating or cooling is required and not continuously available, it is provided by an overall schedule or by any using fan system being turned on. This is appropriate where the usage of the facility is scheduled and there is no need for off-hour services.

2. Temporary override schedules to restore occupied services. This would provide both occupied temperatures and ventilation. It would be appropriate where a qualified operator was available to implement a temporary schedule at the time the need was decided.

3. Occupant override switches to restore a zone to occupied temperatures for a fixed period of time. If the fan system and/or central heating or cooling supply was required, it would also be turned on. This would be appropriate if tenant services were required on demand but no operator was available. This could also be a billable service that would require a log of measured time or quantity of such services.

Building Management Method

The building management method determines the staff requirements and the BAS needs. The range of possibilities go from everything being done by the operating staff to everything being done automatically. The specific actions can be summarized by categories as follows:

1. Operator selection and start–stop of equipment versus automatic selection and scheduling. The starting and stopping of end-use zone equipment was discussed above. The selection and starting and stopping of central supply subsystems becomes a factor on larger systems, where there are many pumps, chillers, and boilers that provide the central supply. There can also be zones of distribution that supply the end-use zone equipment. The selection process involves knowing what is needed and what supply equipment will provide what is needed at the lowest cost. What is needed can be measured as tons of cooling load, MBTUs of heating load, or psi of pumping supply pressure. The capacities of each piece of supply equipment establishes which equip-

ment or combination of equipment can provide what is needed. The part load energy characteristics of each selection will be needed to establish which alternative costs the least.

If the management method is for the operator to do everything, the best approach is with monitored information as to load size and supply pressures to guide the selection of equipment. If this information is not available to the operator, the only indications are when the supply does not keep up with the load. When there is much equipment, instrumentation to measure load and supply pressures is recommended. Also, when there are many chillers, part load energy characteristic information as a function of refrigerant head should be available. This information can then guide the selection of the lowest cost chiller or combination of chillers.

If the management method is to do everything automatically, the measurements of load can be used as inputs to selection programs that automatically select and turn equipment on and off as load levels change. Pumping capacity can be automatically varied as necessary to maintain supply pressure differential at using locations.

2. In-house monitoring and alarm response versus automatic logging and predetermined call out messages. The in-house monitoring can be with or without the help of a BAS. If there is no BAS, the monitoring is by observing or by responding to trouble calls when something goes wrong. If there is a BAS and an on-site operating staff, alarm logging and required alarm acknowledgment are the typical BAS support. If there is a BAS but no on-site operating staff, in addition to alarm logging an automatic call out to an appropriate phone number is recommended.

3. Energy cost and space comfort the responsibility of the operator or handled by management review of BAS reports. The energy cost and space comfort monitoring without the help of a BAS can be accomplished by a review of bills and trouble reports. With the help of a BAS, logs of comfort alarms and summary reports of energy used can be set up to help both operators and management. The content of logs and reports would be defined to best fit the specific facility situation. Comfort alarm limits would be assigned to representative sensing points, and energy summaries would be made of systems or areas with different bills or operating responsibilities. Energy auditor programs are available to help with the task of storing and displaying energy history.

4. Maintenance and repair responsibility by in-house staff or by contracts monitored by management. If there is a BAS, there can be help in managing a maintenance program. This help can be in the form of run-time maintenance alarms, measured condition alarms

such as filter pressure drop, or trend logs. There are also maintenance manager programs that use these capabilities plus other record keeping features. These BAS aids can be used by in-house staff or by maintenance contractors and management.

Total Mechanical System Design

The type of total mechanical system chosen and its design criteria establish the possible modes of operation. Then controls are chosen to implement these modes and control the actual operation. The selection and sizing of the mechanical subsystems must provide for the most extreme loads anticipated, although the actual operation sees these loads only a small percentage of the time. The vast majority of the time the control of the mechanical subsystems is at part load. The part load energy characteristics therefore have a great deal to do with the overall system efficiency and operating costs.

Typically, equipment operates most efficiently above 80% of its design capacity. This is the reason many systems have multiple units of chillers, boilers, and pumps; that is, it allows lower loads to be supplied more efficiently. The selection of the type of subsystems also has a great deal to do with the overall efficiency. For instance, fan systems that can provide heating to some zones and cooling to other zones are inherently inefficient because they mix heating and cooling as a form of control. The load reset control of these types of fan systems was covered in Chapter 10 on air-handling units. This strategy modified the supply temperatures to the minimum required to carry the greatest demand load. The chapters on each type of subsystem gave control recommendations to optimize each operation. This chapter deals with the total system and now considers situations in which interaction between subsystems must be taken into account to achieve optimization of the total system. The common denominator in total system interaction is the load imposed on the system that drives end-use subsystems, which in turn drives distribution and primary supply subsystems. Examples of these interaction situations follow:

1. A minimum energy strategy for variable volume types of systems for both air and water must consider the total system from load back to chiller operation. This consideration trades off changing transport energy (pumps and fans) versus minimizing chiller refrigerant head and operating cost. If the increases in pump and fan energy are greater than the savings in chiller energy when supply temperatures are raised, then load reset of temperature is not an appropriate strat-

egy. This is typically the case when transport energy is more than 30% of the cooling energy.

2. The use of rejected heat from a chiller system should be coordinated with the use of outside air for free cooling. During those times when outside air is a source of free cooling by fan system economizer cycles, rejected heat from the chiller can also be a source of inexpensive heat. The cost of rejected heat is only the decrease in COP resulting from the higher level of condenser water temperature necessary to get a heat source. This cost should be compared to the cost of generated heat. If the cost of generated heat is higher, which it usually is, then the amount of free cooling allowed should be limited to let rejected heat carry as much heating load as it can. This can be accomplished by limiting the maximum outdoor damper air opening on all fan systems with an economizer cycle when the heating circuit is calling for more heat.

3. Alternative energy sources can come from subsystems such as heat pumps, solar collectors, waste heat, and cogeneration of electricity and heat. The availability by quantity and timing are major factors in how these supplementary subsystems are coordinated with the primary supply systems.Usually, the alternative source is lower in cost so it is used in preference to the primary energy source. The coordination is in using what is available, then supplementing that as necessary with the higher cost source. If the primary source is electrically generated, the cost may be dependent on time-of-day demand charges or energy charges. Thus, the lowest cost preferential source may vary with the time of day. For instance, cogeneration energy may cost the least during the day but not at night. The control of sequence of usage must change when the cost relationship changes.

4. Thermal storage is a fixed quantity of energy available for a fixed period of time (normally a daily cycle in which it is charged, then discharged). In this case, the load profile for the 24-h period and for the hours of high electrical cost are the primary factors in coordinating the use of thermal storage. The control involves both charging and discharging storage in a manner that carries loads with a minimum total cost of both demand charges and energy charges.

In its simplest form, control of thermal storage is accomplished by charging during the low electrical cost time of day (off-peak period) and discharging during the high electrical cost time of day (peak period). In control of discharge, there can be two different strategies: A chiller priority strategy is appropriate when the stored energy costs more than the direct cooling energy. An example of this situation is when the stored energy is in the form of ice that was made with low COP at night and the energy cost was the same as daytime energy. In this situation, ice storage is used only as much as is necessary to maintain a lowered demand limit during the peak period. The name of the strategy is chiller priority, because the lower cost of direct cooling is preferred. In the opposite case, or storage priority, the stored energy cost less and is preferred. The savings in the case of storage priority is not only in reduced demand charges but also in reduced energy, so the greatest savings occur when all of the storage capacity is used. The optimal control of storage is therefore a complex problem dependent on cost factors and load profiles. The control involves the interaction of the loads and both the storage subsystem and the chiller subsystems.

An optimal cool storage program was developed for the Electric Power Research Institute (EPRI) as part of a building automation system. This proprietary control scheme is available for licensing by EPRI. Measurement of thermal loads and electrical loads, as well as storage inventory, are important to any type of control of thermal storage. They are essential to the optimal control provided by the EPRI program.

5. Load dynamics are another dimension of total system control that provide an opportunity for improved operations and energy management. The main element of this consideration is the thermal storage in the building structure itself. The opportunities are greatest in situations in which there are daytime cooling loads and cool night temperatures outside. The use of night purge cycles saves daytime cooling costs by precooling the building at night. Another strategy is to move heat stored in the interior of the building out to the skin areas when they need heat. In some cases in which fan systems serve both inside and outside zones, this happens with a normal night cycle operation of fan systems. This can be properly controlled when interior sensors show the presence of heat when the exterior sensors show the need for heat and start the fan. In the use of optimum start and stop programs that adapt to the experience of how long it actually takes to precondition a building, the thermal storage characteristics are automatically used. As the use of measured load profiles becomes more widespread, increased knowledge will lead to better usage of dynamic control. The measurement of loads in zones is becoming a bonus available in many of the zone DDC controllers. This is used for better control at part load and can also present dynamic control opportunities.

Selection of Types of Control Systems

The types of control system can be classified as the following levels: simple local loop control, supervisory control of local loop control, integrated DDC, and building automation systems. The appropriateness of each level of control depends on these factors: size and nature of the facility, type of management philosophy and operating and maintenance staff, and the nature of the total mechanical system. Examples of appropriate selections for each level are as follows:

1. Simple local loop control fits situations in which the mechanical subsystems are simple fan systems and, if there is a central chiller and boiler, there are no multiple units to be controlled. The occupancy of the facility is relatively constant, and there are no unusual requirements for accuracy of control or abnormal ventilation. The options of simple local loop control are pneumatic, analog electric and electronic, and DDC electronic. The choice of a pneumatic control might be exercised if there were an operating and maintenance staff familiar with it. The choice of a DDC electronic control might be exercised if the local energy management programs such as optimum start–stop could be useful or if future growth into a building automation system was a benefit.

2. Supervisory control of local loop control is appropriate in situations in which there are diverse mechanical subsystems and environmental quality and energy costs are important. This presumes that the mechanical subsystems are not complex with a lot of interaction and are adequately controlled by local loop controls. A mixture of existing and new facilities can be well served by a BAS that monitors all systems and provides central scheduling and energy management programs. The BAS can also provide fire and security functions where needed.

3. DDC integrated with a BAS is appropriate in situations in which there are mechanical subsystems with interaction, such as variable volume systems, multiple central chillers, chillers with the use of rejected heat, thermal storage, or cogeneration. This choice is also appropriate when there is a variable occupancy schedule, high standards of environmental quality, and the intention of improving performance and energy costs. The ability to use diverse zone load information to develop advanced control strategies and implement them with custom DDC programs is unique to this level of control system.

Methods of Specifying and Procuring a Control System

Two general approaches to procurement are used for control systems. The first approach is the traditional bid against plans and specifications with evaluation of exceptions or adds to the bid. The second approach is a two-step process. In the first step, a scope plan and a functional specification are issued and technical proposals are requested and evaluated. In the second step, there is further evaluation and negotiation with several qualified proposers to arrive at a decision of contract content and a price with the chosen contractor. There are variations between these two procurement approaches, but the intent is always to evaluate the differences in competing proposals and the value of each proposal.

The traditional method of plans and specifications put out for bid is most appropriate for the simple local loop level of control. This method has typically specified hardware ratings and features plus the required sequence of control for each subsystem. The schedules that defined mechanical equipment sizes have included the valves and dampers installed by the mechanical contractors but sized and furnished by the control contractor.

The plans and specifications for a supervisory control system to monitor and reset local control loops have typically specified which points were to be sensed, which set points of local control loops were to be reset, which start–stop points were to be controlled and monitored, and which sensed or monitored points were to have alarms associated with them. In addition, the general nature of required energy management programs were specified along with which systems were to be controlled by them. The general nature of required logs and reports was also specified.

Some of the differences among proposed systems are as follows:

How the functions were accomplished

How friendly or understandable the presentation was to the operator

How the system responded to device failures

How the system resolved conflicting commands from different programs or sources

What the capacities of panels were

What provisions there were for growth and modification

These evaluations should not only look for proper operation but also for the value to the user of different features in different systems. In that case, a decision could be made on the basis of value received for a price, not just on low price.

The plans and specifications for a control system with DDC integrated into a BAS cover everything needed in the specification of local loop control function and everything needed in the specification of supervisory control, including function and point lists. The specification does not need to deal with the hardware aspects of controllers and accessories that exist only as DDC functions. It does, however, have to include the requirements for software points needed as information or command access by the operator. For example, the control signal from a primary space temperature controller indicates whether the final control signal is coming from the primary controller or from a discharge low-limit controller. This is analogous to reading control signal gauges on both the primary controller and limit controller to understand what is in control at any given time. An example of a software point defined for control access is a discharge low-limit set point. This is analogous to a set point on a hardware controller. Specifying this point to be available to the operator and included in graphic presentations may be desirable in one situation and not in another. Such specification may be best handled by identifying categories of points to be made accessible to the operator. If this method of specification is used, care must be taken not to provide so much information to the operator that the system is difficult to understand. This is an area worthy of discussion with representatives of the operating staff.

A specification defining what is being procured should include a definition of the procedure to be used for submission and approval of contractor plans. It should also include a procedure for arriving at an agreed-upon plan for the commissioning and acceptance of the project. The specification of these procedures should define the level of detail to be given in submission of plans by the contractor. This level of detail should be understandable by the approving engineer so that any misunderstandings or misinterpretations on the part of the contractor can be corrected in the submission stage. For instance, the submitted sequence of control of a DDC program should define all of the output control actions resulting from input changes in each mode of operation. The level of detail in a commissioning and acceptance plan would normally be the same as in the submission. That is, each mode of operation in each subsystem would be checked for proper operation. The operation of the system as a whole is also checked by normal operation of the total system and observing correct interaction between subsystems, for instance, that the chiller system had the correct number of chillers on line for a total load and that optimum start–stop controlled fan systems properly. The specification should define the type of records to be turned over to the owner upon acceptance of the project. Training requirements should be defined that identify who is to be trained and the time period of training.

Index

Absorption chiller solution control, 227
Actuators, electronic, 137
Air balance diagram, 39
Air compressors, 101
  base mounted, 101
  belt guards, 106
  belt tension, 106
  chemical dryers, 109
  diaphragm type, 111
  duplex types, 101, 102, 105
  fan shaft types, 111
  industrial types, 111
  intake filters, 104, 109
  leaks, 104
  lubrication, 101
  motor driven, 102
  oil levels, 106
  oil pumps, 106
  PRV stations, 110, 111
  rings, 104
  run time, 104
  SCIMS, 104
  single, 101
  sizing, 102, 104
  starters, 105
  suction, 105
  tank mounted, 101
  tanks, 101, 102, 104
  valves, 104
  water driven, 101
Air Conditioning and Refrigeration Institute (ARI), 241
Air-handling unit:
  design, 188, 189
  double duct, 161
  dual paths, 161, 162, 172, 178, 179, 180
  induction systems, 177
  industrial types, 178
  maintenance, 187, 188
  make-up air types, 178
    direct fired, 178
  multizone types, 180, 181, 183
  primary air dehumidifier (PAD) types, 177, 178, 205, 206
  single path, 161–163, 166, 171, 172
  trouble shooting, 186, 187
Air Movement and Control Association (AMCA), 35
Air reducing station, 107
American National Standards Institute (ANSI), 56
American Society of Mechanical Engineers (ASME), 300
American Standards Institute (ASI), 53
Analog-to-digital (A/D) convertor, 134, 136
Analog-to-digital interface, 143
Application engineering BAS projects, 287, 288
Arcing, 17, 88, 89, 92
ASHRAE Bacnet proposed standard, 277–280
  object, 277, 279
  services, 279
ASHRAE handbooks, 298, 299
ASHRAE Standards Project SPC 135, 277–230
ASME high pressure vessels, 107

Balancing, 18, 35
BAS documentation, 289–295
Boilers:
  combustion:
    alternate fuels, 216
    primary control, 214
    programming controller, 214
    safeguard controls, 214
    trim control, 214, 215
  electric, 213
  steam, 158
Bristol, 72
Brown Instrument Co., 59, 72
BTU calculations, 83
Building automation systems, 273

Capillaries, 20, 21
  averaging types, 20
  compensated types, 20
  freeze protection types, 21, 26, 166
  limit types, 21
  transmitter, 72
Cast iron raditors, 158, 192
Central panels, 71, 88
Chilled water load reset, 224

Chiller capacity control:
  absorption, 227
  centrifugal, 225
  positive displacement, 227
Chiller optimum selection, 225, 226
Clocks, 97, 98, 121
  astrological, 98
Cogeneration, 212
  engine jacket cooling, 220
  load following control, 221
  prime mover types, 219
  types of cycle, 219
Coil condensate, 67
Coils:
  cast iron, 158
  cooling, 38, 159, 166
  direct expansion types, 166, 180
  evaporative types, 178
  fin tubes, 159
  heating, 166, 181
  preheat, 166, 169, 172
  reheat, 166
  sprayed, 171, 177
  steam, 66, 161, 181
Commissioning of BAS, 289
Communications standards, 277–280
Compressed air, 101
Compressors refrigeration, 159
Condenser evaporative, 159
Constant volume systems (CAV), 35
Constant water flow, 256, 257
Construction, 101
Construction Specification Institute (CSI), 122
Control air systems, 111
Control air tubing, 111
  aluminum tubing, 112
  copper tubing, 111, 112
  Dekabon, 113
  hard copper, 112
  plastic, 113
  PVC types, 113, 114
  soft copper, 112
Control air:
  dryers, 107, 109
  dual PRV systems, 111
  moisture problems, 101
  oil and water problems, 17, 106, 112
  oil indicators, 107
  soldering, 112
  twin tower dryers, 110
Control loops, 143, 144
  closed loop, 276
Controller-damper operator combinations, 26
Controller-valve combinations, 26
Controls, 1
  basic definition, 5
  optimizing principles, 7
Controls (*Cont.*):
  panels, 71
  supervisory definition, 6
Convectors, 191, 212
Cool Storage Surpervisory Controller, 229, 230
Cool storage types:
  eutectic ice storage, 233
    brine ice, 233
    ice harvesting, 233
    ice on coil, 231
  water storage:
    cemented tanks, 230
    empty tank, 231
    stratified, 230
Covers, 14
  room thermostats, 14
Critical alarms, 276
Cumulative energy management, 276
Cumulators, 83, 84, 86, 87

Damper motor, 29, 41
  balancing relays, 47
  bidirectional, 46
  brackets, 44
  crank arms, 42, 44, 46, 48
  diaphragms, 43
  electric, 29, 46, 48, 49
    electric induction, 46
  electronic, 29, 46, 48, 49
  end switches, 48
  feedback potentiometer, 46, 47
  force, 42
  gear trains, 48
  inlet vanes, 45
  materials, 45
  pilot positioners, 29, 43, 46
  pneumatic, 29, 41, 49
  power stroke, 42
  rack and pinion gears, 46
  return stroke, 42
  Series 90 types, 46
  sizing, 42
  slave motors, 48
  Solenoid types, 48
  spring ranges, 42, 43
  spring return types, 46, 47
  stop screws, 42
  torque, 42, 46, 48
  two-position, 47
  two-spring types, 45
  uni-directional, 47
Dampers, 29
  alpha ratio, 37
  approach velocity, 38
  automatic, 29
  balancing, 29, 36
  bearings, 39
  characteristic curves, 29, 31, 37, 38, 40

Dampers (*Cont.*):
- coatings, 29, 36
- compressability, 29
- exhaust air, 37
- face and bypass, 31, 37–39, 41, 166, 169
- face velocities, 37
- fire, 29, 34
- interlinking, 39, 40, 43
- manual, 35
- materials, 39
- opposed blades, 29–31, 36, 38, 40
- outdoor air and return air, 37, 39, 41
- parallel blade, 29, 30, 36, 38, 40
- percent leakage, 34, 41
- pressure drops, 37, 42
- seals, 31, 39–41
- sizing, 36, 37
- smoke, 29, 34
- splitter, 36
- static types, 36, 38
- zone, 41

DDC controllers:
- communicating, 151, 152
- stand-alone, 151
- system level, 151
- zone level, 151

DDC functional flow chart, 144–146, 153
Demand control, 282
Digital-to-analog (D/A) convertor, 134
Digital-to-analog interface, 143
Direct acting, 84
Direct digital controls (DDC), 136, 137
- definition, 143
- functions, 144
- operators, 146
- parameters, 147
- program, 147
- software points, 147

Distributed intelligence, 143, 150
Double bundle condenser, 218
Double-duct mixing boxes, 191, 203, 204
- constant volume types, 204
  - mechanical CV types, 204
  - reheat coil types, 204
  - static pressure controlled, 204
  - system powered, 204

Double-duct variable-air-volume (VAV) boxes, 205
Duty cycle, 283
Dynamic controls, 281

Economizer systems, 163, 172, 218
Electric control motors:
- battery operated, 132
- induction types, 131
- oil immersed, 130
- 135 ohm potentiometer, 131
- spring return types, 130, 131
- slaved types, 132

Electric controls, 11, 18, 26, 93, 125
- floating types, 130
  - Series 90 types, 130, 132

Electronic controls, 11, 18, 26, 93, 125, 136, 185
- analog, 125, 134, 136, 137
- commercial, 130, 135
- digital, 125, 134, 136
- electromechanical controls, 134

EMS functions:
- central, 150
- distributed, 150
- global, 150

Energy efficiency ratio (EER), 135
Energy management system (EMS), 134, 213
Energy monitoring and control systems (EMCS), 273
Enthalphy control, 284
Evaporative cooling towers:
- control, 223
- design approach temperature, 222
- optimizing control, 224

Facility management system (FMS), 273
Factory Mutual Insurance Co., 300
Fan-coil units, 20, 26, 191, 197, 198
- multiple pipes, 199–202
- problems, 198, 199
- unit controls, 200

Fans, 158
- airfoil types, 158
- backward inclined, 158
- centrifugal, 158
- Cfm controls, 161, 173, 175
- Class I, Class II, Class III, 158
- curves and laws, 29, 158, 174
- double inlet double width (DIDW), 39
- forward curved, 158
- horsepower, 37
- propeller, 158
- return air, 166, 171, 172, 175
- squirrel cage, 158
- tube centrifugal, 158
- vane axial, 158

Feedback controls, 17
Filters, air, 17
- charcoal, 107
- coalescent, 106, 107

Fire stats, 26
Fittings:
- barbed, 114, 117
- compression, 114
- crimped, 112, 117
- flared, 114, 117
- flux use, 114
- glued, 114, 118
- screwed, 116
- soldered, 114

Floating controls, 47
Foxboro Co., 59, 72
Free cooling cycle, 228
Fuel-to-air ratio, 215
Furnaces:
  Baso valves, 127
  natural gas, 126, 127
  pilot lites, 127
  safety valves, 127
  thermocouples, 127

Gauges, 20, 49, 71, 97, 98
Global control functions, 280, 281
Grease, 22
Guards, 49

Hardware points, 148
Heat pumps, 235
  absorption types, 245
  air-to-air types, 238–240
  classes of, 238
  closed loop types, 241, 242, 250
  Coefficient of performance (COP), 247
  components, 243, 244
  compressors, 235, 244, 245
  condensers, 235
  cycle chiller, 217, 218
  cycles of, 236
    closed vapor compression, 236, 237
    mechanical recompression, 236, 237
    open vapor compression, 236, 237
    waste heat Rankin, 236, 237
  earth-to-air types, 240
  evaporators, 235
  frost problems, 239, 246
  geothermal, 243
  ground-coupled, 240, 241
  residential split systems, 239
  reversing valves, 245
  solar-assisted, 243
  storage, 247
  water-to-water and air-to-water types, 241
  water-to-water types, 242
  window units, 236
Heat recovery systems, 247–251
  double bundle condensers, 250
  solar collectors, 250
Heat storage, 233, 234
Heating coils, 38, 166, 181
Heating convertors:
  primary to secondary, 217
  steam to hot water, 216
Hierarchial configurations, 275
  management level, 275, 276
  operational level, 275, 276
  system controller level, 275, 276
  zone controller level, 275, 276

High-level language, 149
High-pressure steam, 269
Hot water convertors, 52
Humidification, 169
Humidistats, 11
  duct types, 23
  elements, 24
  high-limit types, 26
  insertion types, 11, 22
  room types, 11, 19
Humidity, sensors, 138
  chilled mirror types, 139
HVAC, 1
  control history, 1–3
  system types, 4, 5

Indicating receivers, 71
Induction units, 20, 191, 205, 206
  controls, 207
  problems, 206
Interface to local loop, 284, 285
Interpretive language, 149

Laminar flow, 18
Leeds Northrup Co., 72
Liquid chillers:
  absorption, 227
  centrifugal, 225, 226
  positive displacement:
    reciprocating, 227
    screw, 227
    scroll, 227
Load reset, 283

Maintaining systems, 297
Master-submaster controls, 24, 25, 85, 179, 183
Menus:
  direct access, 286
  site specific, 286
Microprocessors, 136, 137, 143
Minimum energy strategy, 320
Mixing boxes, 20, 178
Motors, three-phase, 89
Multizone systems, 15, 41

National Electric Code (NEC), 119
National Fire Protection Association (NFPA), 34, 119, 300
Night cycle and night purge, 283, 284

Oil, aerosol, 106
Oil furnaces, 126
Open-loop controls, 24
Open systems, 277–280
Operating systems, 297

Operation and maintainence:
absorption chillers, 304
automation systems, 314, 315
boilers, 298–300
calibrating controls, 312, 313
chillers, 298, 301
hermetic motors, 301
purge systems, 301
vane systems, 301, 302
megging, 302
packages, 302, 303
screw compressor packages, 303
electric controls, 311, 312
heat exchangers, 298
pneumatic controls, 309–311
primary AHU systems, 305–307
primary systems, 293
pumps, 298, 299
steam condensate systems, 305
steam distribution systems, 305
supervisory systems, 297
terminal equipment, 307–309
terminal systems, 297
Operator interface functions, 280, 285–287
Optimizing cool storage, 229–232
Optimizing multiple boilers, 215, 216
Optimizing multiple chillers, 225, 226
Optimum start and stop, 281, 282, 318
Orfice plates, 86

Packaged units, 157, 162
weatherproof types, 162
Peer-to-peer communications, 274, 275
Pitting, 17
Pneumatic transmitters, 71
Poll response protocol, 275
Portable operators terminal (POT), 152
Portable programmers terminal (PPT), 153
Pressure controllers:
differential, 96, 97
high pressure, 96, 97
low pressure, 96, 97
static pressure, 96, 97, 175, 179
velocity pressure, 96, 97, 179
Pressure sensors, 139
Pressure switches, 27
Primary-secondary pumping, 266
secondary zones, 267
throttled interconnect, 268
Primary supply systems, 213
Programming DDC, 147–150
Proportional plus integral (PI) control, 17, 21
Proportional plus integral plus derivative (PID) control, 17, 21, 175, 183
Proprietary systems, 277
Psychometric chart (PSY), 169, 171

Radiators, 191, 192, 212
Receiver controllers, 14, 21, 71, 80, 81
with charts, 81
dual input types, 81
fluidic types, 81, 82
one-pipe, 81
PI control, 81
PID control, 81
pressure types, 97
remote switched, 81
two-pipe, 81
Receiver indicators, 78
multiple ranges, 78
panel-mounted, 80
recorders, 80
circular chart, 80
strip chart, 80
special ranges, 78
Recorders, 18
Refrigerant head, 222
indicated, 222, 223
Register programming, 149
Relays electric, 87, 88
amperage style, 88
coils, 89
dashpot types, 89
hermetic types, 89
magnetic coils of, 88
mechanical latch types, 89
normally closed types, 90
normally open types, 90
pitting contacts, 89
plug-in types, 90
reed types, 90
single pole double throw, 88
single pole single throw, 88
three pole double throw, 88
time delay types, 89
Relays pneumatic, 83, 84
adding and subtracting, 85
averaging, 84
biasing, 85
booster, 85
dividing, 85
highest signal, 84
lowest signal, 84
repeaters, 85
sequencing, 84
signal limiting, 86
square root extracting, 85, 86
two-position, 85, 86
Residential controls, 125
electronic controls, 135
night setback stats, 135
Resistance temperature device (RTD), 13, 18, 72, 92, 136–138
materials, 138
Restrictors, 97
Reverse acting, 84

Self-contained controls, 11
Sensitivity, 16, 21
Sensors, 12–14, 19, 71, 137
  bimetalic, 12
  room 13
Set point indicators, 147
Smoke detectors, 26
Software points, 147, 148, 153
Solar compensators, 97, 98
Specifications, 139
Specifying building automation, 288, 289
Squeeze bulb, 65, 106
Stack gas sensors:
  carbon dioxide, 215
  oxygen, 215
Starters magnetic, 89, 121
Stationary louvers, 38
Steam distribution systems, 269
  one-pipe, 269, 270
  two-pipe gravity, 270
  two-pipe vacumm, 270, 271
  variable vacumm, 271
Steam pressure reduction, 269
Steam traps, 67
Stoker, 158
Stratification, 18, 21, 30, 31
Submaster controllers, 72
Sun shields, 97, 98
Supervisory control, 273
  history, 273–275
Suppliers, 1, 7–9
Switches electric, 83, 92
  bat handle, 92
  explosionproof, 83, 93
  lever, 92
  momentary push-button, 92
  push-button, 92
  rotary, 92
  slide, 92
  stepping, 92
  toggle, 92
  weatherproof, 93
Switches pneumatic, 83, 90
  electric pneumatic (EP), 95, 121
  gradual, 91
  lever, 90
    five-position, 90
    four-position, 90
    six-position, 90
    three-position, 90
    two-position, 90
  minimum position, 166
  panel-mounted, 90
  pressure electric (PE), 83, 95
  push-button, 91
  toggle, 91
System display, 156

Taylor Instrument Co., 59, 72
Terminal units, 20, 191, 192
Thermal storage:
  cool storage, 228–233
  heat storage, 233, 234
  optimized control, 229, 232
Thermistor, 18, 137, 138
Thermocouples, 137
Thermocoupling, 18
Thermometers, 14, 97, 98
Thermostats pneumatic, 11
  capillary, 11, 22, 185
  day-night types, 15
  dead band, 15
  dry bulb, 24
  dual bulb, 23
  explosionproof, 27
  freeze-up, 161, 162
  immersion, 22
  insertion, 11, 22
  leakports, 17
  lid alignment, 17
  lid assemblies, 17
  limiting, 166
  mercury bulb, 12
  one-pipe, 15, 21, 22
  pivot points, 17, 20
  recessed, 14
  room, 11
  safety, 11, 89
  special, 11, 26
  Strap-on, 26
  summer-winter, 15
  unit, 19, 20
  wells, 22
  wet bulb types, 23, 24
    wicks, 24
  wipers, 17
Thermostats residential:
  air-conditioning types, 128
  anticipators, 128
  clock types, 128
  hot water system types, 129
  line voltage types, 129
  night setback types, 128
  room types, 127
Three-way valve controls, 256, 257
Throttling range, 16, 20, 21
Time event programs, 280, 287
Two-way valve controls, 256
Transducers, 83–95, 137
Transformers, 48, 88, 89
Translators, 277
Transmission gage, 76
Transmitters, 13, 14, 21, 72
  calibration of, 76
  capillary, 20, 74
  dials, 74
  differencial-pressure, 77
  high-pressure, 76, 77

Transmitters (*Cont.*):
humidity, 72, 74, 76
insertion, 70
main air, 111
mass, 76
one-pipe, 74
operating range, 74
pressure, 72
room, 14, 74, 76
static pressure, 76, 78
two-pipe, 74
unit types, 74
velocity pressure, 78, 80, 86
Traps condensate, 162
Trend logs, 276
Troubleshooting, 122, 186
ladder diagrams, and point-to-point diagrams, 123
leaks, 122, 123
ohmmeter, 123, 313

Underwriters Laboratories (UL), 34, 119
Unit ventilators, 191, 193, 194
cooling types, 196
cycles, 194–196
electric types, 196
night setback, 197
United States of America Standards Institute (USASI), 119

Valve:
actuators, 52, 60
automatic, 51
balancing, 67
bodies, 53
flanged, 53
flared, 53
screwed, 53
soldered, 53
brackets, 66
butterfly, 51, 67, 69, 70
bearing friction, 69
materials of, 69
sizing of, 69
three-way, 70
torque, 69, 70
cage trim, 56
connections, 54
diaphragms, 60–63, 65
discs, 54, 55
diverting, 51, 54, 55
electric, 51, 65
electronic, 51, 66
flow characteristics, 56
grafite packings, 57
hammering, 53
high-pressure, 56, 57
industrial types, 51, 53, 59
inner valves, 52, 53

Valve (*Cont.*):
maintenance, 63
materials, 55
medium pressure, 56
mixing, 51, 54, 67, 169
motorized type, 65
normally closed, 51
normally open, 51
operators, 51
packings, 51, 57, 58
asbestos, 58
Buna-N, 58
Teflon, 58
U-cup, 58
packless types, 51, 58, 59
plugs, 56
pneumatic, 51
positioners, 62, 63, 65
pressure drops, 66, 67
quick opening, 67
rack and pinion, 60, 66
seats, 54–56
self-contained, 51, 67
sequencing, 52
sizing, 51, 67
Solenoid types, 65
springs, 52, 61
steam, 169
three-way, 51, 54
two position, 56
two way, 51, 52, 169
unions, 51
Variable air volume (VAV) systems:
applications, 86, 96, 97, 171, 173, 175, 177, 179, 208, 211
boxes 191, 192, 207, 208
electric/electronic controls, 211
fan types, 210
pressure dependent, 208, 209
pressure independent, 208, 209
reheat types, 207, 209, 210
static pressure controls, 207
velocity control types, 208, 209
Variable water flow, 256, 257

Water distribution systems:
constant flow, 256
four-pipe, 259, 260
three-pipe, 257–259
two-pipe:
direct return, 256
reverse return, 256
variable flow, 257
Water system pressure regulation, bypass controls:
multiple pumps, 263, 264
pump, 265
system, 265, 266
variable speed pumps, 263

Wheatstone bridge, 136, 138
Wire:
  aluminum solid, 119
    stranded, 119
  Belden, 121
  gauge, 119
  insulation, 119
  material, 119
  metallic shields, 122
  shielding, 119
  structure, 119
Wire wound resistances, 13
Wiring:
  codes, 88
  communications bus, 121
Wiring (*Cont.*):
  for control systems, 118
  DDC systems, 118
  Delta systems, 119
  electric systems, 118, 119
  electronic systems, 118, 119, 121
  line voltage systems, 119, 121
  low voltage, 88, 119, 121
  noisy systems, 121
  Series 90 systems, 120
  three-phase systems, 119
  Wye systems, 119, 120

Zero energy band, 283

ABOUT THE AUTHORS

JOHN LEVENHAGEN is a consultant and retired engineer from Johnson Controls, Inc. He was formerly on the Board of Directors of the American Society of Heating, Refrigeration, and Air Conditioning Engineers (ASHRAE) and past president of the Milwaukee Council of Engineering and Scientific Societies. He has received the Distinguished Service Award of ASHRAE, among others.

DONALD SPETHMANN is a control engineering consultant and spent 40 years as a control engineer with the Honeywell Commercial Buildings Group. His experience includes field, systems, applications, and advanced engineering. He holds more than ten patents, primarily dealing with control for energy conservation. He has been active on ASHRAE technical standards, and handbook committees.